POLYMER NETWORKS
Structure and Mechanical Properties

POLYMER NETWORKS
Structure and Mechanical Properties

Proceedings of the ACS Symposium on Highly Cross-Linked Polymer Networks, held in Chicago, Illinois, September 14-15, 1970

Edited by
A. J. Chompff
and
S. Newman
Polymer Science Department
Scientific Research Staff
Ford Motor Company
Dearborn, Michigan

SPRINGER SCIENCE+BUSINESS MEDIA, LLC 1971

Library of Congress Catalog Card Number 73-163286

ISBN 978-1-4757-6212-9 ISBN 978-1-4757-6210-5 (eBook)
DOI 10.1007/978-1-4757-6210-5

Originally published by Plenum Press, New York in 1971

PREFACE

For several decades, polymer science has sought to rationalize the mechanical and thermodynamic properties of polymer networks largely within the framework of statistical thermodynamics. Much of this effort has been directed toward the rubbery rather than the glassy state. It is generally assumed that networks possess an average composition to which average properties may be assigned; from such a continuum view, a powerful analysis of such properties as modulus, swelling, birefringence and thermoelasticity has emerged.

In the years following the rise of polymer characterization (the late 40's and early 50's), many scientists began to study apparent relations between the properties of linear polymer molecules and the networks obtainable therefrom. This search was also stimulated by the wide range of applications of polymer networks in commercial elastomers, thermosets and coatings. Frequently, these data were confidently matched with curves obtained from statistically describable models of networks of ghost chains, uniformly distributed in space.

More recently, it has become apparent that polymer chains in networks are not as ideal as assumed in the formulation of statistical models, and there has been a shift in emphasis towards the less than ideal, perturbed and possibly inhomogeneous networks which are more frequently encountered in practice. The continuum approach, however, had to be developed before inhomogeneous systems could be described; the present volume, therefore, contains both views.

In practice, most polymer systems are inhomogeneous and contain a variety of defects or inhomogeneities at various structural levels, e.g., loops, pendant chains, wide distributions of chain lengths, fluctuations in crosslink or segment density, anisotropic regions and filler particles. In general, there is a paucity of experimental tools for the study of the defect nature of amorphous polymeric solids. Although light scattering, low angle x-ray diffraction and microscopy may be useful, additional experimental approaches combined with theoretical considerations are needed.

While the properties of polymer networks in the glassy state have attracted many investigators and applications, this area, in our opinion, remains relatively unexplored. Inhomogeneities on the molecular, microscopic and macroscopic levels are particularly important in the glassy state. Unfortunately, despite the importance of glassy network systems, this area is represented by only a few papers in this volume.

In most practical applications, problems which are mechanical in nature dominate. The aim of the present volume is to highlight recent developments in network science which are of prime consideration when polymer networks have to be engineered to certain applications. It is hoped that those actively engaged in polymer applications will find these papers useful and stimulating. Furthermore, the material should be pertinent to other research workers and graduate students as a survey of the present state of the art in polymer networks and might stir their interest in the large variety of materials which fall in this category.

In some cases, network properties depend on the chemical composition of the continuum; in others, on the frequently neglected microstructure of the network. This book attempts to present a balanced view of these two aspects of network properties; and, while incomplete, does arrange presently known facts in a unifying manner. The authors of the present papers have been selected to provide a variety of approaches and methods.

The first ten papers deal with networks as continua. The first paper will be of special interest to chemists working in the field of kinetics of crosslinking processes. The following four papers are equally divided between macro-thermodynamics (Blatz and Shen et al) and micro-statistical mechanics of network chains (Sternstein et al and Edwards). Papers 6, 7 and 8 (Bikerman, Cuthrell and Chompff) consider several aspects of solid/liquid continua and their transitions. Papers 9 and 10 (Williams et al and Landel) deal with the interrelation between various mechanical properties and their structure.

Subsequent papers consider inhomogeneous and heterogeneous networks in various forms. Papers 11 and 12 (Dušek and Rigbi) emphasize the thermodynamics and micromechanics of deliberately or accidentally inhomogeneous networks in the swollen state. Deliberately heterogeneous systems are characterized in papers 13 and 14 (Kotani et al and Picot et al) where the swelling in filled systems has been analyzed. Papers 15 and 16 (VanAartsen and Prins) deal with the characterization of inhomogeneous systems and the possibilities of micro-anisotropy in polymer networks. In papers 17 and 18 (Farris and Hsiao) the mechanical properties of heterogeneous and inhomogeneous systems is considered and new approaches suggested. Papers 19 and 22 (DeVries et al and Labana et al) consider some aspects of network morphology and fracture; papers 20 and 21 (Sperling et al and Frisch et al) present experimental evidence for inhomogeneities formed in interpenetrating polymer networks. Where possible, the discussions following the Symposium have been included in these Proceedings.

Most of the papers in this volume were presented at a Symposium on "Highly Crosslinked Polymer Networks," held in Chicago, September 14 and 15, 1970, at the 160th National Meeting of the American Chemical Society. The program was sponsored by the Division of Polymer Chemistry.

The editors take pleasure in acknowledging the Ford Motor Company and Dr. W. J. Burlant for their interest in, and support of, this project. We are indebted to the authors for their contributions and cooperation and to Dr. H. VanOene for presenting the manuscript of Dr. K. Dušek who was unable to attend the meeting. Finally, we wish to express our appreciation to Miss A. Oslanci for her secretarial assistance and the Polymer Division of the American Chemical Society for sponsoring the Symposium.

A. J. Chompff

S. Newman

Dearborn, Michigan
July 9, 1971

CONTRIBUTORS

P. J. Blatz, Department of Chemical Engineering, Northwestern University, Evanston, Illinois

J. J. Bikerman, 15810 Van Aken Blvd., Cleveland, Ohio

R. Brown, Research Assistant, University of Utah, Salt Lake City, Utah

T. Y. Chen, Department of Chemical Engineering, University of California, Berkeley, California

W. Chen, Department of Aerospace Engineering and Mechanics, University of Minnesota, Minneapolis, Minnesota

A. J. Chompff, Polymer Science Department, Scientific Research Staff, Ford Motor Company, Dearborn, Michigan

C. Chou, Polymer Research Institute and Department of Chemistry, University of Massachusetts, Amherst, Massachusetts

E. H. Cirlin, Science Center, North American Rockwell, Thousand Oaks, California

R. E. Cuthrell, Sandia Laboratories, Albuquerque, New Mexico

K. L. DeVries, Professor of Mechanical Engineering, University of Utah, Salt Lake City, Utah

K. Dušek, Institute of Macromolecular Chemistry, Czechoslovak Academy of Sciences, Prague, Czechoslovakia

S. F. Edwards, Department of Theoretical Physics, The Schuster Laboratory, University of Manchester, Manchester, England

R. J. Farris, College of Engineering, University of Utah, Salt Lake City, Utah

H. L. Frisch, Chemistry Department, State University of New York at Albany, Albany, New York

K. C. Frisch, Polymer Institute, University of Detroit, Detroit, Michigan

M. Fukuda, Central Research Laboratory, Denki Kagaku Kogyo Company, Tokyo, Japan

H. M. Gebhard, Science Center, North American Rockwell, Thousand Oaks, California

M. Gordon, University of Essex, Colchester, England

C. C. Hsiao, Department of Aerospace Engineering and Mechanics, University of Minnesota, Minneapolis, Minnesota

V. Huelck, Materials Research Center, Lehigh University, Bethlehem, Pennsylvania

F. N. Kelley, Air Force Rocket Propulsion Laboratory, Edwards, California

D. Klempner, Polymer Science Department, University of Massachusetts, Amherst, Massachusetts

T. Kotani, Japan Synthetic Rubber Company

T. K. Kwei, Bell Telephone Laboratories, Murray Hill, New Jersey

S. S. Labana, Polymer Science Department, Scientific Research Staff, Ford Motor Company, Dearborn, Michigan

R. F. Landel, Jet Propulsion Laboratory, California Institute of Technology, Pasadena, California

G. M. Lederle, IBM Corporation, Boulder, Colorado

S. Newman, Polymer Science Department, Scientific Research Staff, Ford Motor Company, Dearborn, Michigan

C. Picot, Centre des Recherches Macromoleculaire, Strasbourg, France

W. Prins, Department of Chemistry, Syracuse University, Syracuse, New York

Z. Rigbi, Department of Mechanics, Technion - Israel Institute of Technology, Haifa, Israel

M. Shen, Department of Chemical Engineering, University of California, Berkeley, California

L. H. Sperling, Materials Research Center, Lehigh University, Bethlehem, Pennsylvania

R. S. Stein, Goesmann Laboratory, Polymer Research Institute, University of Massachusetts, Amherst, Massachusetts

S. S. Sternstein, Materials Division, Rensselaer Polytechnic Institute, Troy, New York

D. A. Thomas, Materials Research Center, Lehigh University, Bethlehem, Pennsylvania

J. J. van Aartsen, AKZO Research Laboratories, Arnhem, The Netherlands

T. C. Ward, Virginia Polytechnic Institute, Virginia

R. S. Whitney, University of Essex, Colchester, England

M. L. Williams, University of Utah, Salt Lake City, Utah

CONTENTS

CHEMICAL AND PHYSICAL ASPECTS OF THE THREE STAGES IN FORMING POLYMER NETWORKS

M. Gordon, T. C. Ward* and R. S. Whitney

University of Essex, Colchester, England

*Virginia Polytechnic Institute, Virginia 24061

SUMMARY

Features common to all kinds of network-forming processes are discussed in chronological order along the three stages into which such processes can generally be divided: i) processes without diffusion control, ii) with selective diffusion control of some component steps and iii) with diffusion control of all steps. Apart from the need to study processes in order to understand the resulting structures, and hence their properties, much is to be learnt from chemical kinetics about the properties of the systems in which they occur, e.g. about local molecular mobilities. For instance, polyaddition reactions may show a sudden gel ('Trommsdorf') effect due to the immobilisation of large radicals at the gel point, while polycondensations pass through this point without effects due to diffusion control becoming apparent. The critically branched state of matter (near a gel point) is connected with life and its processes, and is generally rich in quantitative information about network structure and properties. Chemical and physical evidence runs counter to the suggestion that bulk networks tend to shrink in the course of their formation. Curing and aging processes in highly cross-linked systems (stage iii) is governed by physics rather than chemistry, and can be treated in terms of the WLF transform.

1. INTRODUCTION

Properties of network polymers depend on their structure, and therefore on the processes whereby they were synthesised. The study of processes, structure, and properties is best pursued together chronologically along the three stages into which the bond-forming processes are divided[1]: I) processes without diffusion control, II) with selective diffusion control of some component steps and III) with diffusion control of all steps. In polyfunctional additions, I) and II) are often sharply separated by the gel point, but polycondensates made from small monomers remain free from diffusion control until well beyond that point. We describe some models for fitting processes, structures and properties characteristic of the three stages.

2. GENERALISATION OF RING-CHAIN COMPETITION KINETICS TO EMBRACE THE POST-GEL REGION: SMOOTH CONTINUATION OF THE RATE CURVE THROUGH THE GEL POINT.

The accepted model of a gel for purposes of sol-gel analysis or elasticity parameters[2] is the ring-free infinite tree-like molecule. It is successful, even though such a molecule could not be packed[3] into three-dimensional space after the gel point, i.e. when its relative conversion $\alpha/\alpha_c > 1$. (We shall denote this relative conversion by γ^* and refer to it loosely as the cross-linking index. In rubber elasticity theory, our γ^* coincides with the cross-linking index γ for long homodisperse primary chains.) The inability to pack into space arises because, after the gel point, the number of units at a distance of r links from a given unit in the gel would be proportional to $(\gamma^*)^r$ (with $\gamma^* > 1$ after the gel point), while the space available for accommodating them is proportional to r^2: but $(\gamma^*)^r/r^2 \to \infty$. Of course, real gels have rings, and in the melt polyesterification of adipic acid/pentaerythritol (AA/PE), e.g., about one link in ten has been formed intramolecularly already in the sol by the time the gel point is reached.[3] Thereafter, the original model rate-equations for ring-chain competition kinetics,[3,4] cited as equation A1 (with $y = 1$) in Appendix A, fail to apply, because they lead to a spurious divergence of the cyclisation rate at the gel point.[3] It will now be shown that the rate equations can be suitably generalised (by putting $y \neq 1$), so that the rate curve is unaffected up to the gel point but then passes smoothly through the gel point without divergence. The calculation, and the excellent fit (fig.1) of the generalised equations to the data on AA/PE, are taken from the work with G.R. Scantlebury[5], but the theory in his thesis has been reinterpreted.

The ring-chain competition model (Appendix A) postulates one intermolecular rate constant k_u which applies to any pair of suitable free functionalities located on separate molecules, and one intramolecular rate constant k_σ. Any suitable pair of free functionalities, located on the same molecule, reacts with rate proportional to $k_\sigma z^{-3/2}$. Here z is the number of atoms in the ring formed in the reaction, and the factor $z^{-3/2}$ rests on the assumption of Gaussian statistics for any subchain of length z. In general, z ranges over all possible multiples of ν, where ν is the size of the smallest possible ring for the system under consideration, i.e. $\nu = 11$ for AA/PE. The divergence of the original equations ($\gamma^* > 1$, $y = 1$) arises because, for a given free gel functionality, the number of partner functionalities available for cyclisation at a distance of z links within the gel, is proportional to $(\gamma^*)^z$, while the rate "constant" of cyclisation varies as $z^{-3/2}$: but $(\gamma^*)^z z^{-3/2} \to \infty$ as $z \to \infty$. The generalisation in equation A1 of the original rate equations is required to correct the vast overcounting of gel-gel reaction steps (i.e. reaction between two functionalities located on the gel molecule). The existence of a gel molecule to some degree blurs the distinction between intermolecular and intramolecular reaction steps that occur within that molecule. Not only are such gel-gel steps included in the k_σ-term (and to such an extent that they cause this term to give a divergent rate), but they are also included in the k_α-term. In the region just after the gel point, it makes the best physical sense to accept the contribution of gel-gel steps to the k_α-term completely, and to cancel out completely

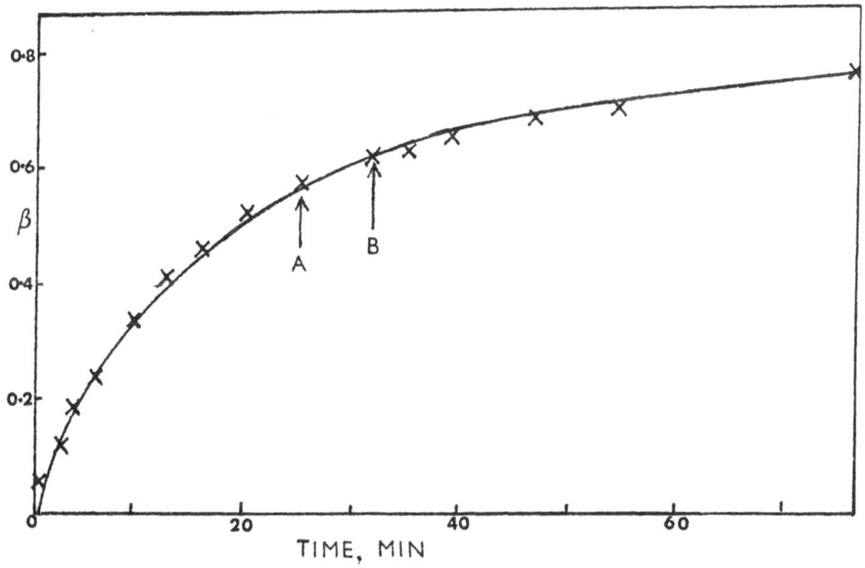

Fig. 1

Fit of rate curve A1 to measurements[5] on stoichiometric AA/PE melt at 175°C, with parameters listed in Appendix A. β denotes total fractional conversion of functionalities ($\beta = \alpha + \sigma$. where α is the fractional conversion by intermolecular, and σ by intramolecular reactions).

their contribution to the k_σ-term, thus removing the divergence. This amounts to treating gel-gel steps as bimolecular collisions governed merely by the law of mass action, and not in terms of Gaussian statistics of the sub-chain linking the two functionalities concerned.

The automatic inclusion of the gel-gel steps in the k_α-term arises as follows. In accord with the mass law, this term treats the chance of a given functionality undergoing reaction as proportional to the total concentration of partner functionalities available for reaction in the system as a whole. If, and only if, the given functionality resides on the gel, a finite fraction of the number of partner functionalities in the system as a whole resides on the same molecule as the given functionality (in this way intramolecular steps creep into the rate term designed for intermolecular steps). The physical plausibility of treating the gel-gel steps in the region just after the gel-point as kinetically equivalent to intermolecular steps may be supported as follows. In this region, the active network chains are very long, and their mean length diverges at the gel point.[2] This represents the limiting case considered by James and Guth "in which all bondable elements take on their positions essentially independently of each other. This limit is approached by a network formed of very long molecular chains bonded together at relatively few points. Each element of such a network will be surrounded for the most part by elements of other chains - elements which can be reached only by a relatively long path through the network. The relation of these chain elements is then almost the same as if they were separate molecules in an ordinary liquid."[8]

The excellent fit in fig. 1 authenticates the theory which, having overcome the divergence trouble at the gel point, provides an accurate model for studying real systems containing rings not only in the sol but in the loose initial gel also. The points shown in fig. 1 give a quite inadequate representation of the experimental accuracy actually attained by means of following the reaction progress with continuous automatic recording.[3] The four parameters[3] fitted are: a (negligible) substitution parameter N_1 for ester groups acting on -COOH in the same repeat unit, a (non-negligible) N_2 parameter for similar effects on -OH, a dimensionless ring-chain competition parameter λ, and a negligible parameter allowing for changes in concentration due to shrinkage during the reaction. The two significant parameters N_2 and λ are amply secured, not only by optimisation of rate curves over five mole ratios AA/PE, but by fitting quite independently the five gel points observed.[3] In addition, the parameter λ agrees well with independent calculations from random-flight statistics.[3] The gel point of the classical random model is shown by arrow A, the gel point observed, and calculated using the substitution and cyclisation parameters mentioned, by arrow B.

3. SOME PHYSICAL ASPECTS OF CRITICALLY BRANCHED POLYMERS

Critically branched systems (i.e. those close to their gel point), besides being useful as stepping stones in understanding the more highly cross-linked networks to which they may lead, are worth study in their own right. This is especially true in Biology. Life occurs in this state of matter, which affords the necessary compromise between a matrix of sufficiently solid consistency to ensure permanence, and fast rates of molecular diffusion through this matrix. Life processes, moreover, generally require passage of a system through its gel point; for instance, fertilisation or the division of a nucleus require the lysis or formation of jelly coats or membranes. The white of a raw egg is critically branched, though presumably by virtue of H-bonds rather than primary valence cross-links. Fig. 2 shows for a model system, how critical the transition is and how well the process is understood in polymer science. The viscosity η and Young's modulus E are plotted for polyesterified decamethylene glycol/benzene 1,3,5 triacetic acid (DMG/BTA). The change from a liquid of the viscosity of castor oil to a gel of the modulus of a jelly-fish is compressed to values of γ^* from 0.99 to 1.01 or 1.02. Nevertheless, plots a and d are one-parameter fittings to theories, with the single parameter in each case in principle calculable; moreover, the absolute position β_c of the transition is predicted theoretically within experimental error. In Biology, the critically branched state is less well understood. A typical statement says about the nucleoplasm that it 'may be gel-like but generally has a fluid consistency'.[6] A perspicacious biochemist, however, remarked in 1963: 'there seems to be a subtle difference between gunk and goo'.[7] The difference is reasonably approximated by equations 1 and B7 (or B8) respectively for our model system.

Each of the four runs plotted in fig. 2 (cf. also fig. 3) was carried out on 3 mg of polycondensate using the magnetic microsphere method [9,10] (see also [11]) for absolute measurements of η and E. Fuller details will be published elsewhere. The observed gel point ($\alpha_c = 0.722 \pm 0.006$ for five experiments) and the chemical kinetics for DMG/BTA agree within experimental error with the classical random model ($\alpha_c = 2^{-1/2} = 0.707$), when allowance is made for the small amount of cyclisation, which in turn agrees with that predicted from Gaussian statistics. The gel point can be measured relatively (in terms of γ^*) to within 0.001 by arrest of the moving sphere, and for each of the three runs styled O,Δ, X has been adjusted to a single value within this small range in fitting both theoretical equations for η and E. The (γ^*)-scale is based on scaling the time measurements to independent measurements by the steam pressure technique[3] of the kinetic rate curve (i.e. third order kinetics), using the observed gel time as the fixed point for the scaling. The error in the (γ^*)-scale factor is unlikely to exceed 3%. (The observed gel times of the runs were

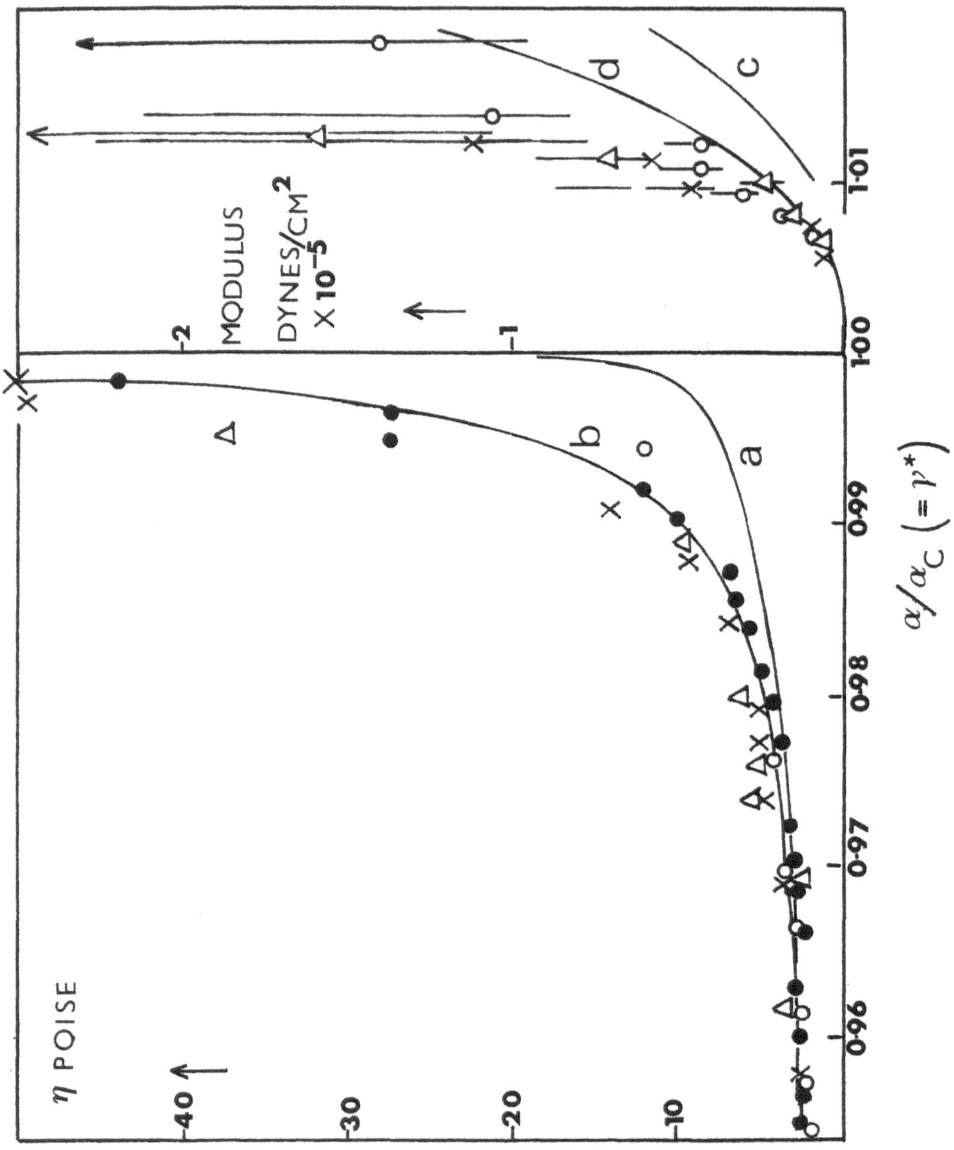

Fig. 2

Viscosity and Young's Modulus for four runs on Stoichiometric DMG/BTA melt at 145°C. See text for details.

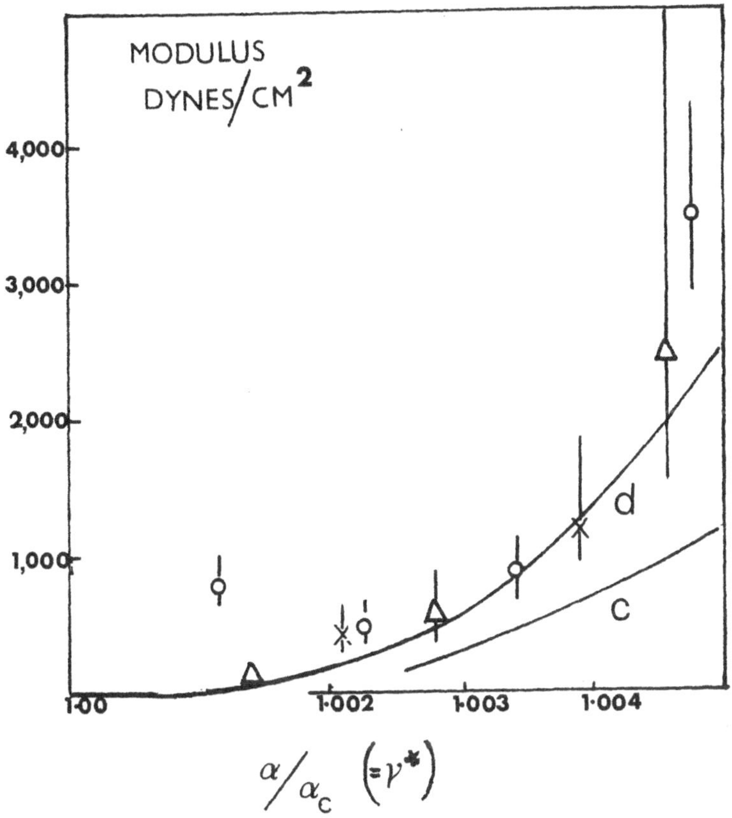

FIG. 3: Enlarged view of initial modulus curve (compare Fig. 2)

Δ = 79.8, O = 81.0, X = 81.0, O = 88.5 min). Estimated error ranges in E are indicated by vertical strokes. The effects due to the small concentration of intramolecular links (cycles) on η and E should be compensated to an excellent approximation since we are comparing the measurements with values calculated for the random ring-free model at corresponding values of $\gamma^* = \alpha/\alpha_c$ (not of α).

3.1 The Viscosity of Critically branched DMG/BTA Polyester

The superposed viscosity measurements of four runs are compared with two such calculated curves. Curve a is a single parameter fitting to $\eta = \text{const.} \langle R^2 \rangle_w$ with const. = 0.133, where the weight-average mean-square radius was calculated assuming Gaussian sub-chain statistics.[12] The equation used is shown in Appendix B. This is appropriate for the free-draining model,[13] and might be expected to fit well at low η, and it does so. However, the measured viscosities rise above the calculated in the range $0.98 < \alpha/\alpha_c < 1$, which could reflect intermolecular interactions or entanglements. The fit to curve b, which is within experimental error, is of much interest. It is a single-parameter fitting to η = const. $\bar{M}_w$ with const. = 1/3000 (See Appendix B). In linear homodisperse linear melts the law $\eta \propto M$ is known empirically to hold for low M, i.e. in the absence of entanglements. For heterodisperse linear polymers, M has often empirically been replaced by $\bar{M}_w$. Long ago, Gordon and Roe[14] fitted a crude dilute-solution theory to their measurements of viscosities of critically branched methacrylate/ethylene dimethacrylate solutions of c ~ 0.01 by weight. They found that η could be treated as a function of $\bar{M}_w$ there also (approximately $\eta \propto \bar{M}_w^{0.8}$). If the relation $\eta \propto \bar{M}_w$ of plot b is found widely applicable to bulk systems - i.e. where coil dimensions are undisturbed and cyclisation is less prevalent - it will be of value for the characterisation of critically branched systems generally.

3.2 Young's Modulus of Critically Branched DMG/BTA Polyester

The system DMG/BTA is thought to furnish the best-characterised synthetic branched and network polymers at present available. It is based on pure crystalline monomers, and kinetic as well as more sensitive statistical tests show that essentially a single 'random' rate process furnishes the intermolecular links, while the small amount of cyclisation can be allowed for to a good approximation. It is therefore of value as a more reliable material than those used previously[15] to test the network theory of the 'tree-like' gel model[2] against measurements of E, and particularly to estimate the controversial 'front factor' in the classical equation of state of rubbers. The plots c and d in figs. 2 and 3 are based on the equations (cf. Dobson and Gordon[2])

$$E = 3g\ RT\ \nu_e/V_{mol}$$

$$= 3g\ \frac{RT}{V_{mol}} \times \frac{3}{5}\left[2^{-\frac{1}{2}}\gamma^*(1-v)\right]^3 \qquad (1)$$

$$v = 1 - 2^{-\frac{1}{2}}\gamma^* + 2^{-\frac{1}{2}}\gamma^*(1 - 2^{-\frac{1}{2}}\gamma^* + 2^{-\frac{1}{2}}\gamma^* v)^2 \qquad (2)$$

for $g = \frac{1}{2}$ (front factor of James and Guth[8], plot c) and g = 1 (front factor of Flory, of Wall and of Kuhn). V_{mol} is the volume of polycondensate per mole repeat units. The elasticity measurements cover a 10^3-fold range of E, and had to be split over two diagrams; fig. 2 shows the E values for more highly cross-linked networks, while the interesting initial region is shown enlarged in fig. 3. It is clear that the results strongly support the value g = 1. Allowing for estimated errors in E and in γ^* we should be able to measure g absolutely to within ± 10%. The fit of eq. 1 and 2 to the initial E values using the classical value g = 1 constitutes a zero-parameter confirmation of the tree-like-gel theory, contributed by statistical mechanics to the science of polymers. It is seen in fig. 2 that in the range $1.01 < \gamma^* < 1.02$ the apparent front-factor rises to somewhere in the region of 2, and such a rise is traditionally explained in terms of physical cross-links due to entanglements.[16] The rationale of our endeavour to measure the behaviour of E as $\gamma^* \to 1 + \varepsilon$ lies in the theoretical prediction[2,17] that effects due to entanglements and to non-Gaussian chain statistics should be asymptotically eliminated. A comment on slow relaxations is added at the end of §4.

Our satisfaction is marred in one respect: we believe that the theory of Imai[18] constitutes a necessary refinement of the classical elasticity theory to allow for coupling of thermal motions between neighbour chain segments (tied to the same junction point). This theory leads incidentally to a front factor different from unity, originally given as 1.82, but now revised to the much lower value 0.55 on grounds stated below. This lies too close to James and Guth's value of ~ 0.5 to be experimentally distinguished with certainty at present. James and Guth derived their value from a totally different theory, which imputes to networks the tendency to shrink during the course of formation in the bulk state. Flory has, we think rightly, rejected the statistical argument underlying the assumption of such shrinkage. The reality or otherwise of such a tendency is of importance to the general theme of this symposium, and fig. 2 and 3 clearly do not support this theory (nor, unfortunately, the value g = 0.54 which is thought to be theoretically sound!).

The following two corrections are required in the original front factor calculation of Imai and Gordon.[18] First, a factor $(1 - (3f/80))$, previously omitted from g, is just significant: since here the functionality f of junction points is 3, it amounts to a reduction of g by a factor 71/80 = 0.8875. More seriously, a further reduction by a factor of 1/3 is required by the realisation that when different equations of state $E = E(\lambda)$ are compared with the classical one, the front factor g must be defined by

$$g = (RT/V_{mol})^{-1} \lim_{\lambda \to 1} (E/\nu_e(\lambda - \lambda^{-2})) \tag{3}$$

The limit is evaluated by de l'Hospital's rule. The factor arising from

$$1/\left[d(\lambda - \lambda^{-2})/d\lambda\right]_{\lambda=1} = 1/3 \tag{4}$$

was regrettably omitted by Imai and Gordon.[18] Thus their theory leads to the revised value g = 0.55.

4. A KINETIC MODEL FOR A STAGE II POST-GEL POLYADDITION REACTION

The second stage of network synthesis, in which some reactions fall under the regime of diffusion control, leads to systems whose long-range structures are much less easily amenable to statistical analysis than those of the first stage. But the chemical kinetics may still be tractable. Methyl methacrylate

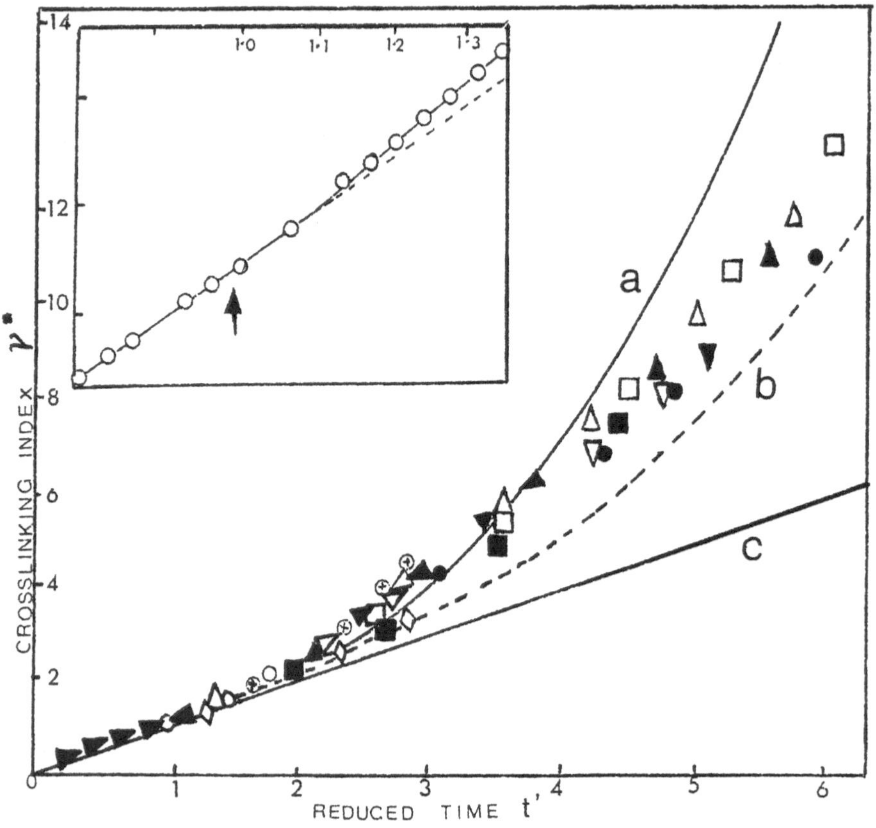

Fig. 4

Superposition of reduced rate curve (eq. 5,6) for copolymerisatic of methyl methacrylate with ethylene glycol dimethacrylate or polyethylene glycol fumarate. Inset: enlarged view of typical results near gel point - ordinate scale arbitrary. a: eq. 5,6 for disproportionation; b: eq. C5, y = 3, for radical combination; c: model without diffusion control. Arrow: observed gel point.

(MMA), when copolymerised with ethylene glycol dimethacrylate (EDMA)[19] or polyethylene fumarate (PEF),[1] but not with styrene,[1] enters stage II sharply at the gel point. A plausible model proposed by Gordon and Roe[17] leads to a parameter-free rate equation (a, fig. 4), which fits extensive measurements on MMA/EDMA and MMA/PEF very well over the range $0 < \gamma^* < 6$. Fig. 4 contains the same data as presented earlier, but the fit to the theory is substantially improved merely as a result of correcting a slip in the value of the statistical factor of the rate constant for termination; the revised derivation is shown in Appendix C. The model postulates that at least one of the two radicals in a termination step after the gel point must belong to a primary chain free from cross-links to other primary chains, as otherwise neither partner is small enough to diffuse sufficiently freely through the gel network. Assuming disproportionation of radicals, the corrected reduced rate curve of $\gamma^*(= \beta/\beta_c)$ versus t^* $(=t/t_{crit})$ reads:

$$t^* = \gamma^* \quad (\gamma \leqslant 1) \tag{5}$$

$$t^* = 1 + 2\left[\ln(\gamma^* + (\gamma^{*2} + 4)^{1/2}/(1 + 5^{1/2})\right] \quad (\gamma^* \geqslant 1) \tag{6}$$

The reader is referred to Gordon and McMillan's[1] fig.2 for details of the various runs superposed in fig. 4. The original fuller range of data showed that the superposition was successful over molar feed ratios $0.2 \leqslant R \leqslant 1.75$ of the unsaturation belonging to the two comonomers (either MMA/EDMA or MMA/PEF). Fig. 4 includes runs ranging in critical conversion β_c from 0.48 to 7.45%.

At the low absolute conversions β (which incidentally include a substantial proportion of intramolecular steps), the reduced curve is linear with unit slope up to the gel time $t^* = 1$. The inset gives a better indication of the experimental accuracy of the dilatometric technique than the main figure 4, and shows the reality of the kink observed in close proximity to the experimental gel point. The ratio of the two slopes at the kink is predicted by eq. 5 and 6 to be 1.12 $(= 5^{1/2}/2)$. The average jump in 28 runs covering widely different conditions was found experimentally as 1.23 ± 0.07. The dotted curve drawn in fig. 4 represents the same model assuming radicals to combine (C5, Appendix C) rather than to disproportionate. This is shown merely for comparison, for it neither fits the data nor independent information on the nature of termination in MMA polymerisation. It is clear from Gordon and McMillan's figure 4 that the systematic fall in rate after $\gamma^* \sim 6$ is due to onset of more general diffusion control (especially of the propagation steps), which leads to stage III (studied also by Barrett and Gordon[20]) and finally to premature arrest of the network synthesis before the unsaturation is exhausted. There is therefore no reason to associate with an incursion of radical combination the fall (beyond $\gamma^* \sim 6$) of the reaction rate below that calculated for radical disproportionation.

Fig. 1 shows that networks synthesised by polycondensation of small molecules are free from diffusion control in the initial ('critically branched') gel; fig. 4 shows that large molecules are hindered in their motions through the initial polyaddition gel synthesised from cross-linked long primary chains. This comparison strengthens the speculation[15] that slow relaxations,[21] which bedevil the approach to mechanical equilibrium in critically branched polyaddition gels, or cross-linked Marlex polyethylene as found by Pechhold (see[15]), are connected with the large $\bar{P}_n$ of the sol fraction. The absence of slow relaxations is important for the interpretation given to measurements of E in terms of equilibrium moduli of polycondensates (fig. 2). Experimentally, for $\gamma^* > 1$, the microsphere attains its presumed equilibrium displacement within < 5 sec of switching on the magnetic field, and no further displacement is discernible over 10 seconds, but the existence of much slower relaxations could not be excluded on experimental grounds. Incidentally, too large a magnetic force must be avoided as it will cause the sphere to tear the weak gel. Displacements up to 0.03d have been found safe, where d is the sphere diameter, typically 7×10^{-3} cm.

5. THE WLF SUPERPOSITION PRINCIPLE FOR STAGE III CROSS-LINKING REACTIONS

In the ultimate, third stage of network synthesis, chemical as well as physical processes are dominated by relaxation of short chain-segments. The long-range structural statistics loose their importance, while they govern networks in stage I, where they are mathematically tractable (fig. 1,2,3), and stage II, where relaxation spectra are difficult to treat and equilibrium is difficult to attain. When cross-link formation has raised the glass transition temperature T_g to the vicinity of the reaction temperature, all component chemical steps of further reactions are likely to be controlled by segmental diffusion rather than by chemistry. Their rates can then be treated[22] in terms of WLF superposition.[23] The basic equation is

$$\log (dT_g/dt) = K - (900/(51.6 + \theta)) \qquad (7)$$

(t = time, K = const., $\theta \equiv T - T_g$). Two simple cases have been verified by the ball rebound technique using systems widely differing in chemistry. In the <u>isoelastic</u>[24] case, θ is kept constant (so that T varies). Then T_g should clearly increase linearly with time. Fig. 5 shows this for a UF resin. The <u>isothermal</u> case leads to

$$\ln\text{-}d\theta/dt \ (= \ln dT_g/dt) = K - (900/(51.6 + \theta)) \qquad (8)$$

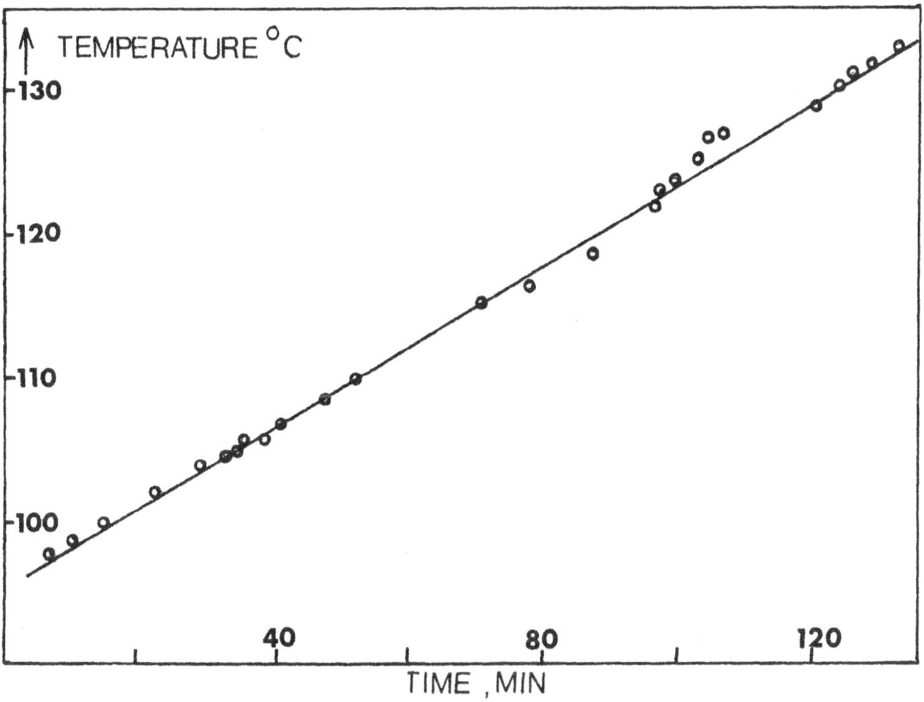

Fig. 5

Typical isoelastic rate curve 24 for urea formaldehyde resin. The temperature required for maintaining constant (74%) energy absorption on ball impact rises linearly with time.

In fig. 6, ball rebound data by Simpson[22] on a silicone varnish and a polyester enamel are plotted according to eq. 3. As predicted, straight lines of slope about 900 are observed, except for some curvature due to a minor superposed reaction component for the silicone, probably non-diffusion controlled oxidation, after the main cross-linking process has been 'frozen out' by the rise of T_g. Many methods for measuring T_g are now available, and the application of the simple eq. 7 or 8 is worth exploring for any industrial resin system whose rates of cure or ageing are thought to be interesting.

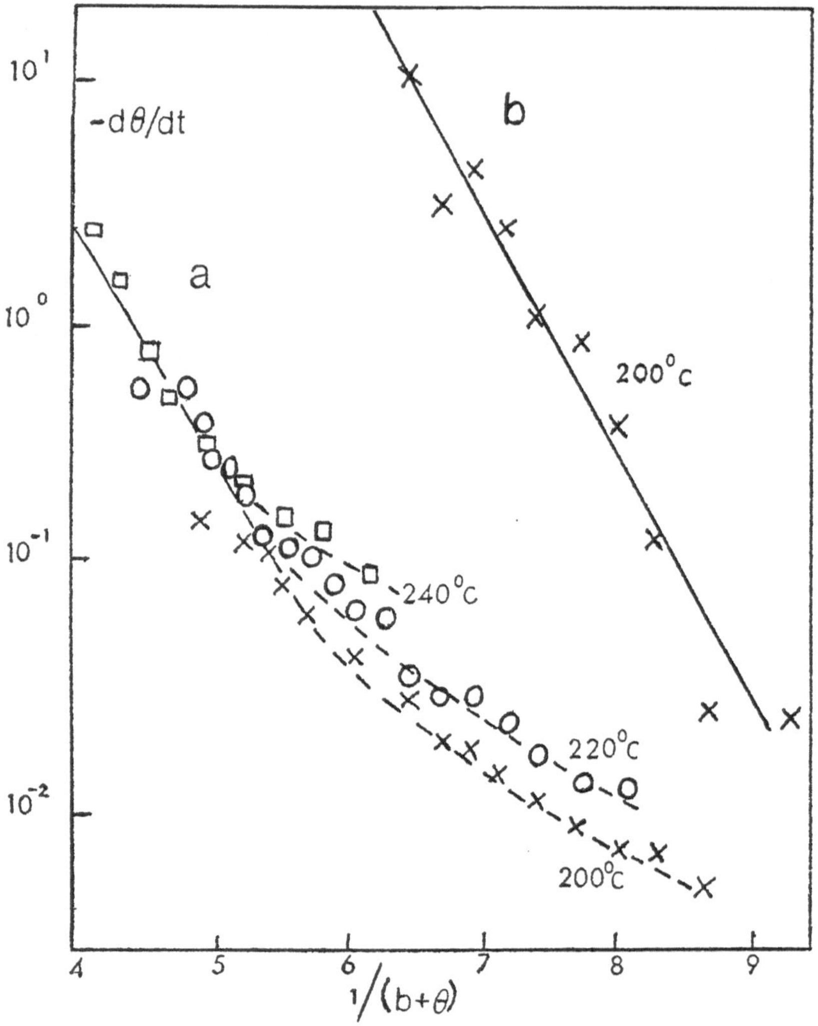

Fig. 6

Rate of cure of silicone (a) and polyester wire enamel (b)22. Slopes of lines drawn: a -987; b -927 (°C); theory: 900°C. b = 51.6, a WLF parameter.

Acknowledgments

T.C.W. thanks the British Petroleum Ltd. for a grant and R.S.W. the Dunlop Company Ltd. and the Science Research Council for a CAPS award.

APPENDIX A

The rate equations for a binary alternating copolymerisation between an f_1-functional unit of type 1 (here AA) and an f_2-functional unit of type 2 (here PE) were derived by Gordon and Scantlebury,[3] in the notation defined in their paper, as the special case y = 1 of the following $[(f_1+1)f_1 + (f_2+1)f_2 + 4]/2$ equations for admissible values of u,v,i,j:

$$-d\,{}_uP_{ij}/dt = k_\alpha\{N_u^{i+j}k\,{}_uP_{i,j} - N_u^{i+j-1}(k+1)\,{}_uP_{i-1,j}\}F_{ov}^{\omega}(N_v,N_v,1)$$

$$F_{01}^{\omega}C_1^0C_v^0 + k_\sigma\{N_u^{i+j}k\,{}_uP_{i,j} - N_u^{i+j-1}(k+1)\,{}_uP_{i,j-1}\}N_vF_{1v}^{\omega}(N_v,N_v,1)$$

$$F_{01}^{\omega}C_1^0\; i\sum_x(F_{11}^{\alpha}F_{12}^{\alpha})^{x-1}(\nu x)^{-3/2}\;y^x \qquad \ldots \text{(A1)}$$

Here u and v represent either the pair of integers 1,2 or 2,1 respectively. (An obvious error in sign of an exponent has been corrected here. Also, all C-terms represent molar concentrations of u-type repeat units, not of functionalities, as stated in error.[3]) The factor $F_{01}^{\omega}C_1^0$ on the right takes care of the catalysis by -COOH groups in AA/PE, but otherwise the equations are quite general and include exactly a linear first-shell substitution effect (through the N_u and N_v-terms), as well as an excellent approximation to the ring-chain competition situation, through the k_σ-term.

The new factor y^x at the end of A1 is introduced as the necessary weighting for reducing to zero the gel-gel reaction contributions to the k_σ-term. The summation over x deals with contributions of rings of size 2x repeat units (of which x are of type 1 and x of type 2 because of the alternation of units). The term y^x is the probability p_s that all $(f_1+f_2-4)x$ functionalities immediately attached to these 2x units lead at most to finite numbers of repeat units (i.e. submolecules which are loose ends). Thus p_s is also the probability that the ring being formed is in the sol fraction, and accordingly gel-gel cyclisation is excluded from the k_σ-term.

We recall that the extinction probability v_i, i.e. the chance that a link to an i-type unit leads to at most a finite submolecule, is found easily from the equations

$$v_1 = F_{11}(v_1, v_2); \quad v_2 = F_{12}(v_1, v_2) \quad \ldots \text{(A2)}$$

(See Butler, Gordon and Malcolm's[12] equation 56). In absence of substitution effects, the fates of all functionalities attached to a ring are independent. In this case, and restricting ourselves to a stoichiometric mixture of 1-type and 2-type units, let α be the fractional conversion of functionalities by intermolecular reactions. Then we have

$$y = (1 - \alpha + \alpha v_2)^{f_1-2} \; (1 - \alpha + \alpha v_1)^{f_2-2} \quad \ldots \quad (A3)$$

This is the equation (with $f_1 = 2$, $f_2 = 4$) used for computing the curve in fig. 1 by integration of A1. The first-shell substitution effect between functionalities attached to the same PE unit in a ring were, for the purposes of calculating y, neglected as a good approximation, though they could be taken into account, using the exact equation

$$y = \frac{F_{1u}^{\alpha}(v_2, v_2, 1)}{F_{1u}^{\alpha}(1,1,1)} \; \frac{F_{1v}^{\alpha}(v_1, v_1, 1)}{F_{1v}^{\alpha}(1,1,1)} \quad \ldots \quad (A4)$$

Of course, up to the gel point the extinction probabilities $v_1 = v_2 = 1$, so that $y = 1$, and the generalisation embodied through the term in y has no effect on A1 before the gel point. The directly measured parameters used in fitting fig. 1 were $C_1^0 = 2C_2^0 = 4.67 \times 10^{-3}$ mol/gm; and the almost negligible shrinkage parameter[3] $k' = 0.45$. The parameters adjusted (except for k_α) by optimising over different kinds of experiments (see main text) were $N_1 = 1.5$, $N_2 = 0.9$ (almost negligible effect), $k_\alpha = 569$ gm^2 (equivalents COOH)$^{-2}$ min^{-1} and $k_\sigma = 36.3$ gm (equivalents COOH)$^{-1}$ min^{-1}.

APPENDIX B

THEORETICAL EQUATIONS FOR $\langle R^2 \rangle_w$ (PLOT a, fig.2), AND $\bar{M}_w$ (PLOT b)

While a three-component vectorial link probability generating function is strictly required for DMG/BTA, and was used (see below) for calculating $\bar{M}_w$ exactly (fig. 2, plot b), the following two-component l.p.g.f. gives a quite negligible error in $\langle R^2 \rangle_w$ (arising from the neglect of mass loss due to the liberation of water):

$$F_o(\theta) = \left[(1 - \alpha + \alpha\theta_2)^2, \; (1 - \alpha + \alpha\theta_1)^3\right] \qquad (B1)$$

and hence the pgf for the first generation of a rooted tree-like molecule

$$F_1(\theta) = \left[(1 - \alpha + \alpha\theta), \; (1 - \alpha + \alpha\theta)^2\right] \qquad (B2)$$

Then

$$u_1 \equiv \theta_1(1 - \alpha + \alpha u_2) \qquad (B3)$$

$$u_2 = \theta_2(1 - \alpha + u_1)^2 \tag{B4}$$

The notation is that of Butler, Gordon and Malcolm.[12] The reader unfamiliar with the power of the 'cascade' approach[25] to polymer statistics will gain a better insight into its fundamentals from a recent paper[26] (cf. also [27]) which generalises the earlier formulae to embrace, e.g., the particle scattering functions through the notion of a trail-weighting function, which is of wide utility.[28]

The relevant equation for $\langle R^2\rangle_w$ reads:

$$\langle R^2\rangle_w = \frac{\bar{P}_n - 1}{\bar{P}_n} \int_0^1 d\theta \left[\theta_n u_j^n(\underset{\sim}{\theta}) b^j c^j\right]^2_{(\theta_1=\theta_2\equiv\theta)} / \theta \tag{B5}$$

with summation over repeated indices, and in the usual notation:

$$u_n^j \equiv \partial u_n / \partial \theta_j \tag{B6}$$

In the case of DMG/BTA, the root mean square distance b between two neighbouring repeat units replaces b^1 and b^2, and $c^i c^j$ must be equated to δ_{ij} (because a randomly chosen link has probability unity of bonding one DMG and one BTA unit). After some calculation, eq. B3, B4 and B5 lead to the equation plotted as a, fig. 2:

$$\langle R^2\rangle_w = \frac{\bar{P}_n - 1}{\bar{P}_n} b^2 \times \tag{B7}$$

$$\int_0^1 d\theta \frac{\theta(\alpha u_2^2 + 2\theta\alpha u_1(1-\alpha+\alpha u_1)(\theta u_2 + u_1)) + u_1 u_2}{1-2\alpha^2\theta^2\left[1-\alpha+\{1-2\alpha^2\theta^2(1-\alpha)-\left[1-4\theta^2\alpha^2+4\theta^2\alpha^3+4\theta^3\alpha^3+4\theta^3\alpha^4\right]^{\frac{1}{2}}\}(2\alpha^3\theta^3)^{-1}\right]}$$

with u_1 and u_2 found from B3 and B4 (with $\theta_1 = \theta_2 = \theta$).
The exact formula (eq. 46 of reference[12]) for $\bar{M}_w$, used in plot b, fig. 2, is

$$(1-2\alpha^2)\bar{M}_w = \left[M_1, M_2, M_3\right] \begin{bmatrix} \frac{1}{2}(5+\alpha-8\alpha^2), & 2(1-\alpha)(1+2\alpha), & 3(1-\alpha^2) \\ (1+2\alpha)/2, & 1+2\alpha^2 & 3\alpha \\ (1+\alpha)/2, & 2\alpha & 1+\alpha^2 \end{bmatrix} \begin{bmatrix} m_1 \\ m_2 \\ m_3 \end{bmatrix} \tag{B8}$$

The molecular weights are $M_1 = 9(=HO_{1/2})$, $M_2 = 156$ (= DMG-$2HO_{1/2}$), $M_3 = 225$(=BTA - $3HO_{1/2}$). The weight fractions of these units are:

$$m_1 = \frac{4n_2(1-\alpha)M_1}{4n_2(1-\alpha)M_1+M_2+(2M_3/3)} \tag{B9}$$

$$m_2 = \frac{M_2}{4n_2(1-\alpha)M_1+M_2+(2M_3/3)} \tag{B10}$$

$$m_3 = \frac{(2M_3\ /3)}{4n_2(1-\alpha)M_1+M_2+(2M_3/3)} \tag{B11}$$

where the number fraction n_2 of units of DMG - $2HO_{1/2}$ is

$$n_2 = 3/[12(1-\alpha) + 5] \tag{B12}$$

APPENDIX C

CORRECTION OF THE RATE EQUATION FOR MMA/EDMA OR MMA/PEF

We denote by c' the concentration of the large (cross-link-bearing) immobile radicals, but otherwise retain the notation of Gordon and Roe[19] whose original uncorrected equations will be denoted by GR. The mistake to be corrected stems from

$$I = k_t c(c+c') \tag{GR9}$$

which should read

$$I = k_t c(c+2c') \tag{C1}$$

Then their

$$c_o/c = S \tag{GR11}$$

becomes

$$c/c_o = S(S - 1)^{\frac{1}{2}} \tag{C2}$$

Similarly, letting z be unity for disproportionation and $\frac{1}{2}$ for combination, their

$$DP_c = DPn_o z/S \tag{GR18, 19}$$

becomes

$$DP_c = DPn_o z\left[S - (S-1)^{\frac{1}{2}}\right] \qquad (C3)$$

Following their derivation through step by step leads to

$$\gamma^* = y(S^2-1)^{1/2} \qquad (C4)$$

where y is 2 for disproportionation and 3 for combination, in lieu of

$$\gamma^* = y(S^2-1)2S \qquad (GR29, 30)$$

After integration this leads to the new and simpler rate law:

$$t^* = 1+y\ \ln\{\left[\gamma^*+(\gamma^{*2}+y^2)^{1/2}\right]/\left[1+(1+y^2)^{1/2}\right]\} \qquad (C5)$$

which for disproportionation (y=2) reduces to eq. 6.

REFERENCES

1. M. Gordon and I.D. McMillan, Makromol. Chem. 23, 188 (1957)
2. G.R. Dobson and M. Gordon, J. Chem. Phys. 43, 705 (1965)
3. M. Gordon and G.R. Scantlebury, J. Chem. Soc. 1 (1967)
4. M. Gordon and G.R. Scantlebury, J. Polymer Sci., C16, 3933 (1968)
5. G.R. Scantlebury, Ph.D. Thesis, Polycondensation Statistics, London University (1965)
6. C.R. Austin, Fertilisation, Prentice-Hall, London, 1965, p.1
7. E. Chargaff, The Origins of Prebiological Systems, Ed. S.W. Fox, Academic Press, New York and London, 1965, 169
8. H.M. James and E. Guth, J. Chem. Phys. 15, 669 (1947)
9. M. Gordon, S.C. Hunter, J.A. Love and T.C. Ward, Nature, (London), 217, 735 (1968)
10. S.C. Hunter, Proc. Edin. Math. Soc. 16, (ser. II), 57 (1968)
11. J. Brooks and H.P. Hale, Biochim. and Biophys. Acta, 32 237 (1959); H. Freundlich and W. Seifniz, Z. Phys. Chem. 104, 233 (1923)
12. D.S. Butler, M. Gordon and G.N. Malcolm, Proc. Roy. Soc. (London), A295, 29 (1966)
13. G. Langhammer, Physik der Kunststoffe, Eds. W. Holzmüller and K. Altenburg, Akademie Verlag, Berlin, 1961, p. 175.
14. M. Gordon and R.-J. Roe, J. Polymer Sci., 21, 39 (1956)
15. M. Gordon, Proc. Int. Rubber Conference, Moscow (1969) in press.
16. B. Meissner, I. Klier and S. Kuchařik, J. Polymer Sci., C16, 793 (1967)
17. M. Gordon, Plaste und Kautschuk, 12, 103 (1965)

18. S. Imai and M. Gordon, J. Chem. Phys. 50, 3889 (1969)
19. M. Gordon and R.-J. Roe, J. Polymer Sci., 21, 57 (1956)
20. R.M. Barrett and M. Gordon, The Physical Properties of Polymers, S.C.I. Monograph, No.5, 183 (1959)
21. J. Janáček and J.D. Ferry, Macromolecules 2, 379 (1969)
22. M. Gordon and W. Simpson, Polymer 2, 383 (1961)
23. M.L. Williams, R.F. Landel and J.D. Ferry, J. Amer. Chem. Soc. 77, 3701 (1955)
24. M. Gordon and B.M. Grieveson, J. Polymer Sci., 29, 9 (1958)
25. K. Dušek and W. Prins, Advances in Polymer Science, 6, 1 (1969)
26. K. Kajiwara, W. Burchard and M. Gordon, Brit. Polymer J., 2, 110 (1970)
27. M. Gordon, Colloquia Mathematica Soc. J. Bolyai (Budapest), 4, 511 (1969)
28. K. Kajiwara, J. Chem. Phys., to appear September (1970)

DISCUSSION

N. Langley (Dow-Corning, Midland): Could you amplify your remark that life processes are connected with the passage of a system through its gel point?

M. Gordon: This is exemplified by fertilization. A spermatozoon cannot swim through a jelly, such as that which surrounds a frog's egg. The spermatozoon carries its acrosome which bursts to release an enzyme which liquefies the jelly. Quite generally, the formation or disappearance of membranes and organelles which accompanies all important cytological events, clearly requires passage through a gel point. The connection of the gel point with the origin of life was appreciated fully by the poet Dante (Purg. 25, 50).

S. Prager (University of Minnesota): You said the chain-end correction concept would not work in your case and that you calculated ν_e in another way. Could you enlarge on that?

M. Gordon: The basic topological idea by Case and by Scanlan is that a network chain is active if at least three links issuing from each of the two terminal junction points lead to paths through the network which can be continued to the surface of the specimen. In the statistical theory, the surface is projected to infinity, and the extinction probability of a link is used in the calculation. The basic theory is reviewed in reference 25 of our paper, and extended in a forthcoming paper, Coll. Czech. Communications, Gordon, Kuchárik and Ward, late 1970.

ON THE THERMOSTATIC BEHAVIOR OF ELASTOMERS

Paul J. Blatz

Department of Chemical Engineering

Northwestern University, Evanston, Ill. 60201

SUMMARY

When a thermally expansible homogeneous isotropic elastic continuum is deformed isothermally, the mechanical work done on the body is partly stored as internal energy, partly converted to heat. In the range of strain up to about 10 percent (which is typical of natural rubber), the internal energy storage exceeds the work done so that heat must be added to the sample. Beyond the strain of 10 percent, heat is increasingly liberated. An equivalent statement conveys the information that, if the sample be stretched adiabatically, there is an initial temperature drop, then subsequent rise.

In the limit of large strain, the ratio of internal energy stored to work done approaches a limiting value which depends on the thermal coefficient of expansion, the absolute temperature, and a parameter m which is related to the temperature coefficient of the shear modulus.

In this paper, general expressions for internal energy, entropy, enthalpy, etc., are presented in terms of the strain energy function and its temperature coefficient.

I. Introduction

The earliest attempts at understanding the behavior of materials have been made under the assumption that the material is a continuum. For example, in an introductory course in physics the concept of density is defined as the ratio of mass to bulk volume. Later, one learns to refine this concept by considering local density fluctuations in terms of the positions of nuclei in some sort of space lattice; and even later on, one refines this concept further by considering density fluctuations which arise from electron orbital populations. It is this sort of development which we have in mind in starting from the assumption that one can approximate bulk elastomeric materials as continua.

From the standpoint of simplicity, it is convenient to restrict this discussion to homogeneous and isotropic continua. The former concept implies that the material properties do not depend on position within the undeformed bulk; and the latter concept implies that the material properties do not depend on orientation around a point within the undeformed bulk. By contrast for example, a solid propellant is inhomogeneous. The density of the undeformed bulk varies from region to region. Wood is anisotropic in the undeformed state; its modulus along the grain is different from that transverse to the grain. Both of these statements may apply to the deformed states as well, but the concepts of homogeneity and isotropy are best defined in the undeformed state.

In nature there are reasonable approximations in the undeformed state to homogeneous isotropic continua. Any elastomer, natural or synthetic, can be prepared reasonably homogeneous, barring a few tenths of a percent of curing agent; reasonably continuous, having pinholes of the order of a micron or less in size; and reasonably isotropic in the undeformed state. Since elastomeric materials evince large deformations at rupture, many without undergoing concatenate crystallization, one is interested in understanding the law by which energy is stored in such a material. In other words, Hooke's law is inapplicable to deformations beyond ten or twenty percent.

A fortiori, in understanding the law by which an elastomer stores energy, one is automatically led by thermostatic arguments to an understanding of how the internal energy of the sample depends upon temperature and deformation. This information, together with a reasonable molecular model, provides a basis for a statistical mechanical exegesis of all thermomechanical behavior.

Depending upon the temperature of the sample relative to the glass transition temperature of the elastomer, depending upon the state of cure of the sample, and depending upon the rate of deformation, the bulk sample will more or less dissipate energy while storing it. It is possible however to cure samples "tight" enough and to carry out deformation in a region of temperature, at least 80°C above the glass transition temperature of the elastomer and at extremely slow rates of deformation, so that the sample is maintained continuously in thermostatic equilibrium. In this paper only such materials and such processes are discussed.

Mechanical energy may be stored in a rubberlike material in a variety of ways -- by stretching, twisting, and dilating or compressing. The form best suited to the description of a general stress-displacement field within an elastomer is that of tensor analysis. However, under the restriction that the only deformations applicable are principal deformations, which means that the stretches are orthogonal and parallel to the rectilinear dimensions of a sample, then the description of the law of energy storage can be tremendously simplified. Such a description has already been presented[1] and will be reviewed here.

In summary, a thermostatic description of the mechanical behavior of homogeneous, isotropic, elastomeric continua, subjected only to large principal stretches, is sought. This information provides a guide for relating phenomenological behavior by the tools of statistical mechanics to a molecular model.

II. The Strain Energy Density

Consider a rectangular parallelopiped, at uniform temperature. T, the lengths of whose sides are denoted by ℓ_i^+, i = 1,2,3. (The superscript $^+$ as used throughout designates the mechanically undeformed state at temperature T.) After applying isothermally a principal deformation to the parallelopiped, the figure is mapped into some new rectangular parallelopiped, the length of whose sides are now denoted by ℓ_i, i = 1,2,3. In carrying out the process, it will be necessary to wait for the heat that arises from the work done on the sample to diffuse out until the sample returns to thermal equilibrium with the environmental reservoir at temperature T. There is no assumption made here that the sample is incompressible.

However in practice, unless extremely high stresses are involved, the product of the ℓ_i will be within a few hundredths of a percent of the product of the ℓ_i^+.

The state of stretch of the sample is denoted by the three ratios:

$$\lambda_i = \ell_i / \ell_i^+ \tag{1}$$

and the state of dilatation by the volume ratio:

$$J = V/V^+ = \prod_{i=1}^{3} (\ell_i / \ell_i^+) = \prod_{i=1}^{3} \lambda_i \tag{2}$$

In order to maintain the sample in this state of stretch, it is necessary to apply loads, f_i, to each of the faces of the sample. Opposite faces, of course, subtend loads of equal magnitude but opposite direction in order to maintain static equilibrium. The so-called <u>stress-on-original-cross-section</u> or simply <u>stress</u> is hereafter given by

$$\sigma_i = f_i / A_i^+ = \ell_i^+ f_i / V^+ \tag{3}$$

Likewise the <u>true stress</u> is denoted by:

$$\bar{\sigma}_i = f_i / A_i = \ell_i f_i / V \tag{4}$$

From (1), (2), (3), and (4), one sees that the true stress is related to the stress by:

$$\bar{\sigma}_i \, J = \sigma_i \, \lambda_i \tag{5}$$

The work done in bringing the undeformed figure into the state of deformation is, de facto:

$$\sum_{i=1}^{3} \int_{\ell_i=\ell_i^+} f_i \, d\ell_i = V^+ \sum_{i=1}^{3} \int_{\lambda_i=1} \sigma_i \, d\lambda_i \equiv V^+ W \qquad (6)^*$$

where the symbol W denotes the work per unit volume of undeformed material, or simply the strain energy density. From the group of homeomorphisms on an isotropic material, it follows that W is an even and symmetric function of the $\{\lambda_i\}$. By the principle of virtual work, it follows immediately from (6) that:

$$\sigma_i = \left(\frac{\partial W}{\partial \lambda_i}\right)_{T,\lambda'} \qquad (7)$$

where the subscript λ' denotes that the remaining λ_i are to be held constant during the differentiation. The existence of such a function W automatically guarantees mechanical compatibility, i.e.:

$$\left(\frac{\partial \sigma_i}{\partial \lambda_j}\right)_{T,\lambda'} = \left(\frac{\partial^2 W}{\partial \lambda_i \partial \lambda_j}\right)_{T,\lambda'} = \left(\frac{\partial \sigma_j}{\partial \lambda_i}\right)_{T,\lambda'} \qquad (8)$$

Since the derivative (7) is evaluated at constant temperature, one can also write:

$$f_i = V^+ \left(\frac{\partial W}{\partial \ell_i}\right)_{T,\ell'} = \frac{\partial (V^+ W)}{\partial \ell_i}\bigg|_{T,\ell'} \qquad (9)$$

and

$$\left(\frac{\partial f_i}{\partial \ell_j}\right)_{T,\ell'} = \left(\frac{\partial^2 (V^+ W)}{\partial \ell_i \partial \ell_j}\right)\bigg|_{T,\ell'} = \left(\frac{\partial f_j}{\partial \ell_i}\right)_{T,\ell'} \qquad (10)$$

Equation (10), as we shall see later on, is a consequence of one of the Maxwell relations. We note also that V^+ and the $\{\ell_i^+\}$ are

---

*Here the terms $(f_i d\ell_i)$ include work done against the atmosphere. The decomposition of these terms into two components, one for each load cell, the other for the atmosphere, will be introduced below.

functions only of temperature, yet to be defined.

III. The Laws of Thermostatics for Large Principal Deformation

A combined statement of the first and second laws of thermostatics, when only mechanical work is involved, may be written in the form:

$$dU = TdS + \sum_{i=1}^{3} f_i \, d\ell_i \tag{11}$$

where all quantities refer to the actual sample, and not to unit mass of material. The reason for this choice will become obvious. By virtue of (9), (11) can be written

$$dU = TdS + \Sigma \left.\frac{\partial(\overset{+}{V}W)}{\partial \ell_i}\right|_{T,\ell'} d\ell_i \tag{12*}$$

In the laboratory, it is convenient to measure independently the temperature and the dimension of the deformed sample, as well as the loads needed to maintain the state of deformation. Since the only states of concern here are homogeneous, the temperature will be homogeneous as well. Thus one can choose the temperature and lengths as independent variables. This implies a need[2] for reformulating (11) in terms of the Helmholtz free energy by the Legendre transform:

$$F = U - TS \tag{13}$$

or

$$dF = - SdT + \Sigma\left.\frac{\partial(\overset{+}{V}W)}{\partial \ell_i}\right|_{T,\ell'} d\ell_i \tag{14}$$

By cross-differentiation, there follows (10), and also:

---

*Henceforth all summations will be understood to be over i from 1 to 3, unless otherwise noted.

$$\left(\frac{\partial S}{\partial \ell_i}\right)_{T,\ell'} = -\left.\frac{\partial^2 (V^+W)}{\partial \ell_i \partial T}\right|_{T,\ell';\underline{\ell}} \tag{15}$$

where the subscripts before the semicolon are associated with the first differentiation and the subscripts after the semicolon with the second differentiation; in this case the subscript $\underline{\ell}$ denotes that all the $\{\ell_i\}$ are to be held constant while varying T.

From (12) and the definition of specific heat, there follows:

$$\left(\frac{\partial U}{\partial T}\right) \equiv C_\ell = T\left(\frac{\partial S}{\partial T}\right)_\ell \tag{16}$$

Together with (15), (16) provides a basis for expanding the entropy as:

$$TdS = C_\ell\, dT - T\,\Sigma \left.\frac{\partial^2 (V^+W)}{\partial \ell_i \partial T}\right|_{T,\ell';\ell} d\ell_i \tag{17}$$

which, when substituted into (12) yields:

$$dU = C_\ell\, dT + \Sigma\left[\left.\frac{\partial (V^+W)}{\partial \ell_i}\right|_{T,\ell'} - T\left.\frac{\partial^2 (V^+W)}{\partial \ell_i \partial T}\right|_{T,\ell';\ell}\right] d\ell_i \tag{18}$$

Eq. (18) can now be partially integrated, holding temperature constant, to yield

$$U = U^+ + V^+W - T\left.\frac{\partial (V^+W)}{\partial T}\right|_\ell \tag{19}$$

$$= U^+ + \left.\frac{\partial\left(\frac{V^+W}{T}\right)}{\partial\left(\frac{1}{T}\right)}\right|_\ell \tag{20}$$

where U^+, the internal energy of the undeformed body depends only on temperature, and where it is understood that, by definition, the strain energy density is zero in the undeformed state.

One can also integrate (17) in like fashion to obtain

$$S = S^+ - \left.\frac{\partial (V^+W)}{\partial T}\right|_\ell \tag{21}$$

After multiplying (21) by T and subtracting from (19), one immediately notes that:

$$\frac{F - F^{+}}{V^{+}} = W \tag{22}$$

which establishes a result that is well known to continuum mechanicians, i.e., the strain energy density is the Helmholtz free energy of deformation per unit volume of undeformed material. Thus, knowing W, and its temperature dependence, one can calculate immediately the internal energy by a Gibbs-Helmholtz relation of the form (20), and the entropy by a relation of the form (21). Furthermore, from (19), we also obtain the specific heat, by differentiation with respect to temperature:

$$C_{\ell} = C_{\ell}^{+} - T \left.\frac{\partial^2 (V^{+}W)}{\partial T^2}\right|_{\ell} \tag{23}$$

where the second term on the right-hand side of (23) expresses the stretch dependence of the specific heat.

Proceeding along classical thermostatic lines, one can also introduce the enthalpy:

$$H = U - \Sigma f_i \ell_i$$

or

$$H = U^{+} + V^{+}W - T\left.\frac{\partial (V^{+}W)}{\partial T}\right|_{\ell} - \Sigma \ell_i \left.\frac{\partial (V^{+}W)}{\partial \ell_i}\right|_{T,\ell'} \tag{24}$$

and the Gibbs free energy, or free enthalpy:

$$G = H - T\,S \tag{25}$$

or

$$G = F^{+} + V^{+}W - \Sigma \ell_i \left.\frac{\partial (V^{+}W)}{\partial \ell_i}\right|_{T,\ell} \tag{26}$$

The specific heat at constant force (analog of C_p) can now be calculated as follows: First one writes the differential of (24)

$$dH = C_{\ell}^{+}\, dT - T\Sigma \left.\frac{\partial^2 (V^{+}W)}{\partial T \partial \ell_i}\right|_{\ell;T,\ell'} d\ell_i - T \left.\frac{\partial^2 (V^{+}W)}{\partial T^2}\right|_{\ell} dT$$

$$- \Sigma \ell_i \left.\frac{\partial^2 (V^+W)}{\partial \ell_i \partial T}\right|_{T,\ell';\ell} dT - \sum_i \sum_j \ell_i \left.\frac{\partial^2 (V^+W)}{\partial \ell_i \partial \ell_j}\right|_{T,\ell'} d\ell_j \qquad (27)$$

Next one introduces the differential of (9)

$$df_i = \left.\frac{\partial^2 (V^+W)}{\partial \ell_i \partial T}\right|_{T,\ell';\ell} dT + \sum_j \left.\frac{\partial^2 (V^+W)}{\partial \ell_i \partial \ell_j}\right|_{T,\ell'} d\ell_j \qquad (28)$$

From (22) and (27) and the definition of specific heat at constant force, there follows:

$$\left(\frac{\partial H}{\partial T}\right)_f \equiv C_f = C_\ell - T \Sigma \left.\frac{\partial^2 (V^+W)}{\partial T \partial \ell_i}\right|_{\ell;T,\ell'} \left(\frac{\partial \ell i}{\partial T}\right)_f$$

$$- \Sigma \ell_i \left.\frac{\partial^2 (V^+W)}{\partial \ell_i \partial T}\right|_{T,\ell';\ell} - \sum_i \sum_j \ell_i \left.\frac{\partial^2 (V^+W)}{\partial \ell_i \partial \ell_j}\right|_{T,\ell'} \left(\frac{\partial \ell_j}{\partial T}\right)_f \qquad (29)$$

From (28), one has

$$0 = \left.\frac{\partial^2 (V^+W)}{\partial \ell_i \partial T}\right|_{T,\ell';\ell} + \sum_j \left.\frac{\partial^2 (V^+W)}{\partial \ell_i \partial \ell_j}\right|_{T,\ell'} \left(\frac{\partial \ell_j}{\partial T}\right)_f \qquad (30)$$

When (30) is multiplied by ℓ_i and summed over i, it implies that the sum of the last two terms on the right-hand side of (29) is zero, so that:

$$C_f = C_\ell - T \left.\frac{\partial^2 (V^+W)}{\partial T \partial \ell_i}\right|_{\ell;T,\ell'} \left(\frac{\partial \ell i}{\partial T}\right)_f \qquad (31)$$

Furthermore (30) is a set of three equations involving the matrix $\left\|\frac{\partial^2 (V^+W)}{\partial \ell_i \partial \ell_j}\right\|$ which can be inverted to yield explicit expressions for the $\{\left(\frac{\partial \ell i}{\partial T}\right)_f\}$, which can then be inserted into (31).

After multiplication by ℓ_i, but no summation, (30) can be rewritten:

$$0 = \left.\frac{\partial^2 (V^+W)}{\partial \ln \ell_i \partial T}\right|_{T,\ell';\ell} + \sum_j \left.\frac{\partial^2 (V^+W)}{\partial \ln \ell_i \partial \ln \ell_j}\right|_{T,\ell'} \left(\frac{\partial \ln \ell_j}{\partial T}\right)_f \qquad (32)$$

The factors in the second term of (32) are the linear expansion coefficients designated by:

$$\alpha_i \equiv \left.\frac{\partial \ln \ell_i}{\partial T}\right|_f \tag{33}$$

In general, these quantities are strongly dependent on force, and slightly dependent on temperature. It is, of course, important to know how these are related to the linear expression coefficients of the undeformed body:

$$\alpha_i^+ \equiv \left.\frac{\partial \ln \ell_i^+}{\partial T}\right|_f \tag{34}$$

Such relations will result formally when the matrix inversion process is carried out, and α^+ will appear by virtue of the differentiation $\left(\frac{dV^+}{dT}\right)$. It is now appropriate to introduce a temperature dependence for $\{V^+, \ell_i^+\}$.

Experimental observations show directly that the lengths of an undeformed body increase linearly with temperature over a fairly broad region of temperature above the glass transition temperature. In fact the linearity can be extended significantly by correlating the log of the dimension with temperature. Thus we write:

$$\ln \ell_i^+ = \ln \ell_{io}^+ + \alpha^+ (T - T_o) \tag{35}$$

or

$$\ell_i^+ = \ell_{io}^+ \exp\left[\alpha^+ (T - T_o)\right] \simeq \ell_{io}^+ [1 + \alpha^+ (T - T_o)] \tag{36}$$

where T_o is a reference temperature, and subscript zero refers to the reference state. Since, in what follows below, the range of temperature over which data is analyzed does not exceed 60°C, and since α^+ is of the order of 10^{-4}, it makes little difference whether the exponential relation or its linear approximation in (36) is used. Contrariwise, with little increase of algebraic complexities, one could make α^+ any function of temperature needed to reproduce the data, and write:

$$\ell_i^+ = \ell_{io}^+ \exp\left[\int_{T_o} \alpha^+ dT\right] \tag{37}$$

We shall need only the form (35), from which there follows:

$$\frac{d\,\ell n\,\ell_i^+}{dT} = \alpha^+ \tag{38}$$

$$\frac{d\,\ell n A_i^+}{dT} = 2\,\alpha^+ \tag{39}$$

$$\frac{d\,\ell n V^+}{dT} = 3\,\alpha^+ \tag{40}$$

where it is now understood that α^+ is a constant.

IV. The Case of Simple Tension With Superimposed Hydrostatic Pressure

In the preceding section there were introduced explicit expressions for the thermostatic quantities $\{f_i, S, U, F, H, G, C_\ell, C_f,$ and $\alpha_i\}$ in terms of the variables $\{T, \ell_i\}$, and the material parameter α^+, and the material function W. It is logical now to ask what a typical form of W is that best represents available data, and particularly what sort of prediction one can make about the form of the internal energy.

Before discussing various candidates for W, we note however that many of the derivatives involved in the preceding formal discussion were to be evaluated holding two or three of the lengths of the sample constant. In practice, in the laboratories, at best do we hold one or two lengths constant, and let the other lengths vary in such a way as to respond to the condition that the pressure is constant. Thus it is necessary to reconstruct expressions which account for this condition.

If the direction of pull is denoted by the subscript 1, and the two transverse directions by the subscripts 2 or 3, then simple tension with superimposed hydrostatic pressure is characterized by the following statement:

$$\lambda_1 = \lambda \tag{41}$$

$$\lambda_2 = \lambda_3 = \sqrt{\frac{J}{\lambda}} \tag{42}$$

$$\sigma_1 \lambda_1 = \sigma\lambda - P_g J \tag{43}$$

$$\sigma_2 \lambda_2 = \sigma_3\lambda_3 = - P_g J \tag{44}$$

Eq. (42) merely expresses the equivalence (2). In (43) are introduced the two components of force acting on the face normal to the direction of pull: one is the force recorded on the load cell, the other is the pressure due to the environment. Since the symbol σ denotes the force on the load cell divided by the cross-sectional area of the undeformed sample, it is necessary to combine the load cell stress with the atmospheric true stress with the help of Eq. (5). In expressing the atmospheric true stress by P_g, the gauge pressure, where

$$P_g = P - P_a \text{ (1 atmosphere)}$$

it has been assumed that the sample is in its reference state of zero deformation (or zero strain energy density) at 1 atmosphere. One could reverse this assumption and choose the reference state of zero deformation to be at zero absolute pressure at the expense of considerable algebraic complexity; from a practical standpoint, this would add nothing to the results we shall obtain.

From (43) and (44), the total force components are given by:

$$f_1 = A_1^+ (\sigma\lambda - P_g J) = \frac{V^+}{\ell_1^+} (\sigma\lambda - P_g J) \tag{45}$$

$$f_2 = A_2^+ (- P_g J) = - \frac{V^+}{\ell_2^+} P_g J = - \frac{P_g V}{\ell_2^+} \tag{46}$$

In computing the linear expansion coefficients by (33), each of these components is to be held constant. At constant pressure, (46) implies that the ratio V/ℓ_2^+ has to be held constant. Likewise (45) implies that the combination $(f\ell_1 - P_g V)/\ell_1^+$ has to be held constant, where $f = \sigma A_1^+$. (47)

Neither of these conditions is easily maintained. In the laboratory it is easier to maintain P_g and one of the lengths (ℓ_1) constant. Thus in order to compute the thermostatic quantities presented in the preceding section, it is necessary to reconstruct these formulae for the case of simple tension with superimposed hydrostatic pressure. We shall not do this in general, but will present the results for one candidate strain energy density.

V. The Compressible Neo-Hookean Strain Energy Density

Consider a modified form of the well-known neo-Hookean strain energy density, namely:

$$W = \frac{G}{2} (J_1 - 3) + (K - \frac{2}{3} G) (J - 1) - (K + \frac{G}{3}) \ell n J \tag{48}$$

where $J_1 = \Sigma \lambda_i^2$ (49)

and G and K are respectively the shear and bulk moduli. This form generates stress components of the form:

$$\sigma_i \lambda_i = G(\lambda_i^2 - 1) + (K - \frac{2}{3}G)(J - 1) \tag{50}$$

In order to motivate the choice of the second term on the right-hand side of (50), one has merely to expand (50) for small strain, i.e., to set

$$\lambda_i = 1 + \epsilon_i \tag{51}$$

where $\epsilon_i \ll 1$. By (2),

$$J = \prod \lambda_i = 1 + \vartheta + \vartheta_2 + \vartheta_3 \tag{52}$$

where

$$\vartheta = \epsilon_i \tag{53}$$

$$\vartheta_2 = \frac{1}{2} \sum_{i \neq j} \epsilon_i \epsilon_j \tag{54}$$

$$\vartheta_3 = \prod \epsilon_i \tag{55}$$

Upon introducing (51)-(55) and retaining only terms of second order, (50) becomes:

$$\sigma_i = [2G\epsilon_i + (K - \frac{2}{3}G)\vartheta] + [-G\epsilon_i^2 - (K - \frac{2}{3})\vartheta\epsilon_i + (K - \frac{2}{3})\vartheta_2] + .. \tag{56}$$

The first bracketed term is merely Hooke's law. Any modification of (48) must be so chosen that the Hookean form is maintained in the limit of vanishing strain. Apart from this restriction, however, there are an infinity of choices that can be made to modify (48) to account for compressibility. From the second bracketed term, we see however, that terms which arise from this modification appear only in the form $(K - \frac{2}{3}G)\vartheta$ or $(K - \frac{2}{3}G)\vartheta_2$ or $(K - \frac{2}{3}G)\vartheta^2$, etc. And since it is known[3] that the hydrostatic pressure is also of the form:

$$P_g = - K \vartheta \tag{57}$$

it follows that if the pressure is small with respect to the shear modulus, then the specific choice of modification introduced into (48) will have no bearing on the results to be presented. This statement is exactly true when the gauge pressure is zero. Contrariwise, if one chooses to analyze data such as were obtained by Bridgman[4] under several thousands of kilobars of pressure, then it is extremely important to define the form which best represents these data. In the following discussion, we shall consider data obtained only for $P_g = 0$. The reason for specifically including P_g in the thermostatic expressions has to do, of course, with the fact that derivatives will be evaluated with respect to P_g, so that P_g can only be set equal to zero after the differentiation.

There is a more important reason for choosing (50) as will now become apparent. For the case of simple tension and superimposed hydrostatic pressure, (50) becomes

$$\sigma\lambda - P_g J = G(\lambda^2 - 1) + (K - \tfrac{2}{3}G)(J - 1) \tag{58}$$

$$- P_g J = G(\tfrac{J}{\lambda} - 1) + (K - \tfrac{2}{3}G)(J - 1) \tag{59}$$

And by difference, one immediately obtains:

$$\sigma = G(\lambda - J/\lambda^2) \tag{60}$$

or

$$f = G\left(\frac{V^+ \ell}{\ell^{+2}} - \frac{V \ell^+}{\ell^2}\right) \tag{61}$$

which is a well-known expression[5] derived from statistical theory. Eq. (59) can be rewritten:

$$P = P_a - G\,\frac{\ell^+}{\ell} - (K - \tfrac{2}{3}G) + (K + \tfrac{G}{3})\,\frac{V^+}{V} \tag{62}$$

Equations (61) and (62) express the two force components in terms of the dimensions of the deformed and undeformed specimen and the material parameters G and K. The temperature also appears implicit in ℓ^+ and V^+ along with the material parameter α^+. The temperature dependence of the material parameters G and K is yet to be discussed.

Proceeding as before, and skipping unnecessary algebraic details, we immediately record the following results:

$$\frac{F-F^+}{V^+} = W = G\left[\frac{\ell^2}{2\ell^{+2}} + \frac{V\ell^+}{V^+\ell} - \frac{3}{2}\right] + \left(K - \frac{2}{3}G\right)\left(\frac{V}{V^+} - 1\right) - \left(K + \frac{G}{3}\right)\ell n\left(\frac{V}{V^+}\right) \quad (63)$$

$$\frac{S-S^+}{V^+} = -\left(\frac{dG}{dT} + G^+\right)\left(\frac{\ell^2}{2\ell^{+2}} + \frac{V\ell^+}{V^+\ell} - \frac{3}{2}\right) - \left(\frac{dK}{dT} - \frac{2}{3}\frac{dG}{dT}\right)\left(\frac{V-V^+}{V^+}\right)$$

$$+ \left[\frac{dK}{dT} + \frac{1}{3}\frac{dG}{dT} + (3K + G)\,\alpha^+\right] \ell n \frac{V}{V^+} \quad (64)$$

$$\frac{U-U^+}{V^+} = \left(G - T\frac{dG}{dT} - G\alpha^+T\right)\left(\frac{\ell^2}{2\ell^{+2}} + \frac{V\ell^+}{V^+\ell} - \frac{3}{2}\right) + \left(K - \frac{2}{3}G - T\frac{dK}{dT}\right.$$

$$\left. + \frac{2}{3}T\frac{dG}{dT}\right)\left(\frac{V}{V^+} - 1\right) - \left[\left(K + \frac{G}{3}\right)\left(1 + 3\,\alpha^+T\right) + T\frac{dK}{dT} + \frac{T}{3}\frac{dG}{dT}\right]\ell n \frac{V}{V^+} \quad (65)$$

Equation (65) suggests that in order for a material to be ideal in the sense that U is independent of (ℓ/ℓ^+) at constant temperature and constant volume,

$$G(1 - \alpha^+T) = T\frac{dG}{dT} \quad (66)$$

or

$$G = G_o \frac{T}{T_o} e^{-\alpha^+(T-T_o)} \quad (67)$$

where T_o is a reference temperature. Equation (67) however is not true in general, and indeed, if we write

$$G = G_o \frac{T}{T_o} \exp[-m\alpha^+(T-T_o)] \tag{68}$$

then the departure of m from unity can be taken to be a measure of the non-ideality of the sample. Note that (68) suggests that, in reducing viscoelastic data to a state of reference temperature one should correct the modulus ratio not by the density ratio ($e^{3\alpha^+(T-T_o)}$), but by a factor containing the appropriate multiple of $\alpha^+(T - T_o)$ as determined from the actual temperature dependence of the equilibrium modulus.

Continuing further, one can derive expressions for H, G, $C_{\ell V}$, C_{fP_g}, and α. Only the last one is of particular interest to this discussion. After eliminating V between (61) and (62), differentiating with respect to temperature holding f and P_g constant, making use of the fact that $G << K$, and then setting $P_g = 0$, one records:

$$\alpha \equiv \left(\frac{\partial \ln \ell}{\partial T}\right)_{f,P_g = 0} = \alpha^+ + \left[(m-2)\alpha^+ - \frac{1}{T}\right]\left(\frac{\lambda^3-1}{\lambda^3+2}\right) \tag{69}$$

where (68) has been used. The behavior of (69) is such that: at $\lambda = 1$, $\alpha = \alpha^+$; as λ increases, α decreases to zero, and then becomes negative, and approaches the limiting value:

$$\alpha|_{\lambda\to\infty} \to (m-1)\,\alpha^+ - \frac{1}{T} \tag{70}$$

The particular value of λ at which α becomes zero is of interest. It is the stretch associated with the so-called thermoelastic inversion and has the value:

$$\lambda = \left[\frac{1 + (4-m)\ \alpha^{+}T}{1 - (m-1)\ \alpha^{+}T}\right] \cong 1 + \alpha^{+}T \tag{71}$$

The linear expansion of (71) reveals that the position of the thermoelastic inversion is insensitive to internal energy effects and depends principally upon thermal expansion.

The calculation of U by means of (65) is straightforward providing the temperature dependences of G and K are known. That of G is adequately described by (68) and will be further documented later on in this discussion. That of K is not as well known, but does seem to be equally well described by a relation of the form (68).

In attempting to evaluate U, previous investigations have chosen a different route, by introducing the internal force analogous to the internal pressure, i.e.:

$$\left(\frac{\partial U}{\partial \ell}\right)_{TV} \equiv f_e = f - T\left(\frac{\partial f}{\partial T}\right)_{\ell V} \tag{72}$$

The value of this quantity is immediately available from (65), namely

$$\left(\frac{\partial U}{\partial \ell}\right)_{TV} = \left(G - T\,\frac{dG}{dT} - G\,\alpha^{+}T\right)\left(\frac{\ell V^{+}}{\ell^{+2}} - \frac{V\ell^{+}}{\ell^{2}}\right) \tag{73}$$

or with (68) and (61):

$$\frac{f_e}{f} = (m-1)\ \alpha^{+}\ T \tag{74}$$

Thus we see that the quantity (f_e/f) is important because it is a direct measure of non-ideality as provided by our parameter (m-1). When m = 1, $f_e = 0$. And more important, (f_e/f), as calculated for a compressible neo-Hookean material, is independent of stretch.

Now other investigators have used an indirect method of evaluating (f_e/f) as follows. It will be shown that (72) is equivalent to:

$$\frac{f_e}{f} = 1 - T\left(\frac{\partial \ell nf}{\partial T}\right)_{\ell P} - T\left(\frac{\partial \ell nf}{\partial P}\right)_{T\ell}\left(\frac{\partial P}{\partial T}\right)_{\ell V} \qquad (75)$$

or

$$\frac{f_e}{f} = 1 - T\left(\frac{\partial \ell nf}{\partial T}\right)_{\ell P} + T\left(\frac{\partial \ell nf}{\partial P}\right)_{T\ell}\frac{\left(\frac{\partial V}{\partial T}\right)_{\ell P}}{\left(\frac{\partial V}{\partial P}\right)_{T\ell}} \qquad (76)$$

or

$$\frac{f_e}{f} = 1 - T\left(\frac{\partial \ell nf}{\partial T}\right)_{\ell P} + \frac{T}{f}\left(\frac{\partial V}{\partial \ell}\right)_{TP}\frac{\left(\frac{\partial V}{\partial T}\right)_{\ell P}}{\left(\frac{\partial V}{\partial T}\right)_{T\ell}} \qquad (77)$$

Expressions (75)-(77) provide an equivalence basis for establishing (f_e/f) in terms of experimentally available data. The quantity $(\partial f/\partial T)_{\ell P}$ is, of course, easily measured, but any of the other quantities are obtained only with difficulties. Furthermore in the limit of vanishing strain, each of these quantities is singular in the strain. For example, after eliminating V between (61) and (62), we can calculate:

$$\left(\frac{\partial \ell nf}{\partial P}\right)_{T\ell} = \frac{1}{K(\lambda^3-1)} \qquad (78)$$

while, from (62) directly, there follows:

$$\left(\frac{\partial P}{\partial T}\right)_{\ell V} = 3K\alpha^{+} \qquad (79)$$

We see that (79) also approaches positive infinity as $\lambda \to 1$. It is readily verified that the product of (78) and (79) added to

$T\left(\frac{\partial \ell n f}{\partial T}\right)_{\ell P}$ and subtracted from (1) yields (75). Thus in attempting to evaluate (f_e/f) by this alternate route, previous investigations have encountered the difficulties attendant upon computing the difference between two large quantities. Apart from the difficulties which arise only at small strain, the suggestion is hereby proffered that, in order to calculate (f_e/f), or better yet $\left(\frac{U-U^+}{V^+}\right)$, one should determine f-ℓ isotherms as well as **P-V** isotherms. Then by representing each isotherm by a suitable equation of state (such as the neo-Hookean one chosen above), one can then calculate the internal energy by means of (65).

VI. An Estimate of the Internal Energy for a Compressible Neo-Hookean Material in Simple Tension

In order to estimate $\left(\frac{U-U^+}{V^+}\right)$, we need merely adduce expressions for dG/dT and dK/dT. We have already noted that

$$\frac{dG}{dT} = G\left(\frac{1}{T} - m\alpha^+\right) \qquad (80)^{(6)}$$

Let us assume, in addition, that

$$\frac{d}{dT}\left(K - \frac{2}{3}G\right) = \left(K - \frac{2}{3}G\right)\left(\frac{1}{T} - p\alpha^+\right) \qquad (81)$$

Then (65) becomes

$$\frac{U-U^+}{V^+\alpha^+T} = (m-1)\frac{G}{2}(J_1-3) - \left[(m-3)G + (p-3)\left(K - \frac{2}{3}G\right)\right]\ell n\, J$$

$$+ p\left(K - \frac{2}{3}G\right)(J-1) \qquad (82)$$

For the case of simple tension (with $J \approx 1$), one obtains

$$\frac{U-U^+}{V^+\alpha^+T} \simeq (m-1)\frac{G}{2}\left(\lambda^2 + \frac{2}{\lambda} - 3\right) + 3G\left(1 - \frac{1}{\lambda}\right) \qquad (83)$$

whereas

$$W = \frac{F-F^+}{V^+} \simeq \frac{G}{2}\left(\lambda^2 + \frac{2}{\lambda} - 3\right) \qquad (84)$$

$$\text{since} \quad \left(\lambda^2 + \frac{2}{\lambda} - 3\right) = O(\varepsilon^2) \tag{85}$$

$$\text{and} \quad 1 - \frac{1}{\lambda} = O(\varepsilon) \tag{86}$$

$$\text{for} \quad \lambda = 1 + \varepsilon \tag{87}$$

We see that initially the ratio of energy stored internally to work done on the body is infinite. This means that heat must diffuse into the body to maintain it isothermal. Contrariwise, if the stretch is performed adiabatically, this means that there is initially a temperature drop prior to a later rise. The stretch at which the temperature is exactly zero for adiabatic stretch is, of course, related to the thermoelastic inversion, and can easily be calculated from (21).

As the stretch increases, U increases monotonically with the stretch for a compressible neo-Hookean material. The rate at which it increases depends on the important parameter (m-1) which measures departure from ideality. According to data presented previously by Shen and Blatz(ibid.), the best current value for m for natural rubber is 2.6. Thus as λ becomes large, we have

$$\frac{U-U^+}{F-F^+} = \alpha^+T \, \frac{(m-1)\,\frac{G}{2}\left(\lambda^2 + \frac{2}{\lambda} - 3\right) + 3G\left(1 - \frac{1}{\lambda}\right)}{\frac{G}{2}\left(\lambda^2 + \frac{2}{\lambda} - 3\right)} \simeq (m-1)\alpha^+T \tag{88}$$

which implies a limiting fraction of about 11 or 12 percent energy stored internally. This value is a direct measure for internal energy, and bypasses the pitfalls that previous authors have staggered into in dealing with (f_e/f).

It is expected that m (more so than α^{+}) should show a strong dependence upon the structure of the polymer chain and the conformation of the network. It is suggested then that a table of internal energy contribution for various polymers can be conveniently provided by tabulating the easily determined experimental quantity

$$m = \frac{1 - \frac{d\ell n\, G}{d\ell n\, T}}{\alpha^{+} T} .$$

References

1) Blatz, P. J., Rheology V, edited by F. Eirich, p. 3, Academic Press, New York City.

2) Callen, H., Thermodynamics, Wiley, New York City.

3) Timoshenko, S., Theory of Elasticity, McGraw-Hill, New York City.

4) Bridgman, P., Proc. National Academy of Sciences (1932-34).

5) Flory, P., J.A.C.S. 73, 5222 (1956).

6) Shen, M. and Blatz, P., J. Appl. Phys. 39, 4937 (1968).

THERMOELASTICITY OF CROSSLINKED RUBBER NETWORKS

M. Shen and T. Y. Chen
Department of Chemical Engineering
University of California
Berkeley, California 94720
and
E. H. Cirlin and H. M. Gebhard
Science Center, North American Rockwell
Thousand Oaks, California 91360

SUMMARY

In recent publications we have demonstrated that by using a new equation, which is based on the temperature coefficient of shear modulus, the relative energy contribution (f_u/f) to rubber elasticity can be shown to be independent of the applied strain. In this work, we have measured f_u/f for natural rubber and poly-cis-1,4-butadiene at a series of degrees of crosslinking. Parallel equilibrium swelling studies in appropriate solvents were also made. It was found that f_u/f is 0.17 for natural rubber and 0.10 for poly-cis-1,4-butadiene. These values are independent of the degree of crosslinking, except for the lowest degree of crosslinking. For the latter rubber networks, f_u/f is 0.26 for natural rubber, and 0.34 for poly-cis-1,4-butadiene. Implications of these findings are discussed.

INTRODUCTION

In the development of statistical theory of rubber elasticity, one of the most important tasks is the search for the free energy function of the rubber network. Heretofore, attention has been centered upon the configurational entropy, while computation of the energetic portion of the free energy on a quantitative basis remains to be a challenge to the polymer physicists. On the other hand, considerable experimental data on the internal energy contrition to rubber elasticity (f_u/f) have now been accrued. It has been demonstrated, for instance, that f_u/f is independent of strain[1] and the presence of diluents.[2,3] However, there are conflicting reports regarding the effect of degree of crosslinking (ν). Mark and Flory[4] found the values of f_u/f to be invariant with ν, while Opshoor and Prins[5] noted a decrease in the energy contribution with increasing degree of crosslinking. In this work, we examined the thermoelastic behavior of natural rubber and polybutadiene at a series of crosslinking densities, in an effort to contribute toward a better understanding of this problem.

THERMOELASTIC EQUATIONS

The internal energy contribution to the elastic force, f_u, is defined as follows:

$$f_u = f - T(\partial f/\partial T)_{V,L} \tag{1}$$

where f is total elastic force. However, due to the difficulty of maintaining constant volume in a thermoelastic experiment, the constant pressure constraint is preferred. The most frequently used equation[6] for this purpose is

$$\frac{f_u}{f} = 1 - \left(\frac{\partial \ell n\, f}{\partial \ell n\, T}\right)_{P,L} - \frac{\alpha_T T}{\lambda^3 - 1} \tag{2}$$

where α_T is the cubic thermal expansion coefficient of the rubber. Equation 2 is derived from the equation of state for rubber elasticity[7,8]

$$f = GA_o\ (\lambda - V/V_o\ \lambda^2) \tag{3}$$

where

$$G = \nu kT \langle r_o^2\rangle / \langle r^2\rangle_o\ / V_o \tag{4}$$

In eqs. 2 to 4, A_o and V_o are the cross-sectional area and volume of the rubber at zero force, zero pressure and temperature T; V is the volume at force f, pressure P and temperature T; λ is the relative extension ratio; ν is the number of network chains in the sample; k is Boltzmann constant; $\langle r_o^2 \rangle$ and $\langle r^2 \rangle_o$ are the mean square end-to-end distances of the chain in volume V_o and in free space respectively. The ratio $\langle r_o^2 \rangle / \langle r^2 \rangle_o$ is commonly referred to as the front factor.

We have demonstrated in previous publications that the stress-temperature coefficient at constant pressure can be recast as[1,2]

$$\left(\frac{\partial f}{\partial T}\right)_{P,L} = \frac{f}{G}\frac{dG}{dT} + \frac{f\alpha_T}{3}\left(\frac{\lambda^3 - 4}{\lambda^3 - 1}\right) \tag{5}$$

on the basis of eq. 3. Combining eqs. 2 and 5, we obtain

$$\frac{f_u}{f} = 1 - \frac{d\,\ell n\, G}{d\ell n\, T} - \alpha^o T \tag{6}$$

where $\alpha^o = \alpha_T/3$ is the linear thermal expansion coefficient of the unstrained rubber. Equations 2 and 6 are equivalent. However, the latter is not as sensitive to the experimental errors normally encountered in thermoelastic measurements. Equation 2 depends on λ to the inverse third power, which magnifies any error in the extension ratios.

Most of the thermoelastic measurements were carried out by keeping the stretched length of the sample constant, and following the changes in the elastic force as the temperature was varied. The alternative procedure is to keep the applied stress constant, and then determine the variations of the sample lengths as a function of temperature. The latter simply corresponds to the condition that the stress-temperature coefficient at constant pressure (but not constant length) vanishes. Differentiation of eq. 3 under these conditions yields[2]:

$$\alpha = \alpha^{o} - \left(\frac{\lambda^{3}-1}{\lambda^{3}+2}\right)\left(\frac{d\ln G}{dT} + 2\alpha^{o}\right) \qquad (7)$$

where α is the linear thermal expansion coefficient of strained rubber. Therefore knowing the linear thermal expansion coefficient of the rubber at given extension ratios, the temperature coefficient of shear modulus of the sample can be obtained directly. The relative energy contribution can then be determined by means of eq. 6.

EXPERIMENTAL

Natural rubber pale crepe was cured in the presence of 0.5, 1.0, 2.0, 3.0, 4.0, 10.0 and 20.0 parts dicumyl peroxide per hundred parts of rubber (p.h.r.) at 145°C for approximately 40 minutes. Poly-cis-1,4-butadiene samples were cured with 0.5, 1.0, 2.0 and 4.0 p.h.r. dicumyl peroxide under similar conditions. Strips of the approximate dimensions of 0.25 x 0.25 x 10 cm^3 were cut by a high-speed rotating blade. Ends of these were glued to two small metal blocks. One end was attached to a pan on which known weights can be placed. The rubber-weight assembly was then suspended inside a glass flask. The flask was repeatedly evacuated and purged with nitrogen. The sample was then permitted to reach its equilibrium length at 60°C under the highest load to be used in the experiment. After no further creep was observed between measurements on successive days, the temperature was decreased by 10°C intervals to 10°C. Two hours were given at each temperature to insure the achievement of thermal equilibrium. Changes in lengths as a function of temperature were determined by a cathetometer accurate to ± 0.1 mm. After this temperature cycle, part of the load was removed. The sample was left standing overnight before the next length-temperature measurements were initiated. From these data linear thermal expansion coefficients at various extension ratios can be readily obtained for use in f_u/f calculations.

The degrees of crosslinking of the rubber samples were determined by equilibrium swelling studies, samples were immersed in solvent for two weeks. The equilibrium degrees of swelling were determined by weighing the samples before and after swelling. For natural rubber, n-hexadecane was used as the solvent; for poly-cis-1,4-butadiene, decane was employed.[10]

Computations were carried out by using the following well-known equation[11]

$$- RT \left[\ln (1-\phi_2) + \phi_2 + \chi\phi_2^2 \right] = G_s v_1 (\phi_2^{1/3} - \phi_2) \tag{8}$$

where ϕ_2 is the equilibrium degree of swelling, v_1 the molar volume of the solvent, χ is the interaction parameter and the shear modulus can be expressed by

$$G_s = \rho RT/\bar{M}_c \tag{9}$$

by assuming the front factor to be unity. In eq. 9, ρ is the density of the rubber, and $\bar{M}_c$ is the molecular weight of the network chain.

RESULTS

Results from the equilibrium swelling measurements are given in Table 1. Values of M_c are seen to vary from about 10^2 to 10^6 gm/mole. For high degrees of crosslinking, or low M_c, there are some questions as to the validity of eq. 8. Mason[12] considered $M_c \approx 4000$ as a lower limit for quantitative treatments. In the last column of Table 1, we include values of shear moduli determined from tensile measurements, G_t. These are crossplotted against G_s in Figure 1. Although theoretically one would expect a linear correlation between these quantities, it has never been experimentally observed. The shapes of these experimental curves are similar to those previously found[13] for other elastomers.

Figure 2 illustrates the length-temperature data for one of the elastomers, poly-cis-1, 4-butadiene cured with 0.5 p.h.r. dicumyl peroxide, at a series of loads. From the slopes of the straight lines in this plot, the linear thermal expansion coefficients of the strained rubber can be readily obtained. In order to compute the relative energy contribution to rubber elasticity by eq. 6, we need the linear thermal expansion coefficients of the unstrained rubber and the temperature coefficient of the shear modulus. These can be determined by plotting α as a function of the quantity $(\lambda^3-1)/(\lambda^3+2)$. If eq. 7 is indeed valid, then one would expect a linear relation. From the intercept and slope of such a plot, values of α^o and $d\ln G/dT$ can be easily obtained.

TABLE 1
Network Characterization of Elastomers

Sample (p.h.r. dicup.)	ϕ_2	M_c (gm/mole)	G_s (10^6 dyn/cm^2)	G_t (10^6 dyn/cm^2)
Poly-cis-1,4-butadiene				
0.5	0.181	96,650	0.256	1.58
1.0	0.294	10,500	2.360	4.00
2.0	0.364	4,250	5.830	6.17
4.0	0.441	1,800	13.500	10.10

ρ = 1.01 gm/c.c.
v_1 = 195 c.c./mole (decane)
χ = 0.546 (polybutadiene - decane[10])
$\alpha^o = 2.1 \times 10^{-4}$/°C

Sample (p.h.r. dicup.)	ϕ_2	M_c (gm/mole)	G_s (10^6 dyn/cm^2)	G_t (10^6 dyn/cm^2)
Natural Rubber				
0.5	0.170	94,850	0.238	0.94
1.0	0.251	22,050	1.020	2.10
2.0	0.300	11,050	2.040	2.91
3.0	0.359	5,400	4.160	5.00
4.0	0.425	2,700	8.390	6.83
10.0	0.625	415	54.300	9.35
20.0	0.744	130	168.500	26.60

ρ = 0.92 gm/c.c.
v_1 = 292 c.c./mole (hexadecane)
χ = 0.530 (natural rubber - hexadecane[9])
$\alpha^o = 2.2 \times 10^{-4}$/°C

Figure 3 summarizes the data for poly-cis-1, 4-butadiene at four crosslinking densities. All experimental points fall on the same straight line within experimental error. For this elastomer, $\alpha^o = 2.1 \times 10^{-4}$/°C and $d\ln G/dT = 2.75 \times 10^{-3}$/°C. Thus from eq. 6, we find $f_u/f = 0.10$. We note, however, that data for the lowest crosslinking density polybutadiene (0.5 p.h.r. dicumyl peroxide) deviate somewhat from those of the other three higher crosslinked samples. For this sample, $d\ln G/dT$ can be shown to be 2.00×10^{-3}/°C with no change in α^o. As a consequence, $f_u/f = 0.34$, which is substantially higher than the balance of the polybutadiene samples.

Similar data for natural rubber sample at seven different crosslinking levels are given in Figure 4. Again all samples appear to fall on the same straight line, giving $\alpha = 2.2 \times 10^{-4}$/°C, $d\ln G/dT = 2.52 \times 10^{-3}$/°C, and $f_u/f = 0.17$. A deviation from the average slope of all natural rubber samples is again observed for the lowest crosslinked sample (0.5 p.h.r. dicumyl peroxide). For this sample, $\alpha = 2.2 \times 10^{-4}$/°C, $d\ln G/dT = 2.20 \times 10^{-3}$/°C and $f_u/f = 0.26$.

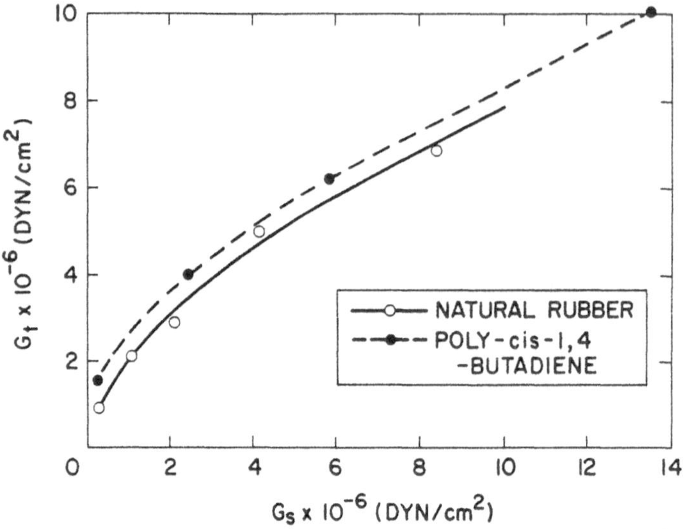

Fig. 1 Shear moduli of natural rubber and poly-cis-1,4-butadiene determined by equilibrium swelling (G_s) and by tensile measurements (G_t).

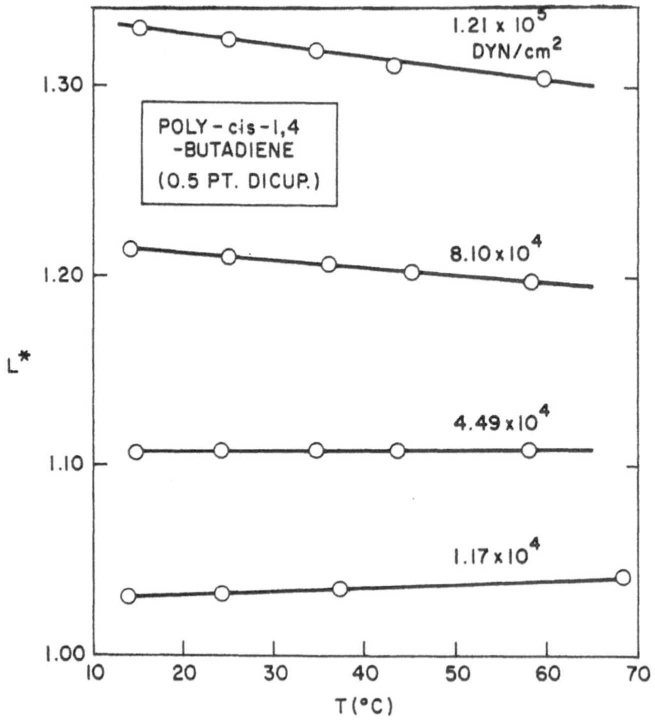

Fig. 2 Length-temperature curves of poly-cis-1,4-butadiene cured with 0.5% dicumyl peroxide at four different stress levels. L^* is the length reduced by the unstrained length at 30°C.

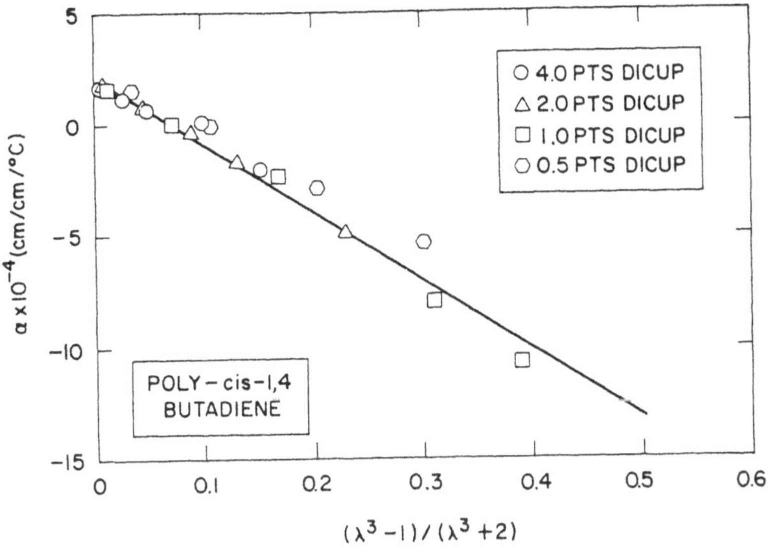

Fig. 3 Linear thermal expansion coefficient of poly-cis1,4-butadiene (cured with 0.5, 1.0, 2.0 and 4.0 parts of dicumyl peroxide per hundred parts of rubber) plotted against $(\lambda^3-1)/(\lambda^3+2)$.

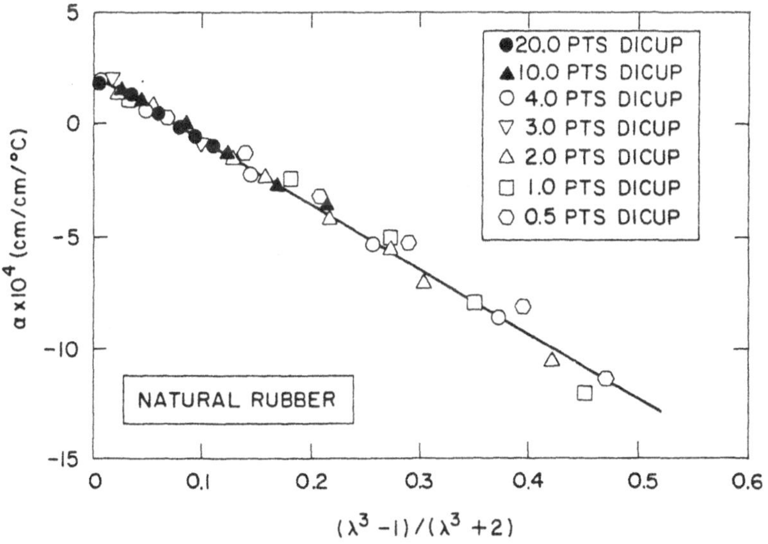

Fig. 4 Linear thermal expansion coefficients of natural rubber (cured with 0.5, 1.0, 2.0, 3.0, 4.0, 10.0 and 20.0 parts of dicumyl peroxide per hundred parts of rubber) plotted against $(\lambda^3-1)/(\lambda^3+2)$.

DISCUSSION

According to the statistical theory of rubber elasticity, as long as the network chain between the crosslink sites are sufficiently long to assume Gaussian behavior, the polymer can be expected to be rubbery at temperatures exceeding its glass transition temperature. In other words, the theory predicts only an increase in modulus (eq. 4) and no other effects as the crosslink density is increased for as long as the sample is a rubber. Thus our observation of the independence of f_u/f of degree of crosslinking is consistent with the theory. The exception is perhaps the lowest crosslinked samples, which exchibits an increased internal energy contribution. The latter fact is consistent with the findings of Opshoor and Prins[5], who reported increases in the values of f_u/f as the crosslinking density decreases.

The preceding observation is a rather surprising one. It has been demonstrated by Katz and Tobolsky[14], that very highly crosslinked polymer networks show no glass transition at all. The moduli of such polymers remain very high for all temperatures up to the decomposition temperature. In other words, highly crosslinked networks are essentially polymeric glasses. For glasses the thermoelastic coefficient is negative, rather than positive as for rubbers. Thus the elasticity of glasses are predominantly energetic in origin, as are most of the ordinary solids. One therefore would expect a very high value of f_u/f (perhaps unity) for very highly crosslinked networks. In fact, we observe a constant energy contribution up to fairly high crosslinking density, and an increased value for the sample with very low crosslinking density. It is intuitively tempting to surmise that if thermoelastic measurements could be carried out for even higher crosslinked rubber samples, an eventual upturn in the value of f_u/f would be found.

The next logical question is just how high a degree of crosslinking one can use and still expect the polymer to behave as a rubber. According to the work of Katz and Tobolsky[14], the statistical theory of rubber elasticity is obeyed by acrylic rubbers up to about $M_c \approx 500$. From our data, the natural rubber remains to be rubbery up to $M_c = 130$ gm/mole, which means that the network chain between crosslinks is only less than two monomers long. This is absurdly short for a polymer chain, and only indicates that eq. 8 is not applicable for this level of crosslinking. Alternatively, we can calculate the same quantity from the shear modulus determined by tensile measurements (Table 1). In this instance, M_c of the natural rubber cured with 20 p.h.r. dicumyl peroxide would have a value of 825, which is about 12 monomer units.

We now compare our f_u/f data with those in the literature[9,15-20]. Figure 5 summarizes these data for natural rubber and poly-cis-1,4-butadiene. In these works, f_u/f was calculated by eq. 2, thus a large strain-dependence is noted in the region of low strains. At high strains, the drop in f_u/f has been attributed to the onset of strain-induced crystallization. In the intermediate region of strains, where f_u/f is relatively constant, the reported values are in agreement with ours calculated by eq. 6. We recall that in the derivation of the statistical theory of rubber elasticity, the configurational entropy of the network is assumed to be the sum of those of the individual chains in the network. In order for this assumption to be valid, it is necessary that configurations of the individual chain are unaffected by the presence of other chains in the network. Thus the deformation of the rubber must not produce any change in interchain interactions. By using eq. 6, this stipulation is seen to be obeyed by natural rubber and polybutadiene within the range of crosslinking examined in this work. Therefore, the principle of free energy additively is vindicated. However, the invariance of the relative energy contribution with strain is required only to preserve the internal consistency of the statistical theory of rubber elasticity. When a strain energy function other than the Gaussian-type free energy expression is used, f_u/f can in fact be shown to vary with strain. The implication is that the Gaussian function is only one of many possible forms, and that in general f_u/f is not necessarily a constant. Quantitative arguments supporting this assertion will be published in the near future.[21]

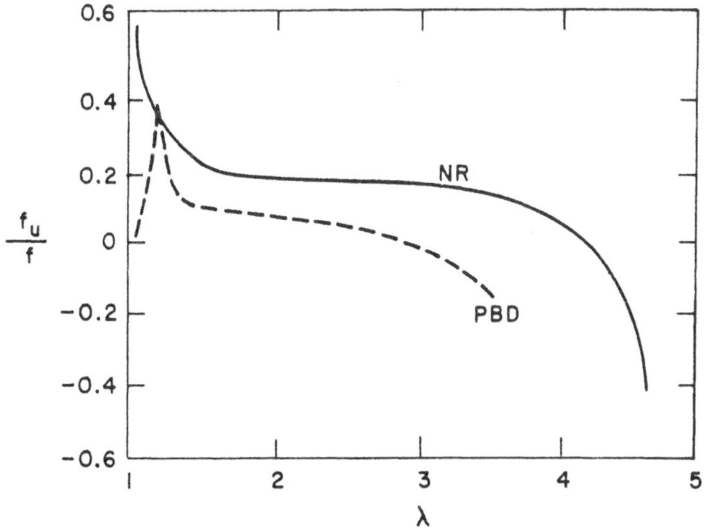

Fig. 5 f_u/f of poly-cis-1,4-butadiene[20] and natural rubber[17] as a function of strain, calculated by eq. 2.

REFERENCES

1. M. Shen and P. J. Blatz, J. Appl. Phys., 39, 4937 (1968).

2. M. Shen, Macromol., 2, 358 (1969).

3. A. Ciferri, C. A. J. Hoeve and P. J. Flory, J. Am. Chem. Soc., 83, 1015 (1961).

4. J. E. Mark and P. J. Flory, J. Am. Chem. Soc., 86, 138 (1964).

5. A. Opschoor and W. Prins, J. Polymer Sci., C16, 1095 (1967).

6. P. J. Flory, C. A. J. Hoeve and A. Ciferri, J. Polymer Sci., 34, 337 (1959).

7. P. J. Flory, J. Am. Chem. Soc., 78, 5222 (1956).

8. A. V. Tobolsky, D. W. Carlson and N. Indictor, J. Polymer Sci., 54, 149 (1961).

9. K. J. Smith, A. Greene and A. Ciferri, Kolleid. Z. u. Z. f. Polymere, 194, 49 (1964).

10. L. D. Loan, J. Appl. Polymer Sci., 7, 2259 (1963).

11. P. J. Flory, Principles of Polymer Chemistry, Cornell Univ. Press, Ithaca, N.Y., 1950.

12. P. Mason, Polymer, 5, 20 (1964).

13. P. J. Flory, N. Rabjohn and M. C. Schaffer, J. Polymer Sci., 4, 225 (1949).

14. D. Katz and A. V. Tobolsky, J. Polymer Sci., A2, 1595 (1964).

15. A. Ciferri, Makromol. Chem., 43, 152 (1961).

16. R. J. Roe and W. R. Krigbaum, J. Polymer Sci., 61, 167 (1962).

17. M. Shen, D. A. McQuarrie and J. L. Jackson, J. Appl. Phys., 38, 791 (1967).

18. G. Crespi and U. Flisi, Makromol. Chem., 60, 191 (1963).

19. G. Moraglio, European Polymer J., 1, 103 (1965).

20. B. M. E. van der Hoff, J. Macromol. Sci., A1, 747 (1967).

21. M. Shen, J. Appl. Phys., in press.

THE MICROMECHANICS OF ELASTOMER NETWORKS

S. S. Sternstein and G. M. Lederle*

Materials Division

Rensselaer Polytechnic Institute, Troy, N. Y. 12181

SUMMARY

Previous formulations of rubber elasticity theory have been based on the assumption that either the deformation, or force field associated with crosslink deformations is the same as in a continuum. There is no apriori justification for such an assumption. In the present theory, consideration is given to a completely arbitrary distribution of crosslink deformations. Generalized equations for the six components of the macroscopic stress tensor and for the stored strain energy function are formulated. The unknown deformation functions of the radius vectors between crosslinks are then found by postulating that the deformation functions are such as to minimize the free energy of the network, subject to the constraints imposed by the macroscopic stress components applied to the body. The problem is solved using the calculus of variations. At vanishing strains, the minimum free energy deformation system for the crosslink network is found to be affine. However, at finite strains the free energy criterion of equilibrium shows the deformation system to be non-affine and a functional of the particular function chosen to describe the macromolecule's conformational behavior. Non-Gaussian and Gaussian chain statistics are examined.

INTRODUCTION

The molecular foundations of the kinetic theory of rubber elasticity were conceived by Kuhn [1] and Meyer [2] and are well established. Subsequent work by James

*Present address: I.B.M. Corp., Boulder, Colorado

and Guth [3], Flory and Rehner [4], and Wall [5] was concerned with the network problem, that is, establishing the relationship between the mechanical behavior of the crosslinked network and the conformational statistics of its constituent elements, the chain molecules. Of major concern in the network problem is the calculation of the crosslink displacements relative to the continuum deformation field. It has been established that this relationship is affine for Gaussian molecular chains [3]. Various three and four chain models have been used for non-Gaussian statistics [4,6] in attempts to calculate the macroscopic behavior of elastomers at large deformations.

The intent of this paper is to present a theory (more correctly, a formalism) that provides a rational, self-consistent scheme for obtaining the macroscopic behavior of a three-dimensional crosslinked elastomeric network in terms of the response of its constituent elements, the macromolecules. The formalism is capable of application to Gaussian and non-Gaussian chain behavior, and to chain behavior that encompasses both energic and entropic force contributions upon change in conformation.

The principal feature of the formalism is the use of variational techniques to obtain the crosslink displacement field that minimizes the free energy of deformation of the macroscopic body, subject to the constraints imposed on this minimum by the externally applied deformation state. In this way, the crosslink displacements are determined as functionals of the molecular chain behavior being considered.

The present formalism is a three dimensional extension of a theory derived by Sternstein to describe the micromechanics of fiber networks [7,8]. The equivalence of formalisms for the elastomer and fiber networks may appear surprising at first glance, and is made possible because of the treatment of both systems as a directed force network composed of a discretely bonded array of elements whose properties are known.

ELEMENTAL DEFORMATIONS

The system to be considered consists of a three-dimensional array of elements (the macromolecules) that are joined together by discrete crosslinks, or junction points, to form a network. The deformation of an element is described conveniently in terms of the vector that connects its two ends. Referring to Figure 1, the junction

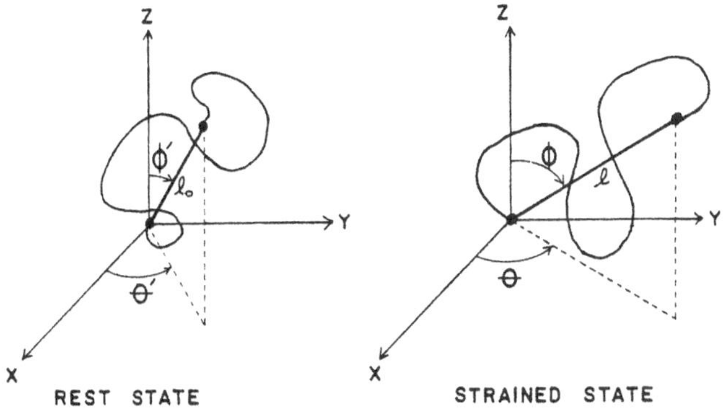

Figure 1: Representation of an element in terms of its junction point vector length and orientation.

point vector of a particular element in the unstrained (or rest) state has length ℓ_o and makes angles φ' and θ' with respect to the reference axis system. The rest length ℓ_o is taken as the root-mean-square end to end distance for the macromolecular chain between the junction points, and is the same for all elements.

In the strained state, the junction point vector deforms to a length ℓ and new orientation angles θ and φ. Choosing the strained state orientation angles as the independent variables, the elemental strain is characterized by three quantities, the extensional deformation function

$$t(\varphi,\theta) = \frac{\ell(\varphi,\theta)}{\ell_o} \tag{1}$$

and two angular deformation functions

$$\chi(\varphi,\theta) = \varphi - \varphi'(\varphi,\theta) \tag{2}$$

$$\Psi(\varphi,\theta) = \theta - \theta'(\varphi,\theta) \tag{3}$$

These functions are written to express the orientation dependence of the elemental deformations for a particular, but arbitrary, network deformation state. The dependence of t, χ, Ψ on the network strain is implicit in eqs. 1-3. The rest state of the network requires that

$$t(\varphi,\theta) \equiv 1 \quad \text{and} \quad \chi(\varphi,\theta) = \Psi(\varphi,\theta) \equiv 0 \tag{4}$$

The functions t, χ, Ψ are taken to represent the average response of those elements at angles φ and θ, but this does not imply that they represent continuum results. The behavior of a continuum line element at angles φ and θ must be obtained by a suitable average of the elemental deformations over all angles of orientation.

The angular deformation functions can be used to relate the strained and unstrained state orientation distributions of elements. Letting δn be the number of elements having rest state orientations between (θ' and $\theta' + d\theta'$, φ' and $\varphi' + d\varphi'$), then in the strained state these elements will be found between angles (θ and $\theta + d\theta$, φ and $\varphi + d\varphi$). The conservation of elements may be expressed as follows:

$$\frac{\delta n}{N} = D^*(\varphi,\theta)\sin\varphi d\varphi d\theta = D(\varphi',\theta')\sin\varphi' d\varphi' d\theta' \quad (5)$$

where N is the number of elements per unit volume, D^* is the strained state orientation distribution function that is to be determined, and D is the known rest state orientation distribution. Making use of eqs. 2 and 3, D^* can be determined and is given by

$$D^*(\varphi,\theta)\sin\varphi = [1-\chi_\varphi]\,[1-\Psi_\theta]D(\varphi-\chi,\theta-\Psi)\sin(\varphi-\chi) \quad (6)$$

where the subscripts on χ and Ψ denote partial derivatives.

The fractional distribution given by eq. 5 requires that D be normalized,

$$\int_0^\pi \int_0^\pi D(\varphi',\theta')\sin\varphi' d\varphi' d\theta' = \text{unity} \quad (7)$$

and from eq. 5, D^* is normalized also. The range of integration in eq. 7 is restricted to half of orientation space because of the indistinguishability of the "head" and "tail" of an element. For an isotropic rest state distribution, D is a constant given by $D(\varphi',\theta') = 1/2\pi$.

The strain energy stored in an element due to network deformation is taken to be a state function of the elemental deformation functions,

$$U = U(t,\chi,\Psi) \quad (8)$$

The elemental potential, U, is a free energy function whose specific functional form and relative energic to entropic character are determined by the particular elements under consideration. The exact differential of

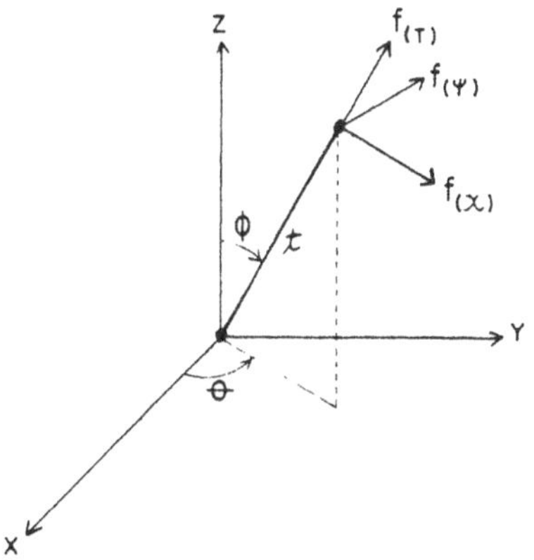

Figure 2: The elemental force components and reduced element length t: $f_{(T)}$ is tangent to the element, $f_{(\chi)}$ and $f_{(\Psi)}$ are perpendicular to the element with the latter parallel to the xy plane. The positive directions of the vectors correspond to the directions of increasing t, φ, and θ.

U can be written as

$$dU = \frac{\partial U}{\partial t}\, dt + \frac{1}{t}\frac{\partial U}{\partial \chi}\, t d\chi + \frac{1}{t}\frac{\partial U}{\partial \Psi}\, t d\Psi \qquad (9)$$

Noting from Figure 1 that $d\ell$ is a tangential displacement of the junction point vector, a tangential force is defined by

$$f_{(T)} = \left(\frac{\partial U}{\partial t}\right)_{\chi,\Psi} = U_t \qquad (10)$$

Similarly, $\ell d\Psi$ and $\ell d\chi$ are transverse displacements and define two normal forces given by

$$f_{(\chi)} = \frac{1}{t}\left(\frac{\partial U}{\partial \chi}\right)_{t,\Psi} = \frac{1}{t}\, U_\chi \qquad (11)$$

$$f_{(\Psi)} = \frac{1}{t}\left(\frac{\partial U}{\partial \Psi}\right)_{t,\chi} = \frac{1}{t}\, U_\Psi \qquad (12)$$

The rest length ℓ_o has been included in eqs. 10-12 so as to define the elemental force components relative to reduced element length t. The directions of the elemental force vectors are shown in Figure 2.

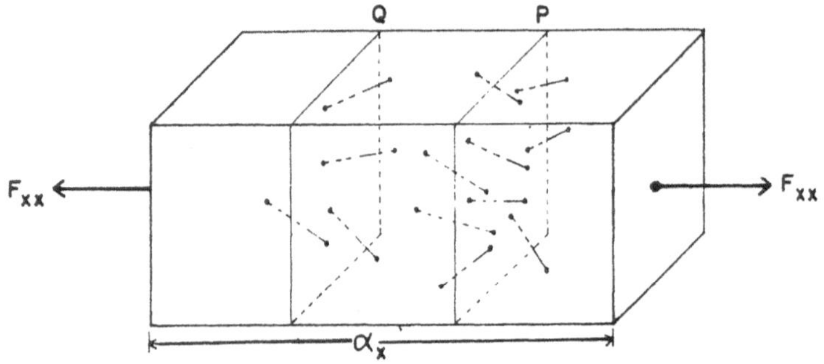

Figure 3: Illustration of the derivation for the macroscopic components of the stress tensor. Planes P and Q are parallel.

The function U is subject to certain mathematical restrictions dictated by the physical situation being described. These restrictions are not obvious and are discussed in a later section. However, it is noted here that U must obey the following rest state conditions:

$$U(1,0,0) = U_t(1,0,0) = U_\chi(1,0,0) = U_\psi(1,0,0) = 0 \quad (13)$$

NETWORK EQUATIONS

A method for determining the macroscopic stress tensor components as integral functionals of the elemental deformations is now developed. The procedure is illustrated in Figure 3 for the xx component of the stress tensor, and consists basically of summing the force contributions of those elements that are "cut" by the plane under consideration. For convenience let the macroscopic body be subjected to a homogeneous stress and deformation field, and let the reference axes x, y, z be both the symmetry axes for the distribution of elements [8] as well as the principal axes of deformation (this restriction is discussed in more detail below).

The fraction of elements per unit volume that have orientations within differentials $d\varphi$ and $d\theta$ of angles φ and θ is given by eqs. 5 and 6. However, any particular plane intersects only a limited number of this fraction. For a homogeneous stress state, parallel planes have equal stress vectors acting on them. Thus, in Figure 3 planes P and Q develop the same tractions although they intersect different elements. The probability that a plane intersects a given element depends, of course, on the projected length of the element perpendicular to the plane. Therefore, the number of elements at orientation angles φ, θ that are intersected by plane P is given by

$$\frac{\ell_x}{\alpha_x} ND^* \ (\varphi,\theta) \sin\varphi \, d\varphi \, d\theta \tag{14}$$

where ℓ_x is the projection of ℓ, and α_x is the macroscopic principal extension ratio, both in the x direction.

Recalling that the elemental forces are defined relative to an element of reduced length t, the general expression for the number of elements at orientation angles φ, θ that are intersected by planes whose normals are x, y, or z is given by

$$\frac{t_i}{\alpha_i} ND^* \ (\varphi,\theta) \sin\varphi \, d\varphi \, d\theta \tag{15}$$

where i = x, y, or z, α refers to the principal extension ratios, and the projections t_i are given by

$$t_x = t(\varphi,\theta) \sin\varphi \cos\theta \tag{16a}$$

$$t_y = t(\varphi,\theta) \sin\varphi \sin\theta \tag{16b}$$

$$t_z = t(\varphi,\theta) \cos\varphi \tag{16c}$$

The elemental forces given by eqs. 10-12 and shown in Figure 2 can be resolved into component forces along the reference axes as follows:

$$f_x = U_t \sin\varphi \cos\theta + \frac{1}{t} U_\chi \cos\varphi \cos\theta - \frac{1}{t} U_\psi \sin\theta \tag{17a}$$

$$f_y = U_t \sin\varphi \sin\theta + \frac{1}{t} U_\chi \cos\varphi \sin\theta + \frac{1}{t} U_\psi \cos\theta \tag{17b}$$

$$f_z = U_t \cos\varphi - \frac{1}{t} U_\chi \sin\varphi \tag{17c}$$

(In eqs. 16 and 17, the subscripts x,y,z refer to components; the subscripts t, χ, Ψ refer to partial derivatives.)

The contribution by those elements at orientation

angles φ, θ to the macroscopic force F_{xx} acting on plane P is given by the product of eq. 15 (with i = x) and eq. 17a. Similarly, the remaining eight combinations of eqs. 15 and 17 define the force contributions in the directions x,y,z on planes whose normals are x,y, or z. The macroscopic stress tensor is obtained by summing the contributions of elements over all of orientation space, and the result is given by

$$\frac{F_{ji}\alpha_i}{N} = \int_0^{\pi} \int_0^{\pi} f_j t_i D^*(\varphi, \theta) \sin \varphi \, d\varphi \, d\theta \qquad (18a)$$

Using eq. 6 to express D^* in terms of D and the elemental deformation functions, the above equation becomes

$$\frac{F_{ji}\alpha_i}{N} = \int_0^{\pi} \int_0^{\pi} f_j t_i [1-\chi_\varphi][1-\Psi_\theta] D(\varphi-\chi, \theta-\Psi) \sin(\varphi-\chi) \, d\varphi \, d\theta \qquad (18b)$$

where i,j = x,y, or z; no sum is implied by the repeated index i; and F_{ji} represents the force acting in the j direction on a plane whose normal is i, and stands for the nine components of the stress tensor referred to unit initial (or rest state) dimensions of the macroscopic body.

The macroscopic work of deformation per unit initial volume $\mathcal{W}$ (for work function) is obtained directly as the scalar summation of $U(t, \chi, \Psi)$ over all elements:

$$\frac{\mathcal{W}}{N} = \int_0^{\pi} \int_0^{\pi} U \, D^* (\varphi, \theta) \sin \varphi \, d\varphi \, d\theta \qquad (19a)$$

or equivalently

$$\frac{\mathcal{W}}{N} = \int_0^{\pi} \int_0^{\pi} U \, [1-\chi_\varphi][1-\Psi_\theta] D(\varphi-\chi, \theta-\Psi) \sin(\varphi-\chi) \, d\varphi \, d\theta \qquad (19b)$$

For reasons to be discussed shortly, an integral representation of the principal extension ratios is required. A measure of the square of the principal extension ratio α_i is taken to be the mean square projection of the element lengths in the i direction relative to the same quantity in the rest state, that is

$$\alpha_i^2 \approx 3 \int_0^{\pi} \int_0^{\pi} t_i^2 \, D^* (\varphi, \theta) \sin \varphi \, d\varphi \, d\theta \qquad (20a)$$

or equivalently

$$\alpha_i^2 \approx 3 \int_0^\pi \int_0^\pi t_i^2\ [1-\chi_\varphi][1-\Psi_\theta]D(\varphi-\chi,\theta-\Psi)\sin(\varphi-\chi)d\varphi\ d\theta \tag{20b}$$

Note that the rest state condition (eq. 4), the normalized character of D and D^* (eqs. 5-7), and the projections of t (eq. 16) combine to give a value of 1/3 for the mean square projection in the rest state; hence the factor 3 in eq. 20.

The stress tensor, eq. 18, has been derived on the basis of an equilibrium between the macroscopic tractions acting on a plane and the elemental forces crossing the plane; however, no such procedure exists for obtaining the macroscopic extensions. Thus, the validity of eq. 20 remains to be established.

The system of equations 18-20 are redundant to the extent that a macroscopic energy balance provides a self-consistency check. For example, suppose that uniaxial tension in the z direction is being considered and that eqs. 18-20 have been evaluated (by procedures to be developed later) and result in sets of numbers $F_{zz}\alpha_z$, $\mathcal{W}$, and α_z^2. A macroscopic strain energy balance requires that

$$d\mathcal{W} = F_{zz}d\alpha_z = (F_{zz}\alpha_z)\ \frac{d\alpha_z}{\alpha_z} \tag{21}$$

or

$$\ell n\ \alpha_z = \int \frac{d\mathcal{W}}{F_{zz}\alpha_z} \tag{22}$$

Clearly, the value of α_z can be obtained from the set $F_{zz}\alpha_z$, $\mathcal{W}$ by numerical procedures and provides a test for the prediction of eq. 20.

This numerical analysis has been performed for a variety of potential functions U of parabolic and non-parabolic form (and therefore different elemental deformation functions, t, χ, Ψ) and for both uniaxial and biaxial deformation states. In all cases the system of equations 18-20 have been found to be self-consistent. Unfortunately, an analytical verification for all admissible forms of U remains to be accomplished.

THE GAUSSIAN NETWORK

For the case where the elemental behavior is that of

a Gaussian chain, the solution to the network equations is obtained without the determination of the elemental deformation functions. The more general case requires solution for the functions, t, χ, Ψ by the method presented in the next section.

The Gaussian element is described by the elemental potential

$$U = \frac{3kT}{2}\left(\frac{\ell}{\ell_o}\right)^2 = \frac{3kT}{2}\, t^2 \qquad (23)$$

where k is Boltzmann's constant, T is absolute temperature, and ℓ_o is taken to be the r.m.s. end-to-end distance for the chain that joins two crosslinks. The potential U is defined uniquely by the extensional deformation t and is not a function of χ and Ψ. This corresponds to the classical viewpoint of the elastomer network, namely, that the conformational behavior of a chain between crosslinks in a bulk solid is the same as that of an isolated chain. Arguments contrary to this viewpoint are presented in a later section.

The elemental forces corresponding to the potential given by eq. 23 are obtained from eqs. 10-12.

$$f_{(T)} = U_t = 3kTt \qquad (24)$$

$$f_{(\chi)} = f_{(\Psi)} = 0 \qquad (25)$$

The force given by eq. 24 acts tangent to the junction point vector, and differs from the classical value $3kTt/\ell_o$ by a factor ℓ_o. This arises from the definition of the elemental forces with respect to reduced length t (see eq. 9), and the incorporation of ℓ_o into eq. 16.

The potential and elemental force defined here do not obey the rest state conditions given by eq. 13 due to the "zero rest length" character of an isolated Gaussian chain. In order to specify $\bar{U}$ as the strain (free) energy of an element relative to its rest state in the network, eq. 23 is rewritten as follows

$$U = \frac{3kT}{2}(t^2 - 1) \qquad (26)$$

which leaves eq. 24 unaltered in form. If eq. 24 is rewritten to express the force exerted by an element relative to the rest state, then $f_{(T)} = 3kT(t - 1)$, in which case $U \propto (t-1)^2$. Thus, it appears that the first two rest state conditions of eq. 13 cannot be met simultaneously without altering the form of either the potential or associated force.

Because the work function defined by eq. 19 must express the actual macroscopic work of deformation relative to the rest state of the network (and not that of isolated chains), we will adopt eq. 26 for the potential, which gives a force, eq. 24, that does not satisfy $U_t(1,0,0) = 0$. However, the presence of rest state forces does not create any difficulty (insofar as the procedure of this section is concerned) provided that incompressible states of deformation are considered. These forces simply contribute to the indeterminacy of the principal stresses by an isotropic stress, and behave as an internal pressure acting against the cohesive forces of the solid. These cohesive forces must exist but are not considered explicitely in a directed force network.

Combining eqs. 24 and 25 with eq. 17 and noting eq. 16, the stress tensor integrals (eq. 18a) obtain the form

$$\frac{F_{ji}\alpha_i}{N} = 3kT \int_0^{\pi} \int_0^{\pi} t_j t_i \; D^*(\varphi,\theta) \sin\varphi \, d\varphi \, d\theta \qquad (27)$$

The work function is obtained by combining eqs. 26 and 19a.

$$\frac{\mathcal{W}}{N} = \frac{3kT}{2} \int_0^{\pi} \int_0^{\pi} (t^2 - 1) D^* \; (\varphi,\theta) \sin\varphi \, d\varphi \, d\theta \qquad (28)$$

The shear stress components of eq. 27 are easily shown to be zero provided that the axes x,y,z are taken to be symmetry axes for the distribution function in the unstrained state (D), as well as the principal axes of deformation. In this case,

$$D^*(\varphi,\theta) = D^*(\pi-\varphi,\theta) = D^*(\varphi,\pi-\theta) \qquad (29)$$

and $$t(\varphi,\theta) = t(\pi-\varphi,\theta) = t(\varphi,\pi-\theta) \qquad (30)$$

Considering either the integral on φ or θ in two stages, from 0 to $\pi/2$ and $\pi/2$ to π, then inspection of eqs. 16, 27, 29, and 30 shows that $F_{ji}\alpha_i$ is zero when $i \neq j$. Thus, the principal axes of stress coincide with the principal axes of deformation.

In the event that the distribution of elements possesses no symmetry axes or that the symmetry axes exist but do not coincide with the principal axes of deformation, then either or both of eqs. 29 and 30 are not valid. The principal axes of stress and deformation would not coincide. For an isotropic distribution, the principal axes of stress and deformation must coincide. In this case, D is a constant and the symmetry axes of D^* coincide with the principal axes of deformation due to eq. 30, regardless of the orientation of the principal axes. A more detailed analysis of symmetry considerations is given by Sternstein [8].

Comparing the principal extension ratio integrals, eq. 20a, with the stress integals, eq. 27, for the three cases i = j = x, y, or z, one obtains by inspection the three principal stresses

$$F_x = NkT\alpha_x,\ F_y = NkT\alpha_y,\ F_z = NkT\alpha_z \tag{31}$$

Similarly, inspection of eq. 28 and eqs. 16, 20a, and the normalized character of D* (eqs. 5 and 7), gives the work function.

$$\mathcal{W} = \frac{NkT}{2}\left(\alpha_x^2 + \alpha_y^2 + \alpha_z^2 - 3\right) \tag{32}$$

If the network is incompressible, then

$$\alpha_x\alpha_y\alpha_z = \text{unity} \tag{33}$$

and eq. 32 is identical to the well-known result for the Gaussian network. The condition of incompressibility implies that the principal stresses, eq. 31, are known only to within an arbitrary additive constant.

These results have been obtained without recourse to, and indeed without knowledge of, the elemental deformation functions t, χ, Ψ. Thus, it would appear that the question of an affine deformation in a Gaussian network need not be answered in order to obtain the work function, eq. 32. Clearly, the procedure used here has succeeded only because of the particular form of potential function given by eq. 26 and the network integrals, eqs. 18, 19, 20.

We now consider elemental potentials of general form, for which the evaluation of the elemental deformation functions is essential to the solution of the network equations.

MINIMUM FREE ENERGY CONSIDERATIONS

The elastic response of an element is described in terms of its potential $U(t,\chi,\Psi)$. For a particular choice of this function, the network behavior is determined by the set of integral equations 18-20, whose evaluation requires that the functions $t(\varphi,\theta)$, $\chi(\varphi,\theta)$, and $\Psi(\varphi,\theta)$ be obtained for each macroscopic state of deformation.

The network can be viewed as an ensemble of elements that are perturbed from their rest states by the imposition of an external force or deformation; the response of the ensemble must be consistent with the average values of stress, deformation, and work that serve to characterize the continuum. If the ensemble is to be in equilibrium with the external perturbation, then the system is in a state of minimum free energy. Thus, a procedure for solution is obtained.

> The elemental deformation functions are to be determined such that they minimize the work function, eq. 19, subject to the constraints imposed on this minimum by the macroscopic state of deformation.

In considering the elemental deformations to be functions of the element's orientation, but independent of the element's environment, the ensemble may be described as uncoupled. Therefore, the formalism used here represents a first approximation to a real network in which additional constraints would act on the various elements of the ensemble.

The minimum free energy approach to a network has been used to describe the stress-strain behavior of hydrogen-bonded solids at small strains [9]. The present forms of the stress tensor and work function (in two dimensions) have been applied to the fiber network problem [7,8], for which the minimizing constraints were taken to be the stress tensor components. In the case of an elastomer network, the stress constraints must be replaced by deformation constraints if the admissable solutions to the problem are to be restricted to constant volume deformations.

The work function, eq. 19, is to be minimized subject to constraints given by the macroscopic extension ratios, eq. 20. The problem so defined can be solved by the calculus of variations, and the constraints are of the isoperimetric type (10,11). A minimizing functional M is defined as follows:

$$M = I + \lambda_x I_x + \lambda_y I_y + \lambda_z I_z \tag{34}$$

where I is the integrand of eq. 19b; I_x, I_y, I_z are the integrands of the three equations designated by eq. 20b; and the λ's are Lagrange multipliers that can be treated as constants for isoperimetric constraints. Noting the integrands of eqs. 19b and 20b and using eq. 16, the previous equation becomes

$$M = \Xi(\theta,\varphi,t,\chi,\Psi)[1-\chi_\varphi][1-\Psi_\theta] \tag{35}$$

where $\Xi = [U(t,\chi,\Psi) + H(\varphi,\theta)t^2]D(\varphi-\chi,\theta-\Psi)\sin(\varphi-\chi)$

and $H = [\lambda_x \cos^2\theta + \lambda_y \sin^2\theta]\sin^2\varphi + \lambda_z \cos^2\varphi$

If the functions t,χ,Ψ are to minimize eq. 19b subject to constraints given by eq. 20b, a necessary condition is that the functional M satisfy the three Euler equations:

$$\frac{\partial M}{\partial t} - \frac{\partial}{\partial\theta}\left(\frac{\partial M}{\partial t_\theta}\right) - \frac{\partial}{\partial\varphi}\left(\frac{\partial M}{\partial t_\varphi}\right) = 0 \tag{36a}$$

$$\frac{\partial M}{\partial \chi} - \frac{\partial}{\partial\theta}\left(\frac{\partial M}{\partial \chi_\theta}\right) - \frac{\partial}{\partial\varphi}\left(\frac{\partial M}{\partial \chi_\varphi}\right) = 0 \tag{36b}$$

$$\frac{\partial M}{\partial \Psi} - \frac{\partial}{\partial\theta}\left(\frac{\partial M}{\partial \Psi_\theta}\right) - \frac{\partial}{\partial\varphi}\left(\frac{\partial M}{\partial \Psi_\varphi}\right) = 0 \tag{36c}$$

where $\partial/\partial\varphi$ and $\partial/\partial\theta$ are total partial derivatives, and all other derivatives are partial partials.

Substituting eq. 35 into eq. 36 and performing the necessary operations results in

$$\Xi_t = 0 \tag{37a}$$

$$[1-\Psi_\theta][\Xi_\chi + \Xi_\varphi + \Xi_\Psi \Psi_\varphi] - \Xi\, \Psi_{\varphi\theta} = 0 \tag{37b}$$

$$[1-\chi_\varphi][\Xi_\Psi + \Xi_\theta + \Xi_\chi \chi_\theta] - \Xi\, \chi_{\varphi\theta} = 0 \tag{37c}$$

where all subscripts on Ξ denote partial partial derivatives

For a given function $U(t,\chi,\Psi)$ and known initial orientation distribution D, eq. 37 results in three simultaneous partial differential equations whose solutions determine the functions $t(\varphi,\theta)$, $\chi(\varphi,\theta)$, and $\Psi(\varphi,\theta)$. These functions (called extremals) minimize the work function (eq. 19b) and satisfy the constraints (eq. 20b), the latter being used to determine the Lagrange multipliers. A less general but more tractable analysis is presented in the next section.

PARTIALLY ISOTROPIC DEFORMATIONS

In order to simplify the variational formulation of the previous section, two restrictions are introduced here. (1) Elastomeric networks of common interest are generally isotropic in the rest state, and the analysis is restricted to such systems. (2) The analysis is restricted to those deformation fields that are isotropic in a plane which, by convention, is taken to be the xy plane. Thus, $\alpha_x = \alpha_y$, $\lambda_x = \lambda_y$, $F_x = F_y$. The z direction constitutes the third principal axis, and F_z and λ_z are not restricted. Any combination of equal biaxial tension or compression in the xy plane with uniaxial tension or compression in the z direction is permitted. Unfortunately, pure shear and torsion are excluded from the restricted analysis.

The first restriction requires that $D = 1/2\pi$ and results in only minor simplification of eqs. 35 and 37. However, the necessity of choosing an analytical description of the rest state distribution is eliminated. Note that this restriction does not imply that the strained state distribution is isotropic (refer to eq. 38). The second restriction results in a reduction of the variational problem from three partial differential equations to two algebraic equations.

Because the xy plane exhibits both structural and deformation isotropy, the angle θ in Figures 1, 2, and 3 is not a variable of the problem. There is no average elemental orientation with respect to θ, and $\Psi \equiv 0$. This is consistent with the coincidence of the symmetry axes of the distribution function with the principal axes of deformation for all rotations of the xy plane about the z axis. Elements that lie parallel to the xy plane ($\varphi = \pi/2$) undergo extension only, while other elements ($\varphi \neq \pi/2$) undergo both extension and orientation (in the φ direction).

If these considerations are applied, then $t = t(\varphi)$, $\chi = \chi(\varphi)$, $U = U(t,\chi)$, and $f(\Psi) \equiv 0$. Restriction (1) and $\Psi \equiv 0$ reduce eq. 6 to

$$D^*(\varphi) \sin \varphi = \frac{[1-\chi']}{2\pi} \sin (\varphi-\chi) \tag{38}$$

where $\chi' = d\chi/d\varphi$. Combining these results with eqs. 16-20 and integrating with respect to θ, gives

$$\frac{F_z\alpha_z}{N} = \int_0^{\pi/2} [tU_t\cos^2\varphi - U_\chi \sin\varphi\cos\varphi][1-\chi']\sin(\varphi-\chi)d\varphi \tag{39a}$$

$$\frac{F_x\alpha_x}{N} = \frac{F_y\alpha_y}{N} = \frac{1}{2}\int_0^{\pi/2} [tU_t\sin^2\varphi + U_\chi \sin\varphi \cos\varphi] [1-\chi']\sin(\varphi-\chi)d\varphi \tag{39b}$$

$$\frac{\mathcal{W}}{N} = \int_0^{\pi/2} U(t,\chi)\, [1-\chi']\, \sin (\varphi-\chi)\, d\varphi \tag{40}$$

$$\alpha_z^2 = 3\int_0^{\pi/2} t^2\cos^2\varphi\, [1-\chi']\, \sin (\varphi-\chi)d\varphi \tag{41a}$$

$$\alpha_x^2 = \alpha_y^2 = 3\int_0^{\pi/2} \frac{t^2}{2} \sin^2\varphi\, [1-\chi']\sin(\varphi-\chi)\, d\varphi \tag{41b}$$

where the integration on φ has been reduced to the interval 0 to $\pi/2$ by symmetry.

The minimizing functional M for eq. 40 with constraints given by eq. 41 is:

$$M = I + \lambda_z I_z + 2\lambda_x I_x$$

or $$M = M(\varphi,t,\chi,\chi') = \Xi(\varphi,t,\chi)\, [1-\chi'] \tag{42}$$

where $$\Xi = [U(t,\chi) + H(\varphi)t^2]\, \sin (\varphi-\chi)$$

and $$H = \lambda_z \cos^2\varphi + \lambda_x \sin^2\varphi$$

The Euler equations to be satisfied by M are

$$\frac{\partial M}{\partial t} - \frac{d}{d\varphi}\frac{\partial M}{\partial t'} = 0 \tag{43a}$$

$$\frac{\partial M}{\partial \chi} - \frac{d}{d\varphi}\frac{\partial M}{\partial \chi'} = 0 \tag{43b}$$

which gives

$$\Xi_t = 0 \text{ and } \Xi_\chi + \Xi_\varphi = 0 \tag{44}$$

where the subscripts in eq. 44 refer to partial partial derivatives. Expressed in terms of $U(t,\chi)$ the previous equation gives

$$U_t + 2Ht = 0 \quad \text{and} \quad U_\chi + 2Pt^2 = 0 \tag{45}$$

where $\quad P(\varphi) = \frac{H'(\varphi)}{2} = (\lambda_x - \lambda_z) \sin\varphi \cos\varphi$

For a given elemental potential $U(t,\chi)$, eq. 45 gives two algebraic equations whose simultaneous solutions yield the associated extremals $t(\varphi)$ and $\chi(\varphi)$.

The Euler equations and associated functions must satisfy certain conditions that are required by symmetry, and are considered in detail elsewhere [8]. These conditions apply to the present analysis and are given without further proof:

If the symmetry axes of the rest state orientation distribution function coincide with the principal axes of deformation (always true for an initially isotropic network), then conditions 1-5 apply for all admissible potentials $U(t,\chi)$:

(1) The Lagrange multipliers are identically zero in the rest state of the network.

(2) The stress tensor is in principal form.

(3) The extensional deformation function is even, $t(\varphi) = t(\pi-\varphi) = t(-\varphi)$.

(4) The angular deformation function is odd, $\chi(\varphi) = -\chi(-\varphi) = -\chi(\pi-\varphi)$, from which it follows that $\chi(0) = \chi(\pi/2) = \chi(\pi) = 0$.

(5) The Euler equations are invariant with respect to changes of reference frame that leave the symmetry unchanged, for example, an inversion of the z axis to form a left-handed coordinate system.

(6) An admissible potential $U(t,\chi)$ must satisfy the following conditions:

(a) $U(t,\chi)$, U_t, U_χ are positive definite for all values of $t > 1$ and $\chi > 0$ experienced by the elements of the network.

(b) The rest state condition requires that $U(1,0) = U_t(1,0) = U_\chi(1,0) = 0$. The second condition was not obeyed in the treatment of the Gaussian potential, but variational techniques were not employed there.

(c) The elemental potential must be even in χ, $U(t,\chi) = U(t,-\chi)$, and therefore $U_\chi(t,\chi) = -U_\chi(t,-\chi)$.

Conditions a and b are consistent with the usual work requirements of an elastic element with finite length rest state, and condition c is required because of the oddness of χ.

A NON-GAUSSIAN EXAMPLE

The end-to-end vector is a symmetry axis for the conformational states of an isolated macromolecule constrained to constant end-to-end distance. This symmetry is a consequence of the environmental isotropy when viewed along axes perpendicular to, and rotated about, the end-to-end vector. The resultant force developed by the conformational entropy of such a macromolecule must be collinear with the end-to-end vector.

When a macromolecular chain is considered as the constituent element of a network, there is some reason to doubt the coincidence of the junction point vector and the axis of time averaged conformations. The deformation of the network by any deviatoric stress field always results in a net reorientation of elements toward the principal axes; and the environment of an element must reflect, on the average, this orientation gradient perpendicular to the junction point vector. Those elements that are oriented along the principal axes of deformation are exceptions to this effect because of the absence of reorientation as discussed in the previous section.

In terms of the elemental potential, symmetry requires that $U = U(t)$, whereas the reorientation induced asymmetry requires $U = U(t,\chi)$. Thus, it may be that the

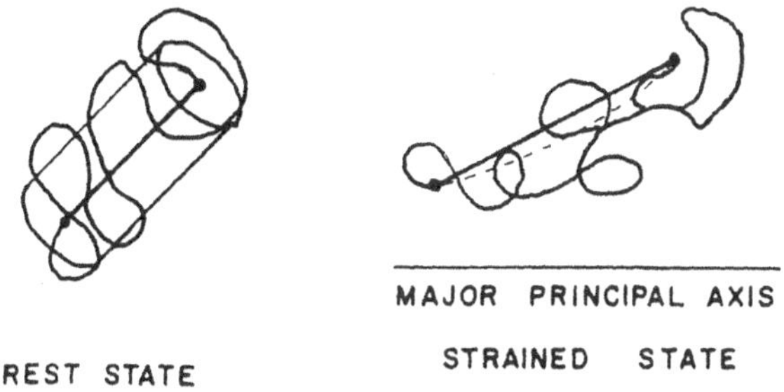

Figure 4: Illustration of asymmetric conformations. In the rest state the symmetry axis and junction point vector coincide, whereas in the strained state the conformations are biased in the direction of elemental orientation.

elemental potential of an isolated chain and that of the same chain in a network subjected to deviatoric deformations are different. The origin of the asymmetric conformations of a macromolecular element may be due to intermolecular entanglements between junction points. This effect is illustrated in Figure 4, where the axis of time-averaged conformations is displaced from the junction point vector in the strained state.

If the elemental potential is to exhibit a stable rest state, then the first approximation to $f_{(\chi)}$ will be linear in χ, as given by eq. 46.

$$f_{(\chi)} = \frac{1}{t} U_{\chi} = q\chi \tag{46}$$

where q is a positive constant. Combining this equation with the Gaussian contribution to $f_{(T)}$ given by eq. 24, then a potential $U(t,\chi)$ is obtained in the form

$$U(t,\chi) = \frac{3}{2} kT \; (t-1)^2 + \frac{q}{2} t\chi^2 \tag{47}$$

In using eq. 24, the net tangential force 3kT(t-1) has been used to express the potential U relative to the rest state (see eq. 13).

A potential of the form $U = k(t-1)^2/2 + p\chi^2/2$ has been used in the description of fiber networks [8], where the factor p characterizes the rigidity of fiber to fiber

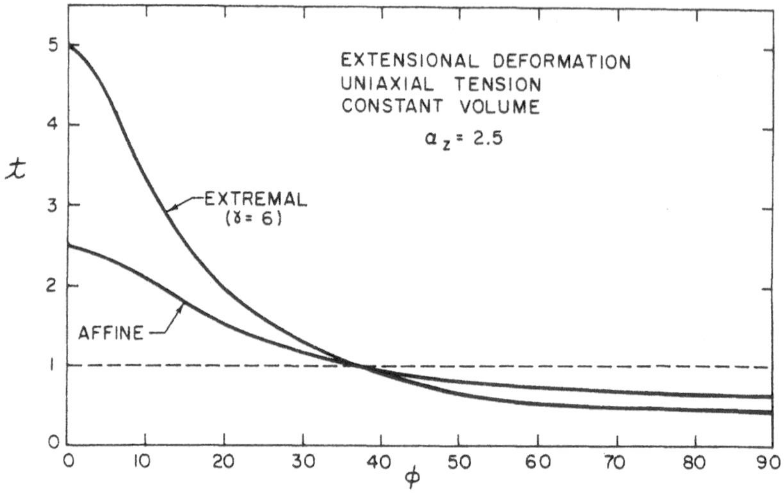

Figure 5: Comparison of the extensional deformation extremal and its affine counterpart for uniaxial tension.

junctions with respect to reorientation. In the case of eq. 47, the parameter q may be interpreted as representing the intensity of transverse coupling (or entanglement) along the contour of the chain element. Thus, the present viewpoint differs from the treatment of an entanglement as an additional (but perhaps labile) crosslink in that a normal force component is generated perpendicular to the junction point vector.

Substituting eq. 47 into the Euler equations (eq.45) and performing the necessary operations, the extremals corresponding to the chosen potential are obtained.

$$t(\varphi) = \frac{-(1 + H) + \sqrt{(1 + H)^2 + 2P^2/\gamma}}{P^2/\gamma} \tag{48}$$

$$\chi(\varphi) = - Pt/\gamma \text{ where } \gamma = q/3kT \tag{49}$$

In eqs. 48 and 49, the functions $H(\varphi)$ and $P(\varphi)$ are as defined in eqs. 42 and 45 but with the Lagrange multipliers multiplied by $2/(3kT)$.

For comparison, the elemental deformations that correspond to an affine deformation are given here:

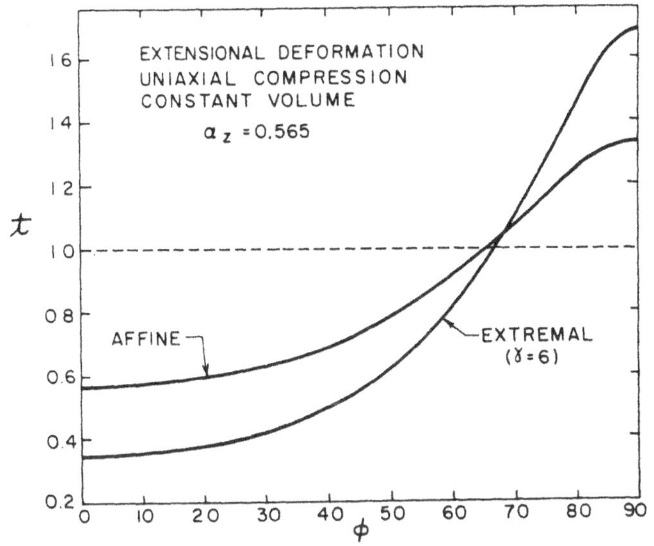

Figure 6: The extensional deformation extremal in uniaxial compression (or equal biaxial tension).

$$t(\varphi) = \frac{\alpha_z}{\cos\varphi}\left[\frac{1}{1 + R^2\tan^2\varphi}\right]^{\frac{1}{2}} \quad (50)$$

$$\chi(\varphi) = \varphi - \arctan\,(R \tan \varphi) \quad (51)$$

where $R = \alpha_z/\alpha'_x$.

Substitution of the extremals into eq. 41 and the use of an iterative numerical integration serve to determine the Lagrange multiplier pairs [L.M.P.(λ_z, λ_x)] that generate the desired network deformations (subject to the restriction of eq. 33). The stress and work function integrals (eqs. 39,40) are evaluated for the same L.M.P., after expressing their integrands in terms of the extremals (eqs. 48, 49 and 47, 10, 11).

In general, the stress field must be corrected by an isotropic stress. For example, if the L.M.P. are determined for uniaxial tension, the $\alpha_z = \alpha$ and $\alpha_x = \alpha_y = 1/\sqrt{\alpha}$ for each pair, with a different pair required for each α. However, these L.M.P. do not reduce eq. 39b to zero. The non-zero value of $F_x (= F_y)$ defines the isotropic stress correction (with due account of the

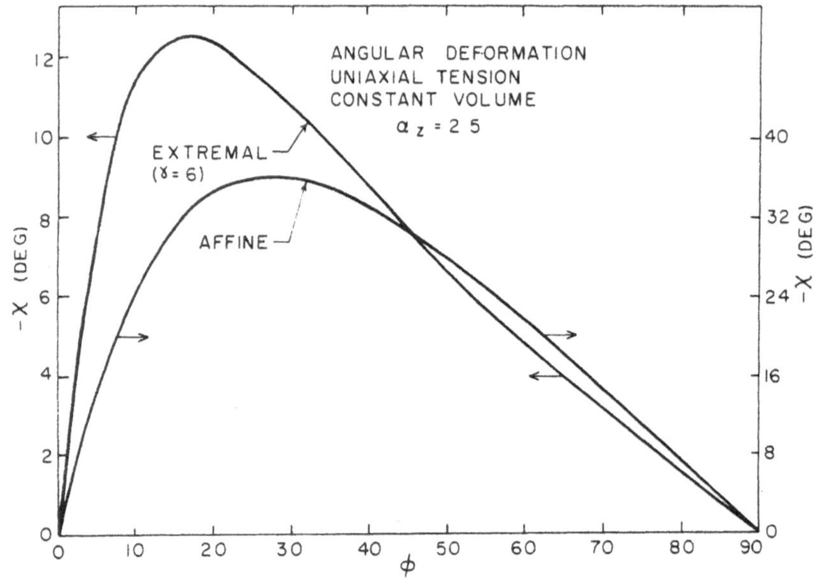

Figure 7: The angular deformation extremal and affine counterpart for uniaxial tension.

strained state areas) required to reduce the x and y principal stresses to zero, and to determine the net stress in the z direction. This procedure is consistent with the condition of incompressibility and implies that the principal stresses are determined only to within an arbitrary isotropic stress.

The extensional deformation extremal (eq. 48) is compared with its affine counterpart (eq. 50) in Figure 5 for uniaxial tension with $\alpha = 2.5$, and a parameter $\gamma = 6$. The same comparison is made in Figure 6 for uniaxial compression with $\alpha = 0.56$ (equivalent to biaxial tension in the xy plane with $\alpha_x = \alpha_y = 1.31$). The corresponding comparisons of the angular deformation extremal (eq. 49) and affine behavior (eq. 51) are given in Figures 7 and 8. Additional values of α have been investigated and it is found that the extremals approach affine behavior as α tends toward unity.

The degree to which the χ extremal deviates from affine response increases with the deviatoric character of the deformation field, a measure of which can be taken as R of eq. 51. In the present case, these deviations are in the order uniaxial tension > biaxial tension (or uni-

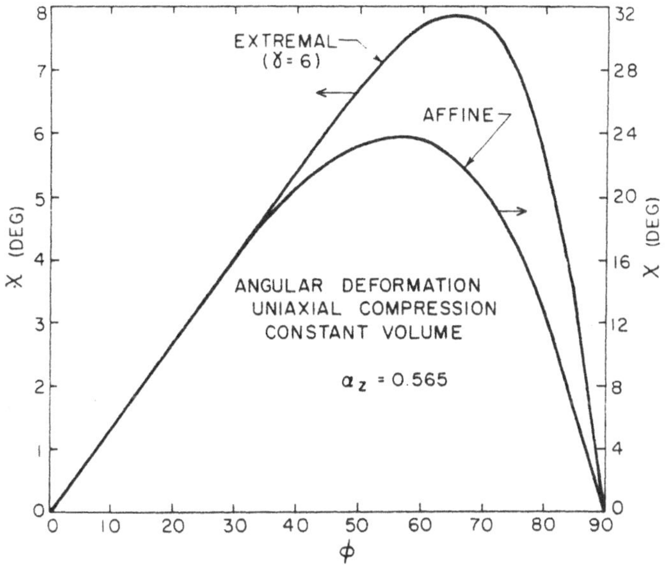

Figure 8: The angular deformation extremal in uniaxial compression (or equal biaxial tension).

axial compression). Unfortunately, other deformation fields require solution of the more general theory (eq.37).

Additional results have been obtained from the two dimensional theory [8] for which it is found that the non-affine character of χ varies in the order uniaxial tension > tension with constrained transverse strain > equal biaxial tension. For the last case, no reorientation of elements occurs ($\chi \equiv 0$) and the t extremal for Hookean elements is identically affine. (It is emphasized that biaxial tension in a three-dimensional network does result in element reorientation.)

From these results, it is apparent that the formalism developed in this paper has the interesting (and perhaps useful) capability of predicting different characteristics of the elemental deformations (associated with a given potential U) as a function of the deformation field. It is known that experimental results on elastomers display deviations from the classical theory that depend on both deformation field and level of strain.

A vanishing strain approximation to the extremals and network integrals (eqs. 48, 49, and 39-41) has been used to obtain Young's modulus. The result (which is valid only for the potential given by eq. 47) is

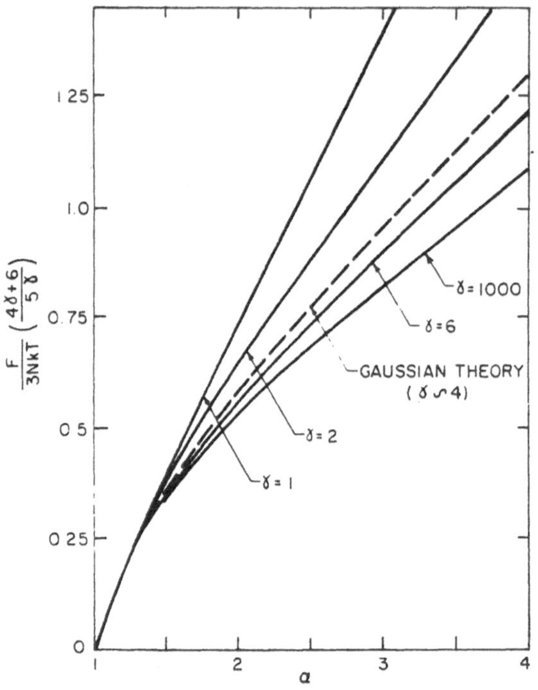

Figure 9: Force-elongation behavior in uniaxial tension for the elemental potential, eq. 47.

$$E = \left(\frac{5\gamma}{4\gamma + 6}\right) 3NkT \quad (52)$$

and differs from the value for the Gaussian theory (eq. 32) by the factor in parentheses.

The force-elongation curves for several values of γ are compared with the Gaussian theory in Figures 9 and 10, where the initial modulii have been scaled to the same value by eq. 52. It is interesting to note that $\gamma = 4$ gives a tensile curve virtually identical to the Gaussian theory. However, the extremals in this case are as non-affine as those shown in Figures 5 and 7 for a γ of 6. This serves to illustrate the non-unique relationship between network behavior and the elemental deformations that exists when different elemental potentials are compared.

For a value of $\gamma = 6$, the factor in eq. 52 is unity, and the non-Gaussian and Gaussian modulii are equal. However, the non-Gaussian curve falls below the class-

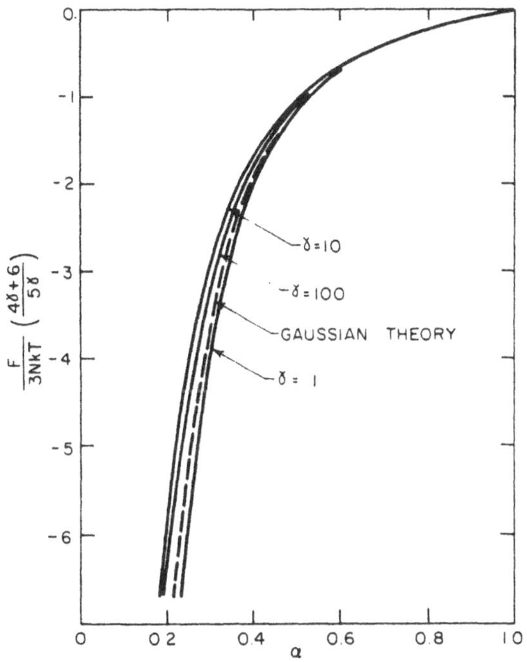

Figure 10: Force-elongation behavior in uniaxial compression for the elemental potential, eq. 47.

ical result at finite deformations in a manner somewhat like the behavior of unswollen elastomers at intermediate deformations (ca. $1.2 < \alpha < 4$). Recent results by Thirion [12] indicate that a Mooney plot for unswollen elastomers may be linear for values of α closer to unity than previously accepted. In any case, the calculations presented here are intended to illustrate the formalism and do not justify detailed comparison with experimental results at present.

The authors are aware of the empirical and rather tenuous nature of the elemental potential given by eq. 47. It is their intention to pursue a more rigorous development of such potentials and the subsequent effects on network response. One incentive for such a study is the obvious application of a more thorough knowledge of the elemental deformations (e.g. Figures 5-8) to the failure of elastomers.

ACKNOWLEDGMENTS

This paper is taken, in part, from a thesis submitted to the Materials Division in partial fulfillment of the requirements for a doctorate [13].

The authors are grateful to the National Aeronautics and Space Administration for a traineeship (to G.M.L.) and a Materials Research Center Grant (to S.S.S.) which supported this work.

REFERENCES

1. W. Kuhn, Kolloid-Z. 68, 2 (1934); 76, 258 (1936).

2. K. H. Meyer, G. von Susich, and E. Valko, Kolloid - Zeitschr., 59, 208 (1932).

3. H. James and E. Guth, J. Chem. Phys. 11, 455 (1943); 15, 669 (1947).

4. P. Flory and J. Rehner, J. Chem. Phys. 11, 512 (1943).

5. F. Wall, J. Chem. Phys. 10, 485 (1942).

6. L. R. G. Treloar, Trans. Faraday Soc., 42, 77, 83, (1946); 50, 881 (1954); The Physics of Rubber Elasticity, 2nd Edition, Oxford Press (1958).

7. S. S. Sternstein, Unpublished work at Rensselaer Polytechnic Institute, 1961-1965.

8. S. S. Sternstein, "The Micromechanics of Fiber Networks", in Cellulose and Cellulose Derivatives, Vol. 5, Bikales and Segal, Eds., J. Wiley, New York (1971).

9. S. S. Sternstein and A. H. Nissan, Trans. of the Oxford Symposium on Formation and Structure of Paper, p. 319, B.P. & B.M.A., London (1961).

10. R. Weinstock, Calculus of Variations, McGraw-Hill, New York (1952).

11. L. Elsgolc, Calculus of Variations, Pergammon Press, London (1962).

12. P. Thirion, Institut Francais du Caoutchouc, Paris, Personal Communication.

13. G. M. Lederle, "A Stress-Strain Law for Rubberlike Materials From Minimum Free Energy Considerations," Ph.D. Dissertation, Rensselaer Polytechnic Institute (1968).

THE STATISTICAL MECHANICS OF RUBBERS

S. F. Edwards

Department of Theoretical Physics, The Schuster Laboratory, University of Manchester, Manchester, England

SUMMARY

The paper concerns the basis of the statistical mechanics of rubber. It is firstly shown that an extension of conventional (i.e. history independent) statistical mechanics can be made for systems with permanent constraints when the weighting of configurations is that of an equilibrium system without the constraints. A method of calculation is then given based on carrying out the usual kind of statistical mechanical calculations with all the coordinates of the ensemble retained i.e. to obtain the effective free energy of a constrained rubber one may average over the constraints of a system with (n+1) members, roughly speaking the rubber in 3(n+1) dimensions. This technique is used to calculate the free energy of rubber firstly in the simplest system, crosslinked interpenetrable chains, then to systems in which the topological constraints are added and finally (but rather incompletely) to the full case with internal energy.

It is possible to determine certain conditions in which a rigorous solution can be found, and for the interpenetrable network a rigorous solution can be obtained which can be shown to be microscopically affine. However the topological constraints can be shown to be not invariant under affine transformations, so that the general microscopic effect of a deformation cannot be affine. The final form of the free energy is a complicated function of the deformation since it does not appear possible to find a simple coupling constant which will lead to a simple answer. The simple $\Sigma\lambda_i^2$ law of the phantom chains is replaced by a more complex law containing all the elastic invariants, when the lack of interpenetrability of chains is allowed for.

1. INTRODUCTION

The goal of predicting the properties of a rubber from its composition and mode of formation is, in general, a difficult one to achieve. This for two reasons: it is not clear what the precise microscopic structure of the rubber is i.e. how it depends on the cross linking cure, and even when properly specified, difficult mathematical problems remain. In this paper I shall explore the theoretical problems from the point of view which has been adopted in the Manchester polymer group, that one should find the circumstances under which one can produce a proper theory and then make laboratory systems which fulfil these conditions as nearly as possible. This is of course a familiar philosophy in polymer science, for example in making studies of polymer solutions near the θ temperature, but it does not seem to have been exploited for the network problem. The work here presented is by no means complete, and the reader is asked to forgive the fact that some parts of this many sided problem are gone into in depth and others shallowly, this is the present state of progress.

The first problem is to formulate the statistical mechanics of a system with permanent constraints.

(Since there are mathematical techniques used in the paper which may be unfamiliar to some readers, a mathematical appendix is added.)

2. STATISTICAL FORMULATION

When a network is formed, the cross links are permanent and they also, when in sufficient density, enforce certain topological constraints between the polymers. Let us label the free energy of a system with some particular cross linkage and topology as F_m . Let the probability of finding this particular system be p_m. Then for a canonical ensemble the effective free energy

$$\mathcal{F} = \sum_m p_m F_m . \tag{2.1}$$

If one makes some change in the system, denote the new values of $\mathcal{F}$ and F_n by $\tilde{\mathcal{F}}$ and $\tilde{F}_n$. Then

$$\tilde{\mathcal{F}} = \sum_m p_m \tilde{F}_m \tag{2.2}$$

where p_m does <u>not</u> change, it having been established during the formation of the rubber. The simplest p_m will be to assume that the configurations have the same weight as that of thermal equilibrium. Thus if the cross links can slip, one has

$$e^{-F_m/\kappa T} = \int_m e^{-H/\kappa T} \tag{2.3}$$

and the free energy of this slipping system is

$$e^{-F/\kappa T} = \sum_m e^{-F_m/\kappa T} \tag{2.4}$$

and

$$p_m = e^{(F-F_m)/\kappa T} \tag{2.5}$$

Hence the network has an effective free energy of

$$\tilde{\mathcal{F}} = \sum_m e^{(F-F_m)/\kappa T} \tilde{F}_m \tag{2.6}$$

(the entropy of the cross linking process is not of interest)

This is a 'history free' formula and represents the <u>simplest</u> rigorous formula which describes a system with constraints. Comments on its attainability will be made later.

Some interesting results can immediately be obtained. Suppose an infinitesimal change is made

$$\tilde{F}_m = F_m + \Delta \tilde{F}_m \tag{2.7}$$

$$\Delta\tilde{\mathcal{F}} = \sum_m e^{(F-F_m)/\kappa T} \Delta\tilde{F}_m \tag{2.8}$$

$$= -\kappa T \sum_m e^{(F-F_m)/\kappa T} e^{\tilde{F}_m/\kappa T} \Delta\left(e^{-\tilde{F}_m/\kappa T}\right)\Big|_{F_m=\tilde{F}_m}$$

$$= -\kappa T e^{F/\kappa T} \Delta \sum_m e^{-\tilde{F}_m/\kappa T}\Big|_{F_m=\tilde{F}_m}$$

$$= -\kappa T e^{F/\kappa T} \Delta\left(e^{-\tilde{F}/\kappa T}\right)\Big|_{F=\tilde{F}} \tag{2.9}$$

$$\therefore \Delta\tilde{\mathcal{F}} = \Delta\tilde{F} \tag{2.10}$$

at $F = \tilde{F}$

So for infinitesimal changes the constraints do not affect the system; but the second order effect gives Hookes Law, for similar algebra gives

$$\Delta^2\tilde{\mathcal{F}} = \Delta^2\tilde{F} + \frac{1}{kT}\sum_m e^{(F-F_m)/kT}\left\{(\Delta\tilde{F}_m)^2-(\Delta\tilde{F})^2\right\} \tag{2.11}$$

For shear, for example, since $\tilde{F}=\tilde{F}(V)$ alone $\left\{\begin{array}{l}\Delta^2\tilde{F}=0\\ \Delta\tilde{F}=0\end{array}\right\}$ and

$$\Delta^2\tilde{\mathcal{F}} = \frac{1}{kT}\sum_m e^{(F-F_m)/kT}(\Delta\tilde{F}_m)^2 \tag{2.12}$$

which is Hookes Law (when evaluated in macroscopic terms).

Edwards and Freed (1) note these formulae and apply them in approximation to gelation problems. In this paper the exact formulation will be carried as far as can be and the formula is adapted to practical calculation by observing that the difficulty lies in the mixture of $e^{-F_m/kT}$ and $\tilde{F}_m$. A characteristic of polymer problems is the central position of gaussian functions, and these have the characteristic that it is as easy to work in n dimensions as in 3 or 1. Thus consider $n+1$ systems all of which have the same constraints m, but n systems are in the present condition, and one in the initial condition. Calculate the free energy $F(n)$ of this system in the normal way

$$e^{-F(n)/kT} = \sum_m e^{-F_m^{(0)}/kT-\tilde{F}_m^{(1)}/kT\,\cdots\cdots\,-\tilde{F}_m^{(n)}/kT} \tag{2.13}$$

$$= \sum_m \int_m e^{-H^{(0)}/kT}\tilde{\int_m} e^{-H^{(1)}/kT}\cdots\tilde{\int_m} e^{-H^{(n)}/kT} \tag{2.14}$$

where $\tilde{\int}$ means integrate in present conditions. Now suppose (as does indeed occur) that $F(n)$ can be expanded in so that

$$F(n) = F + n\tilde{\mathcal{F}} + O(n^2) \tag{2.15}$$

Then $\tilde{\mathcal{F}}$ in (2.15) is indeed the $\tilde{\mathcal{F}}$ of (2.6).

Thus the simplest statistical mechanics of rubber demands the solution of an ensemble problem, but one in which the individuality of each member of the ensemble is preserved till the verv last step of the calculations.

So far the discussion is general, but to proceed further the nature of the constraints must be made explicit. The most powerful mathematics available is that of continuum analysis i.e. one should use differential equations rather than matrices in a first attack. It is therefore convenient (though not at all essential) to represent the polymer molecules by curves $\underset{\sim}{R}(s)$ in space, s being the arc length. Again in the quest for simplicity, if the individual polymers are long, the end corrections are small, so one may treat the whole system as one very long polymer; this eases the notation and can, quite trivially, be returned to the real case as required. The kinetic energy will not be discussed so the Hamiltonian is now $H = H^{(0)} + H^{(1)} + \cdots$

$$H = \sum_{\alpha=0}^{n} \int ds_1 \int ds_2 \, W(\underset{\sim}{R}^{(\alpha)}(s_1) - \underset{\sim}{R}^{(\alpha)}(s_2)) , \tag{2.16}$$

where W is the potential energy.

The constraints firstly contain the fact that it is a polymer chain and inextensible which is expressed in the powerful formulation of Weiner measure i.e. the use of a weight factor

$$\exp\left(\sum_{0}^{n} \int ds \, A([R^{(\alpha)}]) \right). \tag{2.17}$$

For example the usual random flight has the form

$$\exp\left(-\frac{3}{2\ell}\int \dot{\underset{\sim}{R}}^2(s)\,ds\right) \tag{2.18}$$

so that integrating over all $\underset{\sim}{R}(s)$ with boundary conditions

$\underset{\sim}{R}(0) = r'$ $\quad \underset{\sim}{R}(L) = r$ one has

$$G(\underset{\sim}{r}, \underset{\sim}{r}'; L) = \mathcal{N} \int_{\underset{\sim}{R}(0)=\underset{\sim}{r}'}^{\underset{\sim}{R}(L)=\underset{\sim}{r}} \delta \underset{\sim}{R} \, \exp\left\{-\frac{3}{2\ell}\int \dot{R}^2(s)\,ds\right\} = \left(\frac{3}{2L\ell\pi}\right)^{3/2} e^{-\frac{3}{2\ell L}(\underset{\sim}{r}-\underset{\sim}{r}')^2} \tag{2.19}$$

where $\mathcal{N}$ is the usual normalisation, and ℓ the step length. This formalism is readily extended to include curvature and torsion, see for example the forthcoming paper by Freed (2). Readers unfamiliar with this formalism can be assured that it is rigorous, and indeed rather trivial in application. Its great power is in the clarity of notation which permits a direct transcription of physical ideas into mathematics.

A cross linkage at a point S_1, with S_2, now gives

$$\prod_{\alpha=0}^{n} \delta\left(\mathbf{R}^{(\alpha)}(s_1) - \mathbf{R}^{(\alpha)}(s_2)\right) \tag{2.20}$$

This directly expresses the fact that a cross link initially set up, stays in the same place, on the chain.

Finally there are the topological constraints. These express the fact that chains cannot pass through one another, for example

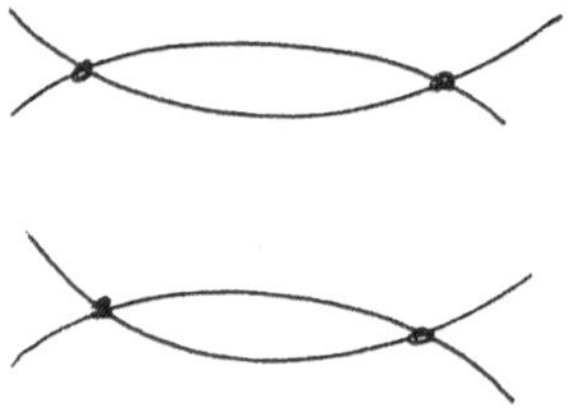

is different, permanently, from

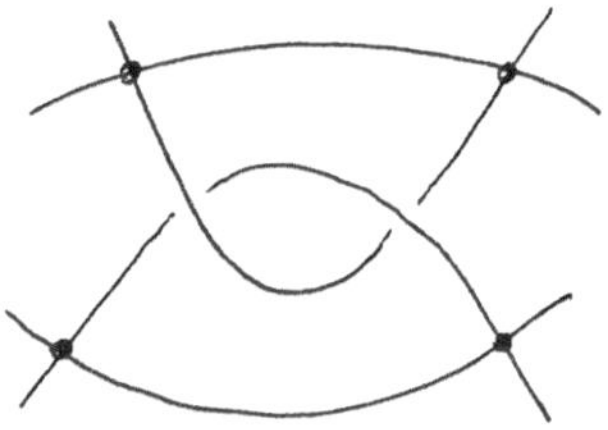

(2.21)

It has been shown (Ref. 4) that this can be expressed by a series of invariant expressions, the simplest of which is

$$I = \iint \left(d\mathbf{R}(s_1) \times d\mathbf{R}(s_2)\right) \cdot \nabla \left|\mathbf{R}(s_1) - \mathbf{R}(s_2)\right|^{-1}.$$

This rather formidable looking expression is just the 3 dimensional generalization of angle swept out, and will be familiar in the theory of Ampere and Gauss concerning the interaction of electric currents via magnetic fields. If for example we consider two curves $\mathbf{A}(S)$, $\mathbf{B}(S)$, and take one to be an infinite straight line, the other a circle

$$I = \iint (d\mathbf{A} \times d\mathbf{B}) . \nabla |\mathbf{A} - \mathbf{B}| \tag{2.22}$$

$= 0$ for

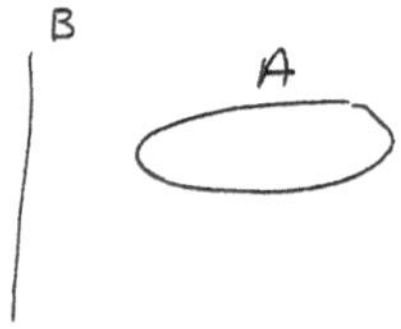

$= \pm 4\pi$ for

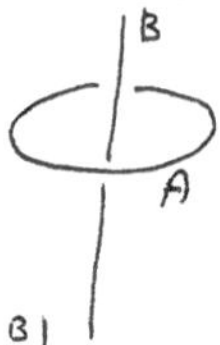

$= \pm 8\pi$ for

Similarly $I = 0$ for

$I = \pm 4\pi$ for

The signs depend on the sense with which $\underset{\sim}{R}(s)$ alters with S i.e. the 'direction of the current', so that arrows should be placed

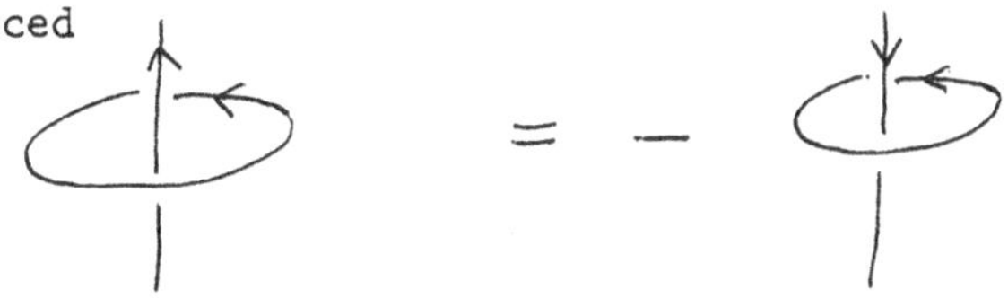

(2.23)

These arrows have no physical significance, but it is impossible to construct the invariants without them. For a single curve, the invariant represents the algebraic number of knots in a curve, but since these contribute + and - contributions to the higher invariants one is required also to say that a particular curve has a particular topology. The problem is rather like defining a function from its Fourier components. In the present paper only the first invariant will be used, and that rather crudely. Nevertheless this invariant can do the vital job of prohibiting a motion from

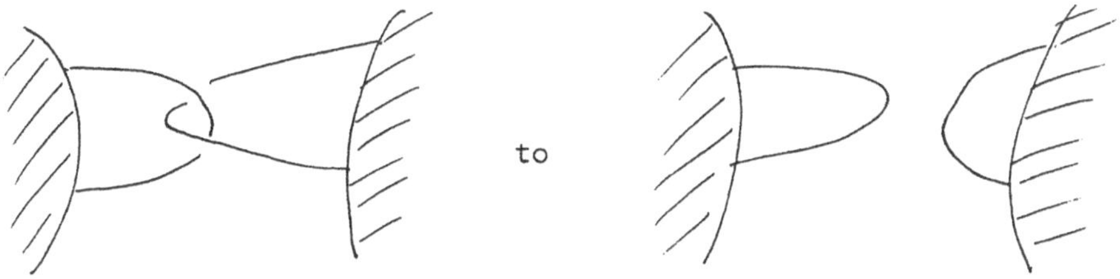

(2.24)

where the shaded regions mean "the rest of the system". Since these invariants are $\pm 4\pi \times$ integers, to say two are the same requires the Kronecker delta, but it will be written as a normal δ function for ease in writing.
Hence the final constraint is that

$$I^{(0)} = I^{(1)} = \ldots\ldots = I^{(n)} \tag{2.25}$$

or

$$\delta(I^{(0)} - I^{(1)})\,\delta(I^{(0)} - I^{(2)}) \ldots\ldots \delta(I^{(0)} - I^{(n)}) \tag{2.26}$$

Finally then

$$F(n) = -\kappa T \log Z(n)$$

$$Z(n) = \mathcal{N} \int \delta R^{(0)} \delta R^{(1)} \ldots \delta R^{(n)}$$
$$\times \exp\left[-\int A^{(0)} - \int A^{(1)} \ldots\right] \left(\int ds_1 \int ds_2 \prod_0^n \delta(R^{(\alpha)}(s_1) - R^{(\alpha)}(s_2))\right)^N$$
$$\times \left(\prod_{\alpha=1}^{n} \delta(I^{(0)} - I^{(\alpha)})\right), \tag{2.27}$$

(where $\mathcal{N}$ is again used for the (uninteresting) normalisation of the Weiner integral $\exp (-\int A$). Here N represents the number of cross links in the system. They are (in accord with the basic formulation) taken at random. This is worth discussing further. In a simple cross linkage process, the links build up in time. One link will affect the neighbouring chain density and therefore affect the position of the next link. In a high density system this is of no account, but in lower density systems, swollen gels etc it is, and quite inhomogeneous situations can arise. There are two courses open here. Either try to produce a theory which allows for this fact, a difficult matter until the simpler case is fully understood; or produce material in which the links really are random. Such a scheme will be described in a forthcoming paper by Allen, Burgess, Edwards and Walsh. Briefly it amounts to cross linking in a 2 stage process. Suppose one puts group A onto a chain at random, then introduces B which links A 's. The positions of the links are determined in the first stage and under careful conditions are truly random. e.g.

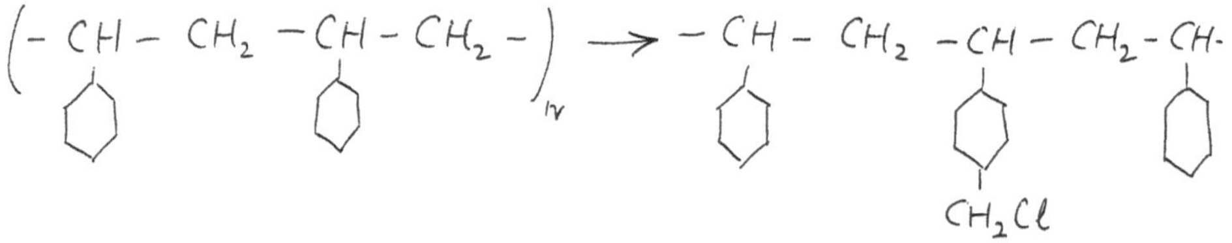

Next react with n-butylamine:

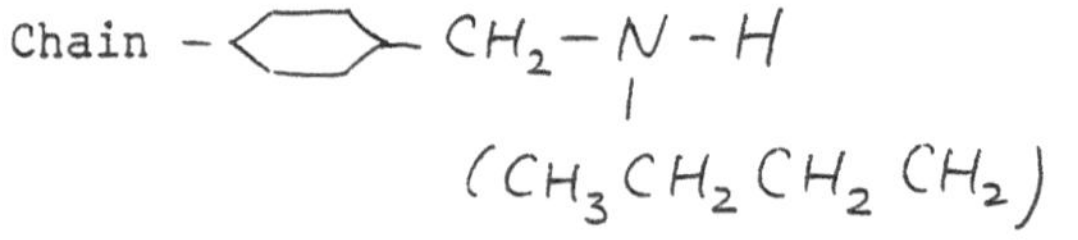

Cross link with a di-isocyanate $(OCN - R - R - NCO)$

to give urethane linkage:

$$(CH_3 CH_2 CH_2 CH_2) \qquad\qquad (CH_3 CH_2 CH_2 CH_2)$$
$$\text{Chain} - \langle \bigcirc \rangle - CH_2 - N - OC - N - R - R - N - CO - N - CH_2 - \langle \bigcirc \rangle - \text{Chain}$$
$$\qquad\qquad\qquad\quad H \qquad\qquad\qquad H$$

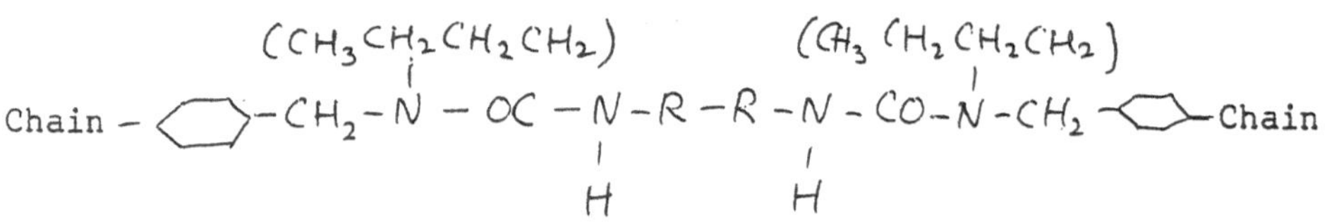

There are of course various other similar reactions.

In the next section the simplest calculation, that of "phantom chains" will be completed. This is a calculation involving cross links alone, neglecting both forces and topology and keeping only pure gaussian chains. It will appear that under certain limits a full solution is possible.

3. THE PHANTOM NETWORK

This critical term was given by Flory to emphasize the unphysical nature of the system. But it is defined mathematically and until it is solved (or shown to have no meaningful solution) the real system cannot be studied. The problem is well defined for gaussian chains, and amounts to the evaluation of

$$\int \delta R^{(0)} \delta R^{(1)} \ldots \int \exp\left[-\frac{3}{2l}\sum_{0}^{n}\int \dot{R}^{(\alpha)2} ds\right]\left\{\iint ds_1 ds_2 \prod_{\beta=1}^{n} \delta\left(R^{(\beta)}(s_1) - R^{(\beta)}(s_2)\right)\right\}^{N} \quad (3.1)$$

where $\int \delta R^{(0)}$ means integrate the chain $R^{(0)}(s)$ through the initial volume V, and $\int \delta R^{(1)}$ etc means integrate the chain $R^{(1)}(s)$ through the new box, of sides $\lambda_1, \lambda_2, \lambda_3 \times old$. Consider firstly

$$G = \mathcal{N}\int \delta R^{(0)} \exp\left[-\frac{3}{2l}\int_0^L \dot{R}^{(0)2}(s)\, ds\right] \quad (3.2)$$

in a box, where L is the entire polymer length which is being used in preference to the true state of many longish chains. This satisfies the differential equation

$$\left(\frac{\partial}{\partial s} + \frac{l}{6}\nabla^2\right) G(\mathbf{r}, \mathbf{r}'; s, s') = \delta(\mathbf{r} - \mathbf{r}')\,\delta(s - s') \quad (3.3)$$

and needs boundary conditions. Now physically the material has a uniform density, so the appropriate boundary conditions will be the cyclic i.e. cosine conditions, and

$$G = \frac{L}{lV} + \frac{8}{V}\sum \cos k_1 x \cos k_1 x' \cos k_2 y \cos k_2 y' \cos k_3 z \cos k_3 z' \times \exp[-l\mathbf{k}^2 (s-s')/6] \quad (3.4)$$

where $\frac{k_1}{\pi}$ runs through the integers 1 to ∞, etc. For the present purposes one may approximate G to be

$$G = \frac{L}{lV} + \left(\frac{3}{2\pi l|s-s'|}\right)^{3/2} \exp\left(-\frac{3}{2l}(\mathbf{r}-\mathbf{r}')^2/|s-s'|\right) \quad (3.5)$$

which is physically obvious: if one has polymer at a point $\mathbf{r}$ what is the probability of finding it at $\mathbf{r}'$? The answer is either L/lV from polymer which is unrelated to that at $\mathbf{r}$, emanating at the boundary, or the normal gaussian chain probability at distance $(s - s')$ along the polymer. (If the many chains had been considered, the first term is 'other chains' second 'same chain'.) A trial calculation will now be made, and its region of validity discussed later. Physically one expects the cross links to pin down the chain to some extent in space so that at any s, $R^{(0)}(s)$, $R^{(1)}(s)$, ... are correlated in same way. However the original $\mathbf{R}^{(0)}(s)$ is free to be anywhere, so that roughly speaking the centre of gravity of the $\underset{\sim}{R}$ is unconstrained, but the remaining n coordinates are constrained in magnitude. For example if one considered just $\underset{\sim}{R}^{(0)}(s_1)$ and $\underset{\sim}{R}^{(1)}(s_1)$ one could expect the effect of the cross links mirrored by a term like

$$\exp\left(-\frac{\omega^2}{2}\left(\underset{\sim}{R}^{(0)}(s_1) - \underset{\sim}{R}^{(1)}(s_1)\right)^2\right) \tag{3.6}$$

and since this argument is true for all s, in all constraints like

$$\exp\left(-\frac{\omega^2}{2}\int\left(\underset{\sim}{R}^{(0)}(s) - \underset{\sim}{R}^{(1)}(s)\right)^2 ds\right). \tag{3.7}$$

There will be no constraint on $\underset{\sim}{R}^{(0)}(s) + \underset{\sim}{R}^{(1)}(s)$. Now all the $\underset{\sim}{R}^{(\alpha)}$ are on a symmetric footing, so it is convenient to generalise $R^{(0)} + R^{(1)}$ and $R^{(0)} - R^{(1)}$ to these coordinates. They are, with $\underset{\sim}{R} = R_1, R_2, R_3$

$$X_1^{(0)} = \frac{R_1^{(0)} + \lambda_1 \sum_1^n R_1^{(\alpha)}}{\sqrt{1+n\lambda_1^2}}, \qquad R_1^{(0)} = \frac{X_1^{(0)} + \lambda_1\sqrt{n}\, X_1^{(1)}}{\sqrt{1+n\lambda_1^2}}$$

$$X_1^{(1)} = \lambda_1\sqrt{n} - \frac{1}{\sqrt{n}}\sum R_1^{(\alpha)}, \tag{3.8}$$

$$R_1^{(\alpha)} = \left(\frac{\lambda_1 X_1^{(0)} - \frac{1}{\sqrt{n}} X_1^{(1)}}{\sqrt{1+n\lambda_1^2}} + \frac{1}{\sqrt{n}}\sum_{-(\frac{n-1}{2})}^{(\frac{n-1}{2})} Y_1^{(\beta)} e^{-2\pi i\beta\alpha/n}\right)$$

$$Y_1^{(\beta)} = \frac{1}{\sqrt{n}}\sum_1^n R_1^{(\alpha)} e^{2\pi i\beta\alpha/n},$$

(β runs from $\frac{n-1}{2}$ to $-\frac{(n-1)}{2}$; there are $n-1$ β's)

and similarly for the 2, 3 cartesian coordinates. These variables have the property that

$$\sum_0^n \underset{\sim}{R}^{(\alpha)2} = \underset{\sim}{X}^{2(0)} + \underset{\sim}{X}^{2(1)} + \sum_{-(\frac{n-1}{2})}^{+(\frac{n-1}{2})} \underset{\sim}{Y}^{(\beta)}\underset{\sim}{Y}^{(-\beta)} \tag{3.9}$$

$$\sum_0^n \dot{R}^{(\alpha)2} = \dot{\underset{\sim}{X}}^{2(0)} + \dot{\underset{\sim}{X}}^{2(1)} + \sum_{-(\frac{n-1}{2})}^{(\frac{n+1}{2})} \dot{Y}^{(\beta)}\dot{Y}^{(-\beta)} \tag{3.10}$$

and n

$$\prod_{\alpha=0} \delta(R^{(\alpha)}(s_1) - R^{(\alpha)}(s_2)) = \delta(X^{(0)}(s_1) - X^{(0)}(s_2))\,\delta(X^{(1)}(s_1) - X^{(1)}(s_2)) \prod \delta(Y^{(\beta)}(s_1) - Y^{(\beta)}(s_2)) \tag{3.11}$$

There are many other possible transformations, this happens to be the simplest and most convenient. One may now model the cross links by a constraint

$$\exp\left(-\frac{\omega^2}{2}\int ds \left(X^{(1)2}(s) + \sum Y^{(\beta)}(s)\, Y^{(-\beta)}(s)\right)^2\right). \tag{3.12}$$

Since the transformation is orthonormal $\prod \delta R \Rightarrow \delta X^{(0)} \delta X^{(1)} \prod \delta Y$ and the boundary conditions are immaterial to $X^{(1)}$ and the Y since they are constrained to be microscopic variables. But whereas $R^{(0)}$ lay inside volume V, one has

$X^{(0)}$ running from 0 to $\sqrt{1+n\lambda_1^2}$ and so on (being the diagonal of an $(n+1)$-tuple cuboid of side $V^{1/3}(1, \lambda_1, \lambda_2 \ldots \lambda_n)$ etc.

It is convenient to represent the constraints as in the grand canonical ensemble

$$\frac{1}{2\pi}\oint \frac{d\mu}{\mu^{N+1}/(N+1)!} \exp\left[-\mu \int_0^L ds_1 \int_0^L ds_2 \prod \delta(R^{(\alpha)}(s_1) - R^{(\alpha)}(s_2))\right]$$

$$= \frac{1}{2\pi}\int \frac{d\mu}{\mu^{N+1}/(N+1)!} \exp\left[-\mu \int_0^L ds_1 \int_0^L ds_2\, \delta(\underset{\sim}{X}^{(0)}(s_1) - \underset{\sim}{X}^{(0)}(s_2)) \times \delta(\underset{\sim}{X}^{(1)}(s_1) - \underset{\sim}{X}^{(1)}(s_2)) \prod \delta(Y^{(\beta)}(s_1) - Y^{(\beta)}(s_2))\right] \tag{3.13}$$

so one can write

$$e^{-F(n)/kT} = \frac{\mathcal{N}(N+1)!}{2\pi}\int \frac{d\mu}{\mu^{N+1}} \times$$

$$\exp\left[-\frac{3}{2l}\int \left(\dot{\underset{\sim}{X}}^{(0)2} + \dot{\underset{\sim}{X}}^{(1)2} + \sum \dot{\underset{\sim}{Y}}^{(\beta)} \dot{\underset{\sim}{Y}}^{(-\beta)}\right) ds - \frac{l}{6}\sum \omega_i^2 \int \left(X_i^{(1)2} + \sum Y_i^{(\beta)} Y_i^{(-\beta)}\right) + C + Q\right] \tag{3.14}$$

(it will turn out that ω is independent of the cartesian index i but it must be kept in at present), where

$$Q = -\mu \iint ds_1\, ds_2\, \delta(X^{(0)}(s_1) - X^{(0)}(s_2))\,\delta(X^{(1)}(s_1) - X^{(1)}(s_2)) \prod \delta(Y^{(\beta)}(s_1) - Y^{(\beta)}(s_2))$$

$$+ \sum_i \frac{l\omega_i^2}{6}\int \left(X_i^{(1)2} + \sum Y^{(\beta)} Y^{(-\beta)}\right) ds - C. \tag{3.15}$$

The chemical potential (evaluated as usual by steepest descent) will turn out real hence one can use the variational principle

$$e^{Q} > 1 + Q \tag{3.16}$$

Hence

$$e^{-F(n)/kT} > \frac{N(N+1)!}{2\pi} \int \frac{d\mu}{\mu^{N+1}} \exp\Big[-\frac{3}{2l}\int \dot{x}^{(0)2}\,ds - \sum_i \frac{l\omega_i^2}{6}\int x_i^{(1)2}\,ds + C\Big]$$
$$= \int \exp \mathcal{A}, \text{ say,} \tag{3.17}$$

where C is chosen to make

$$\int Q \exp \mathcal{A} = 0 \tag{3.18}$$

i.e.

$$C = \int \delta x^{(0)} \delta x^{(1)} \prod \delta y \exp(\mathcal{A}) \times$$
$$\Big[\mu \iint \delta(x^{(0)} - x^{(0)}) \delta(x^{(1)} - x^{(1)}) \prod \delta(y^{(\beta)} - y^{(\beta)}) + \frac{l}{6}\sum \omega_i^2 \int \Big(x_i^{(1)2} + \sum_\beta y_i^{(\beta)} y_i^{(\beta)}\Big)\Big]$$
$$\times \Big(\int \exp(\mathcal{A})\Big)^{-1}. \tag{3.19}$$

Although these formulae seem elaborate they are very simple to evaluate. The weight $\exp(\mathcal{A})$ is equivalent to the differential equation

$$\left(\frac{\partial}{\partial s} + \frac{l}{6}\nabla^{(0)2}\right) G(x^{(0)}, x'^{(0)}; s, s') = \delta(s - s')\,\delta(x^{(0)} - x'^{(0)}) \tag{3.20}$$

for $X^{(0)}$, and has the solution (3.5).
For the others, the differential equation is modified by the constraint to

$$\left(\frac{\partial}{\partial s} + \frac{l}{6}\nabla^{(1)2} + \frac{l}{6}x^{(1)2}\right) G = \delta\,\delta \tag{3.21}$$

and similarly for the Y's. Since L is large only the lowest eigenfunction contributes to G i.e.

$$G \simeq \left(\frac{\omega_1 \omega_2 \omega_3}{\pi^3}\right)^{1/2} \exp\Big[-\frac{1}{2}\sum_i \omega_i x_i^{(1)2} - \frac{1}{2}\sum_i \omega_i x_i'^{(1)2} - \frac{l}{6}\sum \omega_i |s - s'|\Big] \tag{3.22}$$

(this approximation will be further considered later).

Then one has $\int \exp \mathcal{A}$ to be dominated by the ground states and gives

$$\frac{L}{\ell V} \prod_i (1+n\lambda_i^2)^{1/2} \qquad \text{from } \delta x^{(0)}, \text{ times}$$

$$\exp\left[-\frac{n\ell}{6} \sum_i \omega_i L\right] \qquad \text{from the } \delta x^{(1)} \text{ and } \delta \gamma^{(\beta)} .$$

(3.23)

C is now found from (3.19) and gives $\frac{n\ell}{6} \sum \omega_i L$ from the quadratic terms, and the δ functions give: $\prod (1+n\lambda_i^2)^{-1/2} V^{-1}$ from the $x^{(c)}$ δ function, times $(\omega_1\omega_2\omega_3/8\pi^3)^{1/2}$ from the other δ functions.

The steepest descent for μ now is determined by

$$\frac{\partial}{\partial\mu}\left(-(N+1)\log\mu + \mu L^2 \left(\frac{\omega_1\omega_2\omega_3}{8\pi^3}\right)^{1/2} \prod_i (1+n\lambda_i^2)^{-1/2} V^{-1} - \frac{n\ell}{12} \sum \omega_i L\right) = 0 \tag{3.24}$$

and ω_i by $\partial/\partial\omega_i$ of the same expression, $= 0$.
Hence, (dropping terms of order $\frac{1}{N}$),

$$\frac{N}{\mu} = \left(\frac{\omega_1\omega_2\omega_3}{8\pi^3}\right)^{n/2} V^{-1} L^2 \prod (1+n\lambda_i^2)^{-1/2} \tag{3.25}$$

and

$$\frac{n\ell L}{12} = \frac{n}{2}\,\frac{\mu}{\omega_i}\, L^2 \left(\frac{\omega_1\omega_2\omega_3}{8\pi^3}\right)^{n/2} \prod (1+n\lambda_i^2)^{-1/2} V^{-1}. \tag{3.26}$$

Thus $\omega_i = \frac{6}{\ell}\left(\frac{N}{L}\right) = \omega$ say, independent of λ, (3.27)

and $$\mu = \frac{NV}{L^2}\left(\frac{8\pi^3}{\omega^3}\right)^{n/2} \prod (1+n\lambda_i^2)^{-1/2} \tag{3.28}$$

and thus the expansion of $F(n)$ starts with

$$n\left[-3N\log\left(\frac{\omega}{2\pi}\right) + \frac{N}{2}\sum\lambda_i^2 + \frac{3N}{2}\right]\kappa T \tag{3.29}$$

The interesting part is of course

$$\tilde{\mathcal{F}} = \frac{N}{2}\sum \lambda_i^2 (\kappa T). \tag{3.30}$$

This result differs by a factor ½ from that of James and Guth (excluding their modifications due to the cure, which have purposely been avoided here) and of Flory and Wall. The difference appears to come from the fact that the cross links here are not at their affine deformation positions, but are allowed considerable freedom via the term (see also the contribution in this symposium by M. Gordon, whose treatment and answer lies between these discussed) To understand the region of validity of the result consider first that with the variational form (3.16) one can use a more general trial function, for example instead of

$$\exp\left[-\frac{\ell}{6}\sum \omega_i^2 \int X_i^{(1)2} ds - \frac{3}{2\ell}\sum_i \int \dot{X}_i^{(1)2} ds \right] \tag{3.31}$$

one can use the general quadratic form

$$\exp\left[-\iint ds_1 ds_2\, X^{(1)}(s_1) g^{-1}(s_1 - s_2) X^{(1)}(s_2) \right] \tag{3.32}$$

One then finds that, if $\gamma(Z)$ is the inverse of the fourier transform of $g^{-1}(s_1 - s_2)$, γ satisfies

$$\left(z^2 + N\left\{ \iint \frac{d\alpha_1 d\alpha_2}{\left[\int d\zeta g(\zeta) \sin^2 \zeta(\alpha_1 - \alpha_2)\right]^{3/2}} \right\}^{-1} \iint \frac{\{\sin^2 Z(\tau_1 - \tau_2)\} d\tau_1 d\tau_2}{\left[\int d\zeta g(\zeta) \sin^2 \zeta(\tau_1 - \tau_2)\right]^{5/2}} \right) g = 1 \tag{3.33}$$

and in the limit of N large, the new term in g is just a constant, as has been used. If one examines the full series

$$e^Q = 1 + Q + \frac{1}{2}Q^2 + \frac{1}{6}Q^3 + \dots \tag{3.34}$$

one can evaluate every term and recombine as a cluster expansion of the free energy. So far it has been arranged that $\int Q \exp A = 0$. It turns out that in the limit $(N \gg N_{crit})$ all the terms of the series also give zero. This can be explained in this way. Consider firstly the δ functions in say $X^{(1)}$. In Q^P one will have

$$\left[\iint \delta(\underset{\sim}{x}^{(1)}(s_1) - \underset{\sim}{x}^{(1)}(s_2)) \ (\times \text{ other } \delta \text{ functions}) \right]^P$$

This will give rise to

$$\int \begin{matrix} d\underset{\sim}{x}^{(1)} d\underset{\sim}{x}^{(1)\prime} d\underset{\sim}{x}^{(1)\prime\prime} \dots \\ ds\, ds'\, ds'' \dots \end{matrix} \ G(\underset{\sim}{x}^{(1)}, \underset{\sim}{x}^{(1)\prime}; s - s')\, G(\underset{\sim}{x}^{(1)\prime}, \underset{\sim}{x}^{(1)\prime\prime}; s' - s'') \dots \dots \ P \text{ terms} \tag{3.35}$$

where

$$\left(\frac{\partial}{\partial s} + \frac{\ell}{6}\frac{\partial^2}{\partial \underset{\sim}{x}^{(1)2}} + \frac{\ell}{6}\omega^2 \underset{\sim}{x}^{(1)2} \right) G = \delta(x^{(1)} - x^{(1)\prime})\,\delta(s - s') \tag{3.36}$$

and the S lie in any order in $0, L$. The final answer will involve n such terms in the S and a term from $\prod(\delta(X^{(0)}(s_i) - X^{(0)}_{(s_j)}))$. Now G can be expressed in terms of the Hermite polynomials

$$\sum_{\eta_1, \eta_2, \eta_3 = 0}^{\infty}\sum^{\infty}\sum^{\infty} He_{\eta_i}(x_i^{(1)})\, He_{\eta_i}(x_i^{(1)\prime})\, e^{-\Sigma(\eta_i + \frac{1}{2})\frac{\ell\omega_i}{3}(s - s')} \tag{3.37}$$

and the approximation used above is to ignore all but the 'ground state', $\eta = 0$. If only this state is used, the multiple terms in "$\mathcal{G}^A$" become simple products. The higher terms can be neglected provided that $\omega L \gg 1$ (which knocks out $\eta > 0$) since $\exp(-\frac{\ell}{6}(s-s') - \frac{\ell}{6}(s'-s'') \cdots = \exp(-\frac{\ell}{6}\omega L)$, the s being some partition of L i.e. $N \gg L$, which is trivially correct. However there are an infinite number of these Hermite functions and as $\eta \to \infty$, the $\mathcal{G}$ function takes on the structure of the free diffusion equation i.e. at small distances behaves like the free equation. Thus when evaluating the loop  one cannot use $\eta = 0$ alone for small $s-s'$ and one has the difficulty that the free diffusion equation gives a divergence for this contribution. The divergence is removed when the finite size of the cross link is remembered, or when curvature or torsion are invoked, and is simply a constant, so causes no difficulty when cut off. It represents a 'waste' of cross links and can be avoided by an elaboration of the cross link chemistry, e.g. dope half the polymers with -(A), half with (B)-, then cross link with -(C)- by means of -(A)-(C)-(B)-, uniquely. (I have not space to elaborate on this reaction), or just allowed for in an effective N. All remaining terms are finite in the gaussian chain approximation, but proceeding further one finds terms of the type  appearing in the quadratic expansion and so on. An estimate of these terms shows that they are small provided that N is large. When one is precise over this it turns out to be a purely numerical criterion i.e. $N\ell/L > \zeta$, where ζ is a number i.e. there must be a certain number of cross links per length of chain for the system (of phantom chains) to settle into a solid condition. This number is not very significant as a physical quantity, because the effect of real chain forces and entanglements will mean that it has to be rather higher than in a real solid. It will be seen that the mathematics under discussion is identical to that of asking for the shape of an isolated unconfined cross linked phantom chain. If a sufficient number of links are present and a cut off to eliminate the loop divergence is used, such a ball will take up a shape in space and be sufficiently dense that the correlation between links can be neglected. If one does not have this density of links the ball will not form and instead one will get an inhomogeneous substance of much larger dimension i.e. the two pictures are

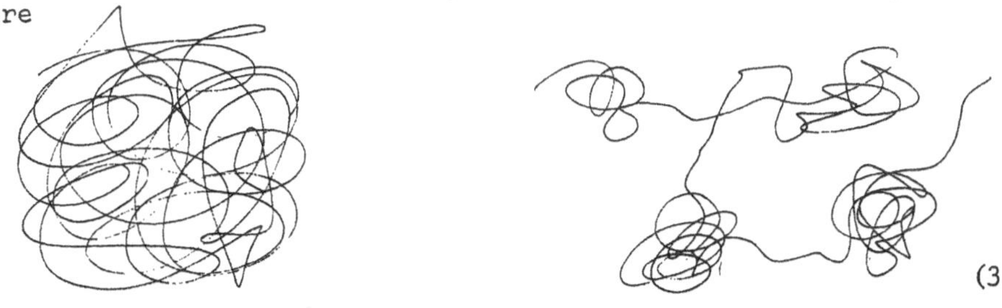

(3.38)

I have estimated N_{crit} very roughly as $1/8\pi$, an absurdly large value for a physical system, but remembering that ℓ is a Kuhn effective step length and all the other factors omitted, the value is not surprising.

Now consider $\underset{\sim}{X}^{(0)}$. Assuming that all the other coordinates have been replaced by averages, one has (3.39)

$$\frac{\mathcal{N}(N+1)!}{2\pi}\oint\frac{d\mu}{\mu^{N+1}}\int\exp\left[-\frac{3}{2\ell}\int\dot{x}^{(0)2}ds-\mu\left(\frac{\omega}{2\pi}\right)^{\frac{3n}{2}}\iint ds_1 ds_2\,\delta(x^{(0)}(s_1)-x^{(0)}(s_2))\right]\delta R.$$

This is however precisely the same integral as arises for the other coordinates, and if evaluated accurately again will collapse into a ball of size $\omega^{-1/2}$. And that is of course just what a phantom chain will do unless the incompressibility due to molecular forces is invoked. If a supplementary condition is imposed that the system is incompressible, this is equivalent to replacing the cross link function $\delta(x^{(0)}(s_1)-x^{(0)}(s_2))$ by its mean value $V^{-1}\prod_i(1+n\lambda_i^2)^{-1/2}$ (or perhaps a little more precisely $cV^{-1}\prod(1+n\lambda_i^2)^{-1/2}$ where c involves the 2-body correlation function at the cross link). Any fuller treatment must invoke a theory of the free energy of unvulcanized rubber. The present work does not therefore comment on the controversy in the literature over the presence of terms like $N\log\lambda_1\lambda_2\lambda_3$ in the free energy. It is hoped to return to this matter when a full study of the unvulcanized rubber is completed: this is not done in the present paper.

It appears then that provided

(a) There are phantom chains

(b) There is an imposed mean density of material

(c) The cross links are formed as in section 2

(d) The density of cross links along the chain exceeds a critical value

(e) The chain statistics are gaussian except possibly at short distances, and the cross links are small but finite in size,

the free energy is $\frac{1}{2}N\kappa T\sum_i\lambda_i^2$. (3.40)

In the next section condition (a) is relaxed, and in the following some discussion given about relaxing (b). As has been mentioned one can arrange the chemistry to accommodate (c), though it may be reasonable anyway. If (d) is violated the cross links will not be homogeneous even though they are randomly spaced according to (c). The final condition (e) can be evaded also by an appropriate chemistry but is not a deep difficulty anyway.

4. ENTANGLEMENTS

The problem of entanglements will prove of such difficulty that a solution will only be attempted for small deformations, though comments on the general solution will be made. The Kronecker delta can be parametrized as

$$\frac{1}{2\pi}\int_{-\pi}^{\pi} e^{iq(I-\tilde{I})}\,dq \tag{4.1}$$

Now I will contain contributions from many chains, so the probability distribution of I will be approximately gaussian, and one can approximate

$$\left\langle \frac{1}{2\pi}\int_{-\pi}^{\pi} e^{iq(I-\tilde{I})}\,dq \right\rangle = \left\langle \frac{1}{2\pi}\int_{-\pi}\left(1+iq(I-\tilde{I})-\frac{q^2}{2}(I-\tilde{I})^2\right.\right. \cdots \Big\rangle \tag{4.2}$$

$$= \left(1-\frac{\pi^2}{6}\langle(I-\tilde{I})^2\rangle + \cdots\right. \tag{4.3}$$

$$\simeq e^{-\frac{\pi^2}{6}\langle(I-\tilde{I})^2\rangle} \tag{4.4}$$

Now perform an expansion both on $\epsilon_i = \lambda_i - 1$, and in the and $X_i^{(1)}Y$ (i.e. in ω^{-1}). Thus

$$I = I([R_i^{(0)}]) = I\left(\frac{[X_i^{(0)}]+\lambda\sqrt{n}[X_i^{(1)}]}{(1+n\lambda_i^2)^{1/2}}\right). \tag{4.5}$$

It will turn out that, to the order of the approximation above the $\tilde{I}$ are independent i.e.

$$\langle \Pi\delta \rangle = e^{-\frac{n\pi^2}{6}\langle(I-\tilde{I})^2\rangle} \tag{4.6}$$

where $\tilde{I}$ represents anyone of the $I([R^{(\alpha)}])$. Hence the factor $\sqrt{1+n\lambda_i^2}$ can be replaced by 1, and

$$I([R^{(0)}]) \simeq I([X^{(0)}]) + \sqrt{n}\sum_i \lambda_i \int X_i^{(1)}\nabla_i^{(0)} I\,ds \tag{4.7}$$

where $\nabla_i^{(0)}$ represents the functional derivative with respect to $X_i^{(0)}$.

similarly

$$I([R_i^{(\alpha)}]) = I\left(\lambda_i[X_i^{(0)}] - \frac{1}{\sqrt{n}}[X_i^{(1)}] + \frac{1}{\sqrt{n}}\sum[Y^\beta]e^{\frac{2\pi i\beta\alpha}{n}}\right.$$

$$= I([X_i^{(0)}]) + \sum \epsilon_i \int [X_i^{(0)}] \nabla_i^{(0)} I\, ds$$

$$+ \frac{1}{\sqrt{n}} \sum_i \int \left([X_i^{(1)} - \sum_\beta [Y_i^\beta] e^{\frac{2\pi i\beta\alpha}{n}} \right) \nabla_i^{(0)} I\, ds .$$

(4.8)

Thus

$$n\langle (I - \tilde{I})^2 \rangle = n \left\langle \left(\sum_i (1+\epsilon_i) \sqrt{n} \int X_i^{(1)} \cdot \nabla_i I \right. \right.$$

$$- \sum_i \epsilon_i \int [X_i^{(0)}] \cdot \nabla_i^{(0)} I\, ds$$

$$\left. \left. + \frac{1}{\sqrt{n}} \sum_i \left([X_i^{(1)}] - \sum_\beta [Y_i^\beta] e^{\frac{2\pi i\beta\alpha}{n}} \nabla^{(0)} I \right)^2 \right\rangle \right.$$

(4.9)

Surviving terms in the average, when terms in n^2 are disregarded, are

$$n \left[2 \sum_i \frac{(1+\epsilon_i)}{3} \frac{1}{2\omega} \left\langle \left\{ \int [\nabla^{(0)} I] \right\}^2 \right\rangle \right.$$

$$+ \sum \epsilon_i^2 \frac{1}{3} \left\langle \left\{ \int [X^{(0)}] [\nabla_j^{(0)} I] \right\}^2 \right\rangle$$

$$\left. + \frac{1}{2\omega} \left\langle \left(\int [\nabla_i^{(0)} I] \right)^2 \right\rangle \right]$$

(4.10)

Defining $\mathcal{B} = \frac{3}{2\rho L} \langle \int (\nabla I)^2 \rangle$, $\mathcal{C} = \frac{1}{3\rho L} \langle \int ((X^{(0)} \cdot \nabla) I)^2 \rangle$

(4.11)

where ρ is the density of polymer $L/\ell V$,

one has

$$nL\rho\left(\frac{\mathcal{B}}{\omega}+\mathcal{C}\sum\epsilon_i^2+\frac{2}{9}\mathcal{B}\sum\epsilon_i\right) \tag{4.12}$$

This result has linear terms in ϵ_i, since the confined chains will exert a pressure, but if the contributions to the rest of the free energy make it incompressible one may set $\epsilon_1+\epsilon_2+\epsilon_3=0$ and the simple result $n\rho L\left(\frac{\mathcal{B}}{\omega}+\mathcal{C}\sum\epsilon_i^2\right)$ follows. The values of $\mathcal{B}$ and $\mathcal{C}$ diverge for purely gaussian chains and their finiteness depends on curvature and torsional effects in the present calculation. Since the structure of the integral to be averaged is invariant under translations the value must be proportional to ρ i.e. I has the structure of a potential (a magnetic potential in fact). The values of $\mathcal{B}$, $\mathcal{C}$ depend on the form taken for the cut off, for example when a weight

$$\frac{3}{2\ell}\int\dot{R}^2ds+\frac{3a^2}{2\ell}\int\ddot{R}^2ds \tag{4.13}$$

is taken $\mathcal{B}\propto 1/a$. In the present study they will be treated as external constants. The value of ω is obtained as before by the variational principle and an additional term is added to the equation (3·27), giving

$$\frac{L\ell}{12}-\frac{N}{2\omega}-\frac{\rho\mathcal{B}L}{\omega^2}=0 \tag{4.14}$$

$$\omega=\frac{6N}{\ell L}\left(\frac{1}{2}+\sqrt{\frac{1}{4}+\rho\frac{\mathcal{B}\ell L^2}{3N^2}}\right). \tag{4.15}$$

Thus for $\rho \gg (N^2/L^2\mathcal{B}\ell)$ one has $\omega\propto\sqrt{\rho}$ (4.16)

but for $\rho \ll (N^2/L^2\mathcal{B}\ell)$ one has $\omega=\frac{6N}{\ell L}$ as before. (4.17)

This formula shows how the cross links and entanglements combine to enclose a chain. The strain energy has now become

$$\frac{1}{2}N\kappa T\sum\epsilon_i^2+L\rho\kappa T\mathcal{B}\sum\epsilon_i^2 \tag{4.18}$$

i.e. the free energy per unit volume is

$$\left(\frac{\rho_x}{2}+\rho_{pol}^2\mathcal{B}\right)\sum\epsilon_i^2 \tag{4.19}$$

where suffixes have been placed to distinguish ρ_x the density of cross links and ρ_{pol} the density of polymer. It is surprising that ω, at least in this approximation, does not enter the free energy. In previous work (3), the author derived (4.14) and assumed that $\mathcal{F}\propto\omega\kappa T\sum\epsilon_i^2$, which is both natural and correct for pure cross linkage. But the contribution of entanglements to the strain energy seems to enter quite independently of the calculation and for a system <u>without</u> cross links, (where strictly a frequency independent strain will cause creep, but for which a very low frequency is sufficient to give good coefficients, for long enough polymer chains), the elastic constants are proportional to ρ_{pol}^{2} and not $\rho_{pol}^{3/2}$ which would come from using ω . I find this a puzzling result which is intuitively surprising. It is of course only correct for small ω^{-1} and small ϵ_i , and only the first terms have been kept. When one looks at the expressions for $\mathcal{I}$, one notices one striking difference as compared with the cross link δ function. The $\mathcal{I}$ expressions do not conform to an affine transformation i.e. under an affine transformation

$$\prod_{\alpha=0}^{n} \delta(R^{(\alpha)}(s_1) - R^{(\alpha)}(s_2)) = \delta(X^{(0)}(s_1) - X^{(0)}(s_1))\,\delta(X^{(1)}(s_1) - X^{(1)}(s_2)) \prod_{\beta} \delta(Y^{(\beta)}(s_1) - Y^{(\beta)}(s_2))$$

but $\prod_{1}^{n} \delta(\mathcal{I}([R^{0}]) - \mathcal{I}([R^{(\alpha)}]))$ goes to nothing simple. This is reasonable because an affine transformation does not conserve length. If one took a set of arcs and affinely deformed each (thus stretching them) their topology is of course invariant and $\tilde{\mathcal{I}} = \mathcal{I}$, but when they are inextensible, the topological constraints imply that although the system may deform affinely on a macroscopic scale, it does not on a microscopic scale. By expanding in the deformation and constraint, this difficulty has been dodged, but to get to large λ it will be essential to handle this local non-affine nature. In any treatment no very simple function of λ will result. I have made some exploration of these forms but the work is still in progress. In two dimensions these problems can be largely solved, but in three dimensions very unpleasant equations are involved, as given for example for the entanglement of a pair of polymers by the author (4).

5. THE INTERNAL ENERGY

This section will only contain a few general ideas on the interaction of the normal forces with cross linkages. As was shown earlier the pressure when links are frozen is the same as when they are free. To get the pressure we may then calculate

$$\int \frac{d\mu}{\mu^{N+1}} \int \delta R \exp\left[-\frac{3}{2l}\int \dot{R}^2 ds - H([R])/kT + \mu \iint \delta(R - R')\right]. \tag{5.1}$$

Proceeding as before one has

$$e^{-F/kT} = \oint e^{-F_0/kT + N\log\mu - \mu C}\, d\mu \tag{5.2}$$

$$F = F_0 + N\log C \tag{5.3}$$

where F_0 is the free energy of unlinked material and

$$C = \langle \iint \delta(R - R') \rangle \tag{5.4}$$

which is the value of the 2-body correlation function at zero argument (times (L/ℓ)). Very roughly one can say C will be $(N/V)(1 + \gamma \rho (\partial P/\partial p)_T)$ where γ is a constant depending on the fine structure of the system. Hence

$$P = P_0 - \frac{N}{V} + O\left(\frac{\partial P}{\partial p}\right)_T . \tag{5.5}$$

If the material is rather incompressible, the primitive result $P = P_0 - N/V$ is obtained. However from (Ref. 1) one can develop a good cluster expansion and get an expansion for (5.1) in terms of 2, 3, 4-body correlation functions. When one turns to the full expression this is no longer possible since one cannot treat the quadratic constraint as a perturbation, and the affine transformation is not useful on $\sum H^{\alpha}/kT$. Just as with the entanglements, one does not expect it to be. But unlike there one can argue that $\omega^{-1} \gg$ range of forces, so that it may be possible to justify writing

$$F = (n+1)F_0 + N \log \prod_i (1 + n\lambda_i^2)^{-1/2} \omega^{-3n/2} (VC)^{n+1} \tag{5.6}$$

i.e. adopting the volume V for one coordinate and $\omega^{-3/2}$ for the other n . This in effect is what is always assumed. What is needed is a direct evaluation of the integral (2.27) using only the correlation functions of the unvulcanized material, and work is in progress on this problem.

6. CONCLUSION

This paper is very much an account of work in progress and has not yet progressed very much further in direct results than the existing theories. Nevertheless the formulation is rigorously founded and it is hoped to complete the various programmes here suggested, coupled with experiments on novel cross linked systems in which the various pieces of theoretical optimism can be directly checked.

7. ACKNOWLEDGEMENTS

I should like to thank Professors Allen and Gee for helpful discussions and to Dr. R. Alexander-Katz for many discussions during his Ph.D. days at Manchester.

MATHEMATICAL APPENDIX

Since the mathematical manipulations of this paper may be unfamiliar to some readers, Dr. Chompff suggested that a brief appendix could be of value.

The basic problem is to see under what conditions these problems can be reduced to differential equations, from which point ordinary analysis can be performed. It is well known that a freely hinged chain can be seen to satisfy a differential equation provided that its long range effects are solely studied. Suppose the qth link of a chain is at $\underset{\sim}{R}_q$, and that the number of ways the $(q+1)$th link can arrive at $\underset{\sim}{R}_{q+1}$ is $\Gamma(\underset{\sim}{R}_q, \underset{\sim}{R}_{q+1})$. Then if the number of configurations spanning from $\underset{\sim}{R}_0$ to $\underset{\sim}{R}_q$ is $G(\underset{\sim}{R}_0, \underset{\sim}{R}_q)$

$$G(R_0, R_{q+1}) = \int \Gamma(\underset{\sim}{R}_q, \underset{\sim}{R}_{q+1}) G(\underset{\sim}{R}_0, \underset{\sim}{R}_q)\, d^3R_q \tag{A.1}$$

$$= \int \cdots \int \Gamma(\underset{\sim}{R}_0, \underset{\sim}{R}_1)\Gamma(\underset{\sim}{R}_1, \underset{\sim}{R}_2) \cdots \tag{A.2}$$

$$\cdots \Gamma(\underset{\sim}{R}_{q-1}, \underset{\sim}{R}_q)\Gamma(\underset{\sim}{R}_q, \underset{\sim}{R}_{q+1}) \prod dR$$

a typical Markov chain. If Γ is $\Gamma(\underset{\sim}{R}_q - \underset{\sim}{R}_{q-1})$ with a fourier transform of $\gamma(\underset{\sim}{k})$, clearly $(A\cdot 2)$ implies

$$G(\underset{\sim}{k}) = \gamma^{q+1}(\underset{\sim}{k}) \tag{A.3}$$

where $G(\underset{\sim}{k})$ is the fourier transform of G.

For large $\underset{\sim}{R}_0 - \underset{\sim}{R}_{q+1}$ one needs small $\underset{\sim}{k}$ and

$$G(\underset{\sim}{k}) = e^{(1+q)\log\gamma(\underset{\sim}{k})} = e^{(1+q)\log\gamma(0) + \underset{\sim}{k}^2 \frac{(q+1)}{\gamma(0)}\left(\frac{\partial\gamma}{\partial k^2}\right)_{k^2=0} + O(k^4)} \tag{A.4}$$

If this is carried out explicitly for rods of length ℓ, freely hinged, then $\frac{(q+1)}{\gamma(0)}\left(\frac{\partial\gamma}{\partial k^2}\right)_{k^2=0} \underset{\sim}{k}^2$ is $-\frac{(q+1)\ell}{6}\, \underset{\sim}{k}^2$ and

$$G(\underset{\sim}{R}_0 - \underset{\sim}{R}_q) = \left\{\int G(\underset{\sim}{z})\, d^3z\right\}^q \left(\frac{3}{2\ell\pi}\right)^{3/2} e^{-\frac{3}{2\ell L}(R_0 - R_q)^2} \tag{A.5}$$

where $L = \ell q$, and

$\left\{\int G(\underset{\sim}{z})\, d^3z\right\}^q$ represents the total number of configurations.

In this approximation (of dropping the k^4 terms) one can break the length L up <u>in any way</u> and still retrieve G, and one can also notice that within the constant $\int \mathcal{G} d^3 z$, Γ is $\mathcal{G}$ for one step. Thus

$$\mathcal{G}(\underset{\sim}{R}_0, \underset{\sim}{R}_L; L) = \int \cdots \int \mathcal{G}(\underset{\sim}{R}_0, \underset{\sim}{R}_1; s_1)\, \mathcal{G}(\underset{\sim}{R}_1, \underset{\sim}{R}_2; s_1 - s_2) \cdots$$
$$\cdots \mathcal{G}(R_{L-1}, R_L; s_{L-1} - L)\, d^3R_1 \ldots d^3R_{L-1}, \tag{A.6}$$

where $s_1, s_2 \ldots$ is <u>any</u> ordered decomposition of $(0, L)$, not necessarily $s_i - s_{i-1} = l$.

Then the set of points

$$s_1, s_2 \quad \ldots \quad s_{L-1}$$

can be made more and more numerous until they constitute a whole continuous arc $\underset{\sim}{R}(s)$, $0 < s < L$

and

$$\mathcal{G}(\underset{\sim}{R}, \underset{\sim}{R}'; L) = \int_{\underset{\sim}{R}(0) = \underset{\sim}{R}'}^{\underset{\sim}{R}(L) = \underset{\sim}{R}} \prod_s \mathcal{G}(\underset{\sim}{R}(s) - \underset{\sim}{R}(s+ds), ds)\, \delta R \tag{A.7}$$

where $\delta \underset{\sim}{R}$ means integrate over all curves such that $\underset{\sim}{R}(0) = \underset{\sim}{R}'$, $\underset{\sim}{R}(L) = \underset{\sim}{R}$

But for a small distance ds,

$$\mathcal{G} \text{ is just } \left(\frac{3}{2 l\, ds\, \pi}\right)^{3/2} e^{-3 \frac{(R(s) - R(s+ds))^2}{2 l\, ds}}$$
$$\to \left(\frac{3}{2 l \pi\, ds}\right)^{3/2} e^{-\frac{3}{2} \dot{R}^2(s)\, ds} \tag{A.8}$$

and

$$\mathcal{G} = \mathcal{N} \int_{R(0) = R'}^{R(L) = R} \exp\left[-\frac{3}{2l} \int_0^L \dot{\underset{\sim}{R}}^2(s)\, ds\right] \delta \underset{\sim}{R} \tag{A.9}$$

where $\mathcal{N}$ is the normalization $\prod_s (3/2 l \pi\, ds)^{3/2}$. This term is completely inert and need not worry us.

That Brownian motion can be treated this way was first suggested by Wiener, and by now has a considerable literature. The key point is that the mathematicians are happy that by their standards these expressions are meaningful, so we may press on using them without worries. (Since all these operations are a very direct representation of what happens physically i.e. in a diffusion equation the fine structure of the Brownian path doesn't matter, if something did go wrong, it would be for mathematicians to be worried.)

It follows now that since we know the $\mathcal{G}$ satisfies

$$\left(\frac{\partial}{\partial s} + \frac{l}{6}\nabla^2\right)\mathcal{G} = \delta(r-r')\delta(s-s') \tag{A.10}$$

the integral on the l.h.s of (A.9) must satisfy the differential equation. If now a constraint is added on each element ds of the chain, which depends on $\underset{\sim}{R}(s)$ only, say as exp $(V(R(s))ds)$ the total effect is to add in exp $(\int V(R(s))ds)$ into (A.9) and the new

$$\mathcal{G}' = \mathcal{N}\int \exp\left[-\frac{3}{2l}\int_0^L \dot{R}^2 ds + \int_0^L V(R(s))ds\right]\delta R \tag{A.11}$$

satisfies

$$\left(\frac{\partial}{\partial s} + \frac{l}{6}\nabla^2 + V(r)\right)\mathcal{G}(r,r';s,s') = \delta(r-r')\delta(s-s') \tag{A.12}$$

since from (A.7) one only needs the $\mathcal{G}$ of infinitesimal ds to get the whole $\mathcal{G}'$, and over small ds this factor clearly does the job.

At this point we can return to Γ and ask about the terms in k^4, or, more generally about terms in curvature and torsion which have been omitted. Such terms are included if one used a Γ with $\Gamma(\underset{\sim}{R}_q, \underset{\sim}{R}_{q-1}, \underset{\sim}{R}_{q-2})$ (curvature) or $\Gamma(\underset{\sim}{R}_q, \underset{\sim}{R}_{q-1}, \underset{\sim}{R}_{q-2}, \underset{\sim}{R}_{q-3})$ (torsion). It is only at this latter level that one could for example describe a polyalkane correctly. Following the same lines of analysis as above one will end up with

$$\exp\left[\int P(\dddot{\underset{\sim}{R}}, \ddot{\underset{\sim}{R}}, \dot{\underset{\sim}{R}}, \underset{\sim}{R})ds\right] \tag{A.13}$$

as against $\int P(\dot{\underset{\sim}{R}}, \underset{\sim}{R})ds$ of the first analysis. In fact quite apart from obtaining a form for P from a polymer chain one gets something like this from the statistical mechanics of a wire. The curvature energy is proportional to $\ddot{R}^2$ and the torsional energy to $\mathrm{Det}|\dot{R}\,\ddot{R},\dddot{R}|$ i.e. one gets

$$\exp\left[-\frac{\xi}{kT}\int[(\dot{R}\times\ddot{R})\cdot\dddot{R}]^2 ds - \frac{\eta}{kT}\int \ddot{R}^2 ds - \frac{3}{2l}\int \dot{R}^2 ds\right] \tag{A.14}$$

as a special case of (A.13). This also can be cast into the form of a differential equation, for example the case with $\ddot{\underset{\sim}{R}}$ alone gives

$$\left(\frac{\partial}{\partial s} + \underset{\sim}{v}\frac{\partial}{\partial \underset{\sim}{r}} + \frac{kT}{4\eta}\frac{\partial^2}{\partial \underset{\sim}{v}^2} + \frac{3}{2l}v^2\right)\mathcal{G}(\underset{\sim}{r},\underset{\sim}{r}';\underset{\sim}{v},\underset{\sim}{v}';s,s') = \delta\delta\delta$$

where $\underset{\sim}{v}, \underset{\sim}{v}'$ are the values of $\dot{\underset{\sim}{R}}$ at the end points which now need to be specified to give the integral a proper definition. The torsion takes us up to one more variable $\ddot{\underset{\sim}{R}}$ needing to be defined at the end points.

As has been shown, the physical constraints of cross linkage and topology, and indeed ordinary forces, lead to expressions in the exponent which are not just single integrals and it is not possible to transform them into differential equations, and as a result approximate techniques are required.

The Variational Principle

The proof that $e^Q > 1+Q$ is easily seen from their graphs

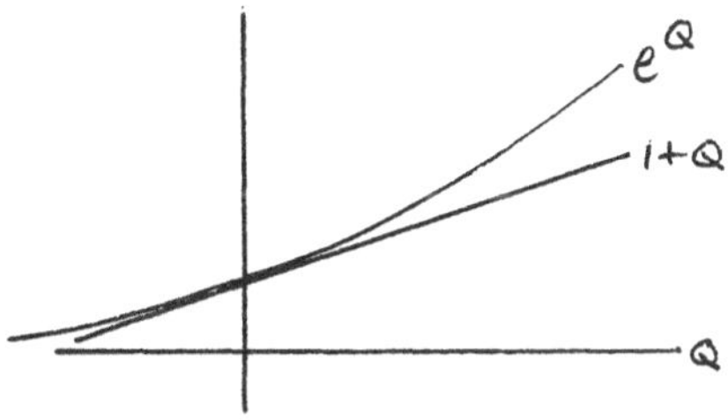

Since the free energy is defined in terms of exp $(-F/\kappa T)$ it follows that

$$F_{\text{true}} < F_{\text{found here}}.$$

It will be noted that the free energy found here is ½ that of pervious work so can be claimed to be closer to F_{true}.

The Transformation

This is based on taking the $X^{(1)}$ system to be the centre of mass of the n additional coordinates and $n-1$ others, then taking this into $c.m.$ system with the '0' or principal system. It is very much simpler to set up collective coordinates when their number is odd for the complex labelling can then be employed. So the transformation goes this way

$$R^{(1)} \ldots\ldots R^{(n)} \longrightarrow \underbrace{\frac{1}{\sqrt{n}}\sum_1^n R^{(i)}, \quad \text{the } Y\text{'s}}_{<\!-\,c\cdot m\,-\!>}$$

$$R^{(0)}, \frac{1}{\sqrt{n}}\sum_1^n R^{(i)} \longrightarrow \underbrace{X^{(0)}, \quad X^{(1)}}_{<\!-\,c\cdot m\,-\!>}$$

The Topological Invariants

This subject is sufficiently technical that it is doubtful whether the effort is justified by the modest results so far obtained, so the reader is referred to the authors papers (4).

REFERENCES

1) Edwards S. F. and Freed K. F. J. Phys. C. 1970, 3, L31; J. Phys. C 1970, 3, 739, 750, 760.

2) Freed K. F. to be published in J. Chem. Phys.

3) Edwards S. F. J. Phys. C 1969, 2, 1.

4) Edwards S. F. Proc. Phys. Soc. 1967, 91, 513; J. Phys. A 1968, 1, 15.

DISCUSSION

M. Gordon (University of Essex): We have found experimentally (see our paper) that, as expected, the effects of entanglements on the initial Young's modulus, vanish as the concentration of active network chains goes to zero at the gel point. Does your theory predict this, or are effects of crosslinks and entanglements additive independently of the concentration of active chains?

S. F. Edwards: The calculations given here have assumed one is well past the gel point. The method of attack will give a gel point due to entanglements (i.e. high enough density) or crosslinks, or both. The two effects are apparently not additive, because I would take the system as having gelled when the constraint $w \neq 0$, and (as 4.15 shows for example even though this is valid only in well gelled situations) the equation for w can be expected to be complicated.

S. Prager (University of Minnesota): I would like to return to a point you made at the beginning of your talk, namely that crosslinks and entanglements in a network constitute microscopic constraints on the topology, constraints which cannot be relaxed when the network is subsequently deformed. Now in regard to entanglements, such constraints work both ways: two entangled network chains cannot become disentangled, and two disentangled chains cannot form an entanglement. Thus if we treat an entanglement as some sort of additional crosslink, must we not also take into account a corresponding repulsion between chains which are not entangled with one another? Might not these two effects largely cancel out in the end, so that ideal Gaussian network model, which ignores them both, is perhaps rather better than it deserves to be?

S. F. Edwards: This point is clearly seen by means of diagrams. If one has two configurations of equal probability say

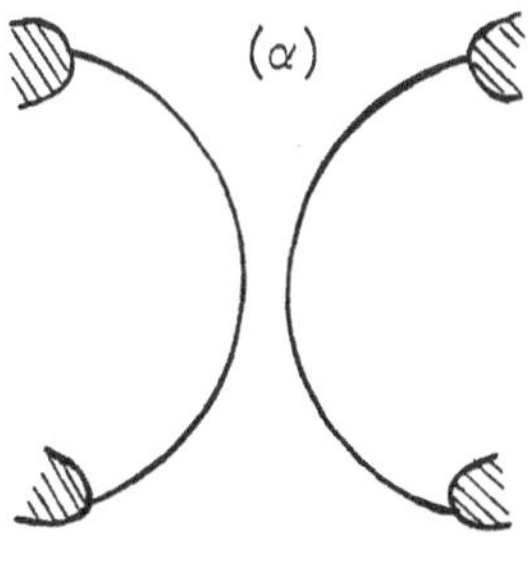

and

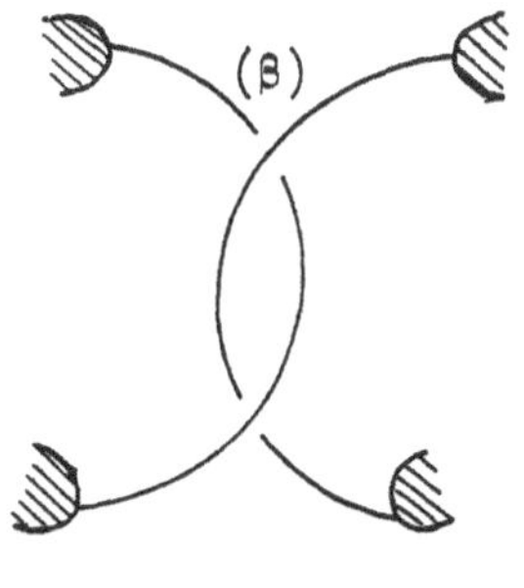

then under (horizontal) compression (α) resists whilst (β) relaxes whilst under tension the reverse is the case. This shows in that I has different signs for the two situations and in $\langle (I-\tilde{I})^2 \rangle$ there are a large number of contributions to the integrals, both positive and negative. The final number is a kind of r.m.s. value and is much smaller than if one took $\iint \left| (\underline{dR} \times d\underline{R}') \cdot \nabla \frac{1}{|R-R'|} \right|$

for which I would have much the same effect as looking at only (β) in extension and only (α) in compression. The formulae do however take these terms in correctly and the effect is allowed for.

THE RUPTURE WORK OF CROSSLINKED POLYMERS

J. J. Bikerman

15810 VanAken Blvd., Cleveland, Ohio 44120

SUMMARY

The work "γ" of fracture was identified by Dupré and Griffith with the surface energy of solids. In reality work is spent on the deformation leading to rupture rather than on the break itself. In simple instances it is equal to the work needed to extend a column of material (of unit cross-section) in front of the growing crack until the column snaps. This extension may be purely elastic. The new concept, contrary to the old, (a) indicates the analogy between the breaks of a liquid and a solid, (b) accounts for the absolute value of "γ," (c) agrees with the effect of crosslinking on "γ," (d) is compatible with the rate dependence of "γ," and (e) explains the liberation of heat during fracture. The absence of any clear effect of macroscopical plasticity on "γ" contradicts the hypothesis attributing the high values of "γ" to plastic deformation.

If a Hookean cylindrical bar of length l_0 and radius r is extended to the length l_m and then breaks, and if the tensile force F at every moment is only insignificantly greater than the elastic resistance of the bar, then the work (which has achieved rupture) is $0.5\ F_m(l_m - l_0)$, if F_m is the highest value of F (i.e., in the moment of fracture). This product obviously has no fundamental importance and is not a material constant as it depends on the absolute extension $l_m - l_0$ which is greater the greater the initial length l_0. It also depends on the value of r but this complication usually can be successfully avoided by substituting the external (average) stress $f_m (= F_m/\pi r^2)$ for F_m and referring the work of rupture to unit cross-section.

Three attempts are known to define, and to account for, a work of rupture which is independent of the sample dimensions and thus may represent a property of the material studied. The most common mode of rupturing a solid involves the growth of a crack. In many instances it is possible, more or less convincingly, to separate the work W (ergs or joules) necessary for this growth from the work done on the simultaneous gross deformation of the test sample (and the instrument employed). Usually it is found that this W is proportional to the area A of the new crack surfaces, so that $W/A = "\gamma"$ (g/sec^2 or kg/sec^2) remains constant when the crack expands; it is moreover independent of the initial length l_0. The physical meaning of the quantity "γ" is the subject of the present discussion.

Rupture Work and Surface Energy

The "γ" was identified by Dupré (1866) and Griffith (1921) with the specific surface energy of the solid. This concept is still popular but at least six weighty arguments may be marshalled against it.

(a) The hypothesis attributes great importance to the air - solid interface in the crack but completely neglects the identical air - solid interface along the cylindrical surface of the bar. As a simple calculation shows, the total air - solid area of the bar decreases, rather than increases, during fracture whenever $\varepsilon - \nu_p \varepsilon - \nu_p \varepsilon^2 > (r/l_0)$; ν_p is the Poisson ratio and ε is the maximum strain of the bar material. If surface energy really were decisive, then bars thin and long enough for this inequality to be valid would rupture spontaneously; no wire or filament would be stable.

(b) The hypothesis exaggerates the difference between liquids and solids. When a liquid drop is gradually extended, constriction (or "necking-in") starts as soon as the length of the liquid cylinder exceeds its circumference (Plateau 1869). This phenomenon takes place because the area of the two resulting drops is smaller than that of the cylinder before splitting. Thus, in liquids, rupture is associated with a decrease in area and is caused by the ensuing lowering of the free surface energy. We are asked to believe that, in solids, rupture is associated with, and resisted by, an increase in area.

(c) Although the surface energy of solids has never been convincingly measured or calculated [1,2], it is universally believed that its order of magnitude is 100 - 1000 ergs/cm^2, the lower limit being valid for organic solids and the upper, for high-melting metals. The values of "γ" for many polymers range between 10^5 and 10^7 ergs/cm^2. Thus, the experimental results frequently exceed the theoretical expectation by factors reaching 100 or even 10000.

(d) The theoretical cohesion of a polymer generally is greater when the degree of cross-linking increases. According to Dupré, surface energy is the work required to separate two rows of atoms or molecules (of unit area) against their cohesional attraction. Thus, also the surface energy would be expected to be raised by cross-linking. Experimentally, the opposite behavior is observed in many instances. A butadiene - styrene rubber was vulcanized with one, three, or nine parts of a peroxide (for 100 parts of the rubber), and the value of "γ" decreased in this series from 20 to 6 and to 2 megaergs per square centimeter [3] . In a recent study [4] , the "γ" of an epoxy resin hardened with phthalic anhydride was almost independent of the ratio of hardener to resin within the range of 0.12 to 0.30.

The surface energy of water is not altered by pebbles in the latter; hence, the "γ" of polymers would be expected to be unaffected by inorganic fillers. In reality [4], incorporation of 1 part of silica in 1 part of cross-linked epoxy resin raised the "γ" by a factor of nearly 10.

(e) Surface energy, as a rule, is independent of the rate of formation of a surface [2]. Consequently, "γ" would be expected to be independent of the rate of rupture. In reality, "γ" usually increases with this rate [3,5,6].

(f) If the work of rupture really remained in the resulting fragments as their surface energy, no energy would have been left to be liberated as heat. Contrary to this conclusion, the major part of the work spent on fracture usually is released as heat to the surroundings [7].

Rupture Work and Plastic Deformation

A later concept (Orowan 1948, Irwin 1948) supplies a reasonable explanation for the discrepancy indicated in item (c) above. The "γ" is supposed to be the sum of the free surface energy and the work expended on plastic deformation of the solid near the front of the advancing crack. There is no doubt that plastic deformation in many instances precedes rupture and that it may be extensive enough to account for the high experimental values of "γ"; the colored fracture surfaces of methacrylates ("crazing") and the milk-white surfaces of poly(tetrafluoroethylene) bent to break are visible proofs of such deformations. However, when different materials are placed in the order of increasing "γ" values, this order is not identical with that of increasing (macroscopic) plasticity. Also observation (e) of the preceding section lends no support to the plasticity mechanism. As a rule, a more profound plastic deformation would occur during a slow than during a rapid operation. Hence, "γ" would be expected to decrease on an increase in the rate of rupture. The opposite behavior seems to be more common.

Plastic Work and Elastic Deformation

A third theory [7] appears to be the most trustworthy at present; it does not exclude the plasticity mechanism but claims that the main experimental observations can be accounted for also when no plastic flow is present.

The unbroken material right in front of the growing crack is extended by the local tensile stress until its maximum strain is reached; then the "column" of this material snaps, which is equivalent to an advance of the crack. The "γ" is the work required to stretch this column to its maximum elongation, referred to unit cross-section. Thus, the work is spent not on rupture as such but on the deformation leading to rupture.

The theory can be expressed in mathematical terms. Since the column in the front of the crack is very short and is not likely to contain serious flaws, it may be assumed to possess the strength of a perfect material; hence, its tensile strength is equal to its molecular cohesion ξ g/cm.sec^2. If Δ is the maximum absolute extension (measured, for instance, in centimeters), then the work of stretching (per unit area) is

$$"\gamma" = o\xi\Delta; \qquad (1)$$

o is a number between 0.5 and 1.0 depending on the conditions of stretching. The breaking stress σ_m of the specimen, that is the external force (causing fracture) divided by the cross-section of the bar, naturally is different from ξ because of the stress concentrations at the crack. If 2a is the length of the crack in the direction perpendicular to the external force and if R is its radius of curvature at the advancing tip, then

$$\xi \approx \left(\frac{4a}{R}\right)^{0.5} \sigma_m . \qquad (2)$$

If the above column is a Hookean solid of length l and modulus of elasticity E, then

$$\Delta = l\,\xi/E. \qquad (3)$$

Introduction of equations (2) and (3) into (1) results in the formula $"\gamma" \approx 4\,oal\,\sigma_m^2/RE$. It is usually assumed that $l \approx R$. Hence,

$$"\gamma" \approx 4\;oa\,\sigma_m^2/E. \qquad (4)$$

Equation (4) is similar to that derived by Griffith for calculating the magnitude believed by him to be the free surface energy of the solid. The specific work of rupture "γ". as given by (4), is of course no surface property at all.

The advantages of the new concept may be listed following the scheme used for pointing out the limitations of the surface energy hypothesis.

(a) The interfaces in the crack and the cylindrical surface of the bar are treated equally.

(b) The difference between the behaviors of solids

and liquids vanishes. In both forms of matter, the work is spent on the deformation which precedes rupture.

(c) The absolute value of "γ" derived from the experiment is in a reasonable agreement with equation (1). The molecular cohesion ξ of many classes of solids is of the order of magnitude of 10^{11} g/cm.sec^2. The value of o is constant as far as the order of magnitude is concerned. Thus, "γ" depends above all on the extension Δ. If the maximum strain of a solid is high, also its "γ" is likely to be considerable. The relation between "γ" and Δ, at $\xi = 10^{11}$, is illustrated in the table:

Δ	10^{-8}	10^{-7}	10^{-6}	10^{-5}	10^{-4}	cm
"γ"	10^{3}	10^{4}	10^{5}	10^{6}	10^{7}	g/sec^2

The experimental values of "γ" are near 3 x 10^3 g/sec^2 for silicate glasses whose maximum strain usually is less than 0.01. Their Δ ought to be very small; and equation (1) predicts Δ of a few ångströms for these solids. Values near 7 x 10^3 g/sec^2 were observed [4] for epoxy resins hardened with 15 - 30% of phthalic anhydride; unfortunately, the stress - strain curves of these materials have not been determined. For polyesters "in which are many cross-links, tending to cause glassy behavior, and also residual polystyrene chains which are free to contribute" viscoelasticity [8], the "γ" was 2 x 10^5g/sec^2, corresponding to a Δ of about 200 ångströms. The "γ" of a deeply vulcanized butadiene - styrene rubber is quoted above as 2 x 10^6 g/sec^2. Since the ξ of this vulcanizate presumably was smaller that 10^{11} g/cm.sec^2, the Δ would be near to 5 x 10^{-5} cm. When a polymer is only a little cross-linked, its "γ" often is in the range between 10^6 and 10^7 g/sec^2, corresponding to an extension of 1 micron or so; and the maximum strain of such polymers usually is high (of the order of 1.0). It is seen that the total relative elongation rather than the tendency for plastic flow increases, on the whole, when "γ" increases.

(d) Only one publication [3] was found by the author, in which both "γ" and the maximum relative elongation ε_m were determined on identical samples. The following table shows the percentage (p%) of the peroxide employed in vulcanizing a butadiene - styrene rubber, the "γ", the ε_m, and the estimated Δ, calculated under the assumptions that $\xi = 6 \times 10^6$ g/cm.sec^2 for all vulcanizates and that o was equal to 0.5.

p%	1	3	9
"γ"	2×10^7	6×10^6	2×10^6 g/sec^2
ε_m	6	3	1.5
Δ	7×10^{-4}	2×10^{-4}	1×10^{-4} cm

It is interesting to notice that the ratio Δ/ε_m does not vary much when the nature of the rubber varies. This lends credence to the above reasoning in which conclusions on the probability of a definite value of Δ are derived from the knowledge of ε_m. To test this reasoning in a more convincing manner, additional measurements of both "γ" and ε_m on identical materials would be highly welcome.

(e) The effect of the rate u of rupture on "γ" cannot be unambiguously predicted from the above equations (1) or (4) but it is clear that the existence of this effect is in agreement with the theory. Suppose that the rate u affects the work of fracture because the sample acquires a higher temperature when it is deformed in a shorter time. A higher temperature would lower the true surface energy so that "γ", if it were equal to this energy, would decrease when u increases, contrary to the usual observations. On the other hand, the maximum strain ε_m may rise with temperature more rapidly than ξ decreases; hence, the product $\xi\Delta$ may be expected to increase when u is greater.

At a constant temperature, usually a greater breaking stress σ_m is required when a very rapid break is aimed at. Equation (4) shows then that, as long as a and E are independent of the rate u, "γ" must increase with σ_m, that is with u. Unfortunately, the mechanism of the correlation between σ_m and u is not well understood, and the above consideration must be treated as a preliminary suggestion only. It should be remembered also that equation (3) is valid solely for Hookean solids, while the measurements of the relation between "γ" and u generally are performed on viscoelastic materials.

No systematic information could be found on the effect of cross-linking on the relationship between "γ" and u.

(f) The extension of the material column in front of the crack may be purely elastic; only in these instances can equation (3) be correct. When this column snaps, the work spent on the stretching would, ideally, be completely transformed into heat. It is sometimes stated that no heat is liberated in purely elastic deformations. This statement is erroneous. Every process which is irreversible in the sense of thermodynamics must be associated with evolution of heat. If, for instance, a perfect gas is extremely slowly compressed and the piston is rapidly and suddenly removed, then the compressed gas expands without doing any work, and the work spent by the experimenter during the compression half-cycle is now present as heat.

The work of rupture is not transformed into heat quantitatively because the strata next to the rupture surfaces, as a rule, are in a more disordered state than the bulk of the two solid fragments; see the above section on Rupture Work and Plastic Deformation. The energy residing in the disturbed surface layers received the name "cuticular". Thus, fracture work = heat + cuticular energy [2].

References

1. J. J. Bikerman, Physica Status Solidi **10**, 3 (1965).
2. J. J. Bikerman, "Physical Surfaces". Academic Press, New York 1970.
3. H. W. Greensmith and A. G. Thomas, J. Polymer Sci. **18**, 189 (1955).
4. R. Griffith and D. G. Holloway, J. Materials Sci. **5**, 302 (1970).
5. P. I. Vincent and K. V. Gotham, Nature **210**, 1254 (1966).
6. S. J. Bennett, G. P. Anderson and M. L. Williams, J. Appl. Polymer Sci. **14**, 735 (1970).
7. J. J. Bikerman, SPE Trans. **4**, 290 (1964).
8. B. Harris and E. M. de Ferran, J. Materials Sci. **4**, 1023 (1969).

DISCUSSION

A. Peterlin (Research Triangle Institute, N. C.): The difference between the surface energy γ and the fracture work per unit area of new surface "γ" is usually explained by the deformation of a layer of finite thickness δ which is many times larger than the length $\Delta\ell$ needed for separation of adjacent molecules ($\Delta\ell \approx 1$Å). The ratio $\delta/\Delta\ell$ is practically identical with "γ"/γ. The deformed layer shows up in glassy polymers as craze .

To what extent does your treatment agree with that model?

J. J. Bikerman: As mentioned in my paper, plastic flow has been observed near many fracture surfaces, and in these instances the work required to cause the flow undoubtedly was a part of the experimental "γ." My point is that, even when there is no plastic flow, the "γ" is not related to the surface energy γ of the solid (if this γ exists). It may be asked what part of the "γ" is likely to be caused by effects differing from plasticity; perhaps this part is negligibly small in nearly all solid materials. If this were so, then the series of solids arranged in the order of their "γ" values would be identical with that arranged according to their plasticities. In reality, the highest "γ" is observed for rubber vulcanizates which possess very little plasticity but a very high maximum strain. Hence, the new theory appears to have an important range of application.

INTERMOLECULAR FORCES IN POLYMERS AND LIQUIDS*

Robert E. Cuthrell

Sandia Laboratories

Albuquerque, New Mexico

SUMMARY

A variety of physical properties of several liquids and polymers can be quantitatively described by considering a crystalline arrangement of molecules or other volume elements bound by non-directional forces. In some cases we propose that the intermolecular forces in the liquid or solid under question are simply due to van der Waals interactions. In those cases we demonstrate that physical properties such as surface energy, cohesive strength, compressibility, thermal expansion, and work of vaporization can be calculated from atomic constants and related to one another by the proposed model. In the case of other liquids and polymers, intermolecular forces cannot be described in terms of van der Waals binding alone and other (directional) forces such as dipole-dipole binding must be included. It will be shown that a variety of the calculated properties favorably compare with experimental results.

INTRODUCTION

A number of authors have previously demonstrated that some of the properties of liquids and polymeric materials may be described in terms of models in which a material is bound primarily by van der Waals forces. An early example of this is the calculation by de Boer indicating that the theoretical strength of phenol-formaldehyde and m-cresol-formaldehyde polymers may be qualitatively described by considering dispersion interactions alone.[1] Other examples of such treatments are those of Pastine,[2] Ulbrich,[3] and

* This work was supported by the U. S. Atomic Energy Commission.

Muller[4] in which they attempted to describe the theoretical strength, surface tension, compressibility, and heat of sublimation in terms of dispersion interactions.

This paper is essentially an extension of those treatments in which a somewhat more detailed model of a liquid or an amorphous polymer is considered and a variety of properties are calculated on the basis of that model. In particular, we consider that the liquid or polymer may be described by a crystalline array of molecules interacting purely by van der Waals forces and then calculate the theoretical tensile strength, surface energy, work of vaporization, Young's modulus, compressibility, and coefficient of thermal expansion on the basis of that model. These results are compared to ex perimentally observed values for a variety of liquids and polymers. We ignore the effects of "bond breaking" mechanisms just as previous authors and consider only a highly idealized model.

THE THEORETICAL MODEL

We will consider that each molecule of a liquid or an amorphous organic polymer is centered on a normal site in a face centered cubic lattice and that all interactions may be described in terms of van der Waals interactions. The total potential of a molecule will be calculated by summing over a large number of neighbors. Similarly, calculations of various properties of such a system will be performed in terms of the fcc lattice array. There is, of course, no physical basis for the selection of any one of the isotropic crystalline models except from the standpoint of simplifying the subsequent calculations. One could also perform calculations using a radial distribution function and integrating rather than performing lattice sums. Clearly these calculations are of an approximate nature and are used only to demonstrate that several properties of liquids and polymers may be described in terms of van der Waals interactions.

The potential energy for the i-th molecule at a lattice site interacting with any other molecule, designated by index j, is given by

$$\phi_{ij} = C_r\, r_{ij}^{-12} - C_d\, r_{ij}^{-6} . \qquad (1)$$

It is assumed here that all interactions are identically expressed in terms of the intermolecular spacing, r_{ij}, and that a simple pairwise summation is an accurate description. The constants C_r and C_d are the repulsion and attraction constants in the London expression and will be discussed in somewhat greater detail later. The total interactions of the i-th molecule with molecules at all other sites in the lattice are given by,

$$\phi_i = \sum_j \phi_{ij} = \phi_{i1} + \phi_{i2} + \phi_{i3} + \cdots, \tag{2}$$

It is convenient to introduce another expression for r_{ij},

$$r_{ij} = P_{ij} R, \tag{3}$$

where P_{ij} is a dimensionless factor dependent upon the lattice coordinates and R is the near-neighbor spacing. Then,

$$\begin{aligned} \phi_i &= \sum_j \left(C_r P_{ij}^{-12} R^{-12} - C_d P_{ij}^{-6} R^{-6} \right) \\ &= M_{12} C_r R^{-12} - M_6 C_d R^{-6}, \end{aligned} \tag{4}$$

and

$$\frac{d\phi_i}{dR} = 6M_6 C_d R^{-7} - 12 M_{12} C_r R^{-13}, \tag{5}$$

where M_n are Madelung-like constants for the fcc lattice. Their values are $M_6 = 14.45392$, and $M_{12} = 12.13188$.[5] At equilibrium, $R = R_o$ and $d\phi_i/dR = 0$, hence,

$$C_r = \frac{1}{2} \frac{M_6}{M_{12}} C_d R_o^6, \tag{6}$$

$$\phi_i = M_6 C_d \left(- R^{-6} + \frac{1}{2} R_o^6 R^{-12} \right), \tag{7}$$

and

$$\frac{d\phi_i}{dR} = 6 M_6 C_d \left(R^{-7} - R_o^6 R^{-13} \right). \tag{8}$$

The total lattice energy for a lattice of N molecules is

$$U = \frac{1}{2} N \phi_i, \tag{9}$$

where the factor $\frac{1}{2}$ has been included so that all interactions are counted only once.

With these general formulations we may now calculate a variety of associated properties for the lattice at 0°K.

Pressure-Volume Calculations

When an isotropic material is subjected to isothermal triaxial compression, the pressure, P, can be calculated in terms of $d\phi_1/dR$, the number of moledules in each unit cell n, and the unit cell cross sectional area a,

$$- P = (n/a)\, \frac{d\phi_1}{d\,R}\,, \tag{10}$$

where $a = 2R^2$. The equilibrium near-neighbor spacing, R_o, is decreased incrementally, hence,

$$R_o - x = K\,R_o = R, \tag{11}$$

$$- P = 12\, M_6\, C_d\, R_o^{-9} \left(K^{-9} - K^{-15}\right), \tag{12}$$

and

$$V/V_o = K^3, \tag{13}$$

where V/V_o is the volume ratio.

Compressibility

The Compressibility, k, is defined as

$$k = -V^{-1}\, dV/dP, \tag{14}$$

where P and V are the pressure and volume, respectively.[6] The unit cell volumie is $(R\sqrt{2})^3$, hence

$$k^{-1} = 12 M_6\, C_d\, R_o^{-9}\left(5K^{-15} - 3K^{-9}\right). \tag{15}$$

Young's Modulus

If the lattice is extended along a single direction while leaving the dimensions unchanged in the plane perpendicular to the extension, Young's modulus, Y, may be calculated directly from the compressibility. In this case Poisson's ratio is assumed to be zero and Y will be

$$Y = 3/k \tag{16}$$

Work of Vaporization

The isothermal, reversible work of expansion, ΔF, of the lattice from the liquid density to the gas density for the Avogadro number of molecules, N, is given by

$$\Delta F = N\{\phi(\rho_2) - \phi(\rho_1)\}, \tag{17}$$

where the two near-neighbor separations are calculated from the gas and liquid densities, ρ, respectively, the molecular weight, M, and the Avogadro number. The near-neighbor separations are given by

$$R_o = \{4M/(\rho N)\}^{1/3}/\sqrt{2}. \tag{18}$$

When the process involves pressure-volume work only, as in this lattice expansion, the free energy change and lattice energy are equivalent.[7]

Surface Energy

The dispersion surface energy, γ^d, is the difference between the energies of bulk and surface vacancy formation for the appropriate number of molecules, n, to make up unit area, a, and is given by

$$\gamma^d = (n/a)\,\phi_S - (2\,n/a)\,\phi_B. \tag{19}$$

The factor 2, in the second term, has been included since a plane passing through the bulk contains two surfaces, and, therefore, twice as many molecules as the (1,1,1) surface plane.

Coefficient of Thermal Expansion

The coefficient of thermal expansion may be calculated by expanding Equation (4) in binomial series in terms of a small displacement, x, of the lattice from its equilibrium spacing. This calculation may be performed using a method outlined by Kittel[8] in which we express the lattice energy as

$$\phi(x) = cx^2 - gx^3 - fx^4. \tag{20}$$

Using the Boltzmann distribution function, which weights the possible values of x according to their thermodynamic probability, Kittel calculated the average displacement $\bar{x}$,

$$\bar{x} = 3k\,Tg/4c^2, \tag{21}$$

where k is the Boltzmann constant and T is the absolute temperature. The coefficient of linear thermal expansion $\alpha = \bar{x}/TR$ and the coefficient of volume expansion $\beta = 3\alpha$, and is expressed as

$$\beta = \frac{9}{4}\,k\,\frac{\left(-56\,M_6\,C_d\,R^{-6} + 364\,M_{12}\,C_r\,R^{-12}\right)}{\left(-78\,M_{12}\,C_r\,R^{-12} + 21\,M_6\,C_d\,R^{-6}\right)^2}. \tag{22}$$

Theoretical Tensile Strength

The theoretical tensile strength is calculated using an idealized concept in which the lattice is assumed to separate along a single plane with no defects participating in the process. The sum of all pairwise interactions, $\sum_j d\phi_{ij}/dR$, between the i-th molecule and its neighbors across the fracture plane was multiplied by the appropriate number of molecules, n, to make up unit area, a. This calculation was performed as a function of the separation of the two bodies at the fracture plane to give the tensile stress, σ,

$$\sigma = (n/a) \sum_j d\phi_{ij}/dR. \tag{23}$$

The theoretical tensile strength is given by the stress at the maximum in the resultant calculated stress-strain curve.[9]

Force Constants

Dispersion forces are due to induced electronic polarizations and are expressed in the second-order perturbation terms in the quantum mechanical calculation of the interaction potential. First interpreted by London,[10] the fundamental theory of dispersion forces is discussed by Margenau[11] and Pitzer.[12] The attraction force constant C_d, defined by London, is

$$C_d = \frac{3}{4} h \nu_o \alpha^2. \tag{24}$$

where h is Plank's constant, ν_o is the frequency for the electronic excitation, and α is the molecular electronic polarizability. The repulsion force constant C_r is empirically determined in terms of C_d using Equation (6) and the condition $d\phi_i/dR = 0$ when $R = R_o$.

Empirical Equations

We are aware of several alternate methods for calculating physical properties. For instance, Equation (10) can be written, from the first law of thermodynamics,

$$-P = dU/dV = (dU/dR)(dR/dV). \tag{25}$$

Young's modulus, Equation (16), can also be formulated in terms of Equation (23). We obtain the same dependence upon C_d and R, but slightly different numerical coefficients, using each of these methods. The correct method is not obvious. Therefore, we retain the formal dependence upon C_d and R and determine the numerical coefficients empirically using the experimental data for carbon tetrachloride. The attraction constant, C_d, was calculated using Equation (24), a spectral cut-off wavelength of 2770 Å from which the frequency was calculated, and an electronic polarizability of 10.5 Å^3.[13] The experimental data in Table I were used in determining the numerical coefficients in Equations (26) through (32).

TABLE I

Experimental Properties for Carbon Tetrachloride

Property	Experimental
Tensile Strength (psi)	4057[14]
Surface Tension (dyn/cm)	26.4[15]
Work of Vaporization (KCal/mole)	1.10[16]
Compressibility (psi^{-1}) x 10^6	4.32[17]*
Cubical Coefficient of Thermal Expansion (°C^{-1}) x 10^3	1.24[18]
Near-Neighbor Spacing, R_o, (Å)	6.46[19]

*V/V_o = 0.924

The equations describing the physical properties, in terms of the force constant C_d and the near-neighbor spacing R_o, are given by

$$-P = 1.457 \times 10^2\, C_d\, R_o^{-9}\, (K^{-9} - K^{-15})\ \text{dyn/cm}^2, \tag{26}$$

$V/V_o = K^3$, for pressure volume calculations;

$$k^{-1} = 1.457 \times 10^2\, C_d\, R_o^{-9}\, (5K^{-15} - 3K^{-9})\ \text{dyn/cm}^2 \tag{27}$$

for compressibility, k;

$$Y = 8.742 \times 10^2\, C_d\, R_o^{-9}\ \text{dyn/cm}^2, \tag{28}$$

for Young's modulus, Y;

$$\Delta F = 1.349 \times 10^{14}\ C_d\ R_o^{-6}\ \text{KCal/mole}, \tag{29}$$

for the work of vaporization, ΔF;

$$\gamma^d = 13.51\ C_d\ R_o^{-8}\ \text{dyn/cm}, \tag{30}$$

for the dispersion surface tension, γ^d;

$$\beta = 1.011 \times 10^{-17}\ R_o^6/C_d\ {}^\circ C^{-1}, \tag{31}$$

for the coefficient of thermal expansion, β;

and

$$TS = 9.245\ C_d\ R_o^{-9}\ \text{dyn/cm}^2, \tag{32}$$

for the tensile strength, TS.

Figures 1 through 4 show that these equations are useful over an extensive range of temperatures and pressures.

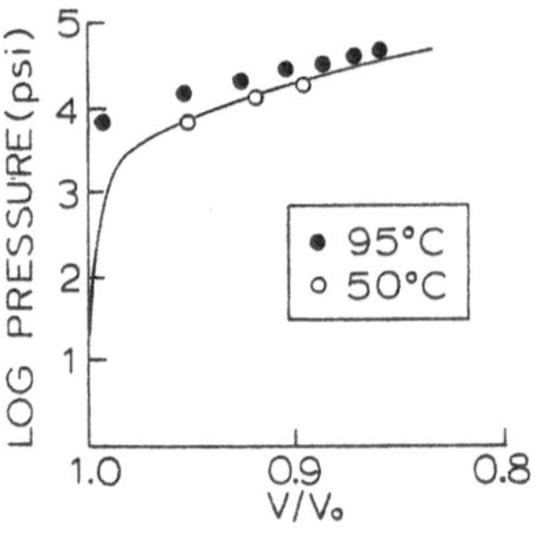

Figure 1. The theoretical compression curve for carbon tetrachloride compared with the experimental data points.[17]

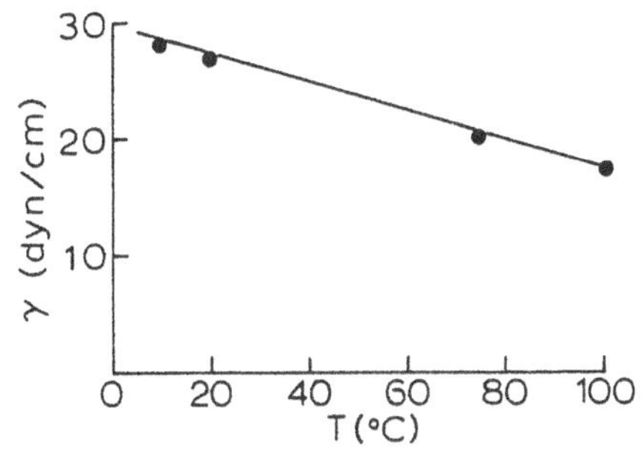

Figure 2. The theoretical temperature dependence for the surface tension of carbon tetrachloride compared with experimental data.[15]

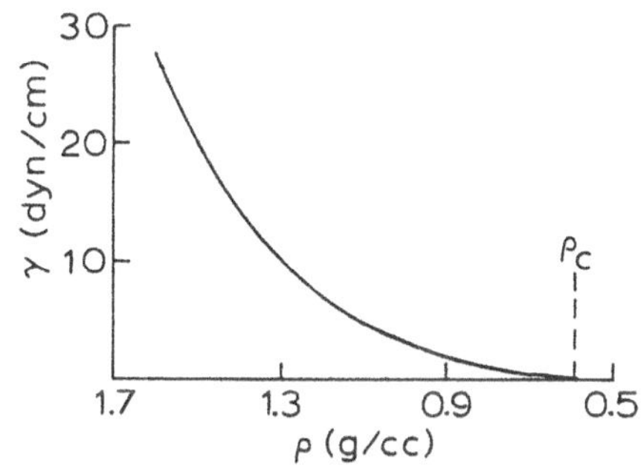

Figure 3. The theoretical relation between surface tension and density through the critical point for carbon tetrachloride.

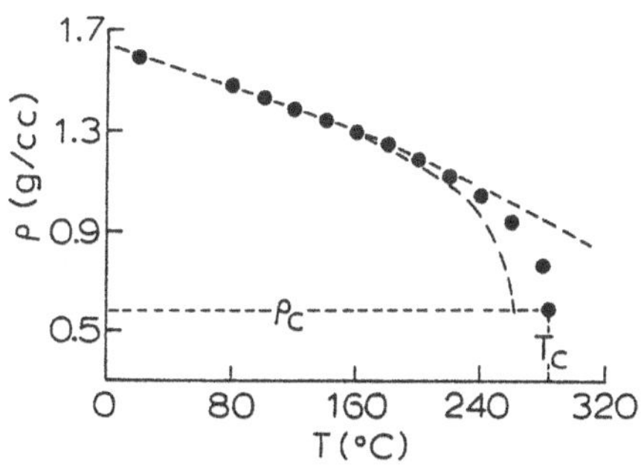

Figure 4. The theoretical temperature dependence of the density through the critical point for carbon tetrachloride compared with experimental orthobaric data.[20]

The results shown in Figures 2 and 3 were calculated using Equations (30) and (31). Figure 3 shows that the calculated surface tension decreases to zero at the critical density. The theoretical critical density (0.574 g/cc) is within 2.4% of the measured value (0.588 g/cc).[20] Equations (18) and (31) were used to calculate the temperature dependence of the density shown in Figure 4. Equilibrium conditions were assumed at each temperature such that $d\phi_1/dR = 0$. Since the lattice vibration contributions to the energy are ignored in this theory, the equilibrium conditions require C_r to increase with increasing temperature (upper dashed line in Figure 4). Linder derived expressions which indicate that the attraction constant C_d decreases with increasing temperature.[21] The lower dashed line in Figure 4 was calculated for $C_d = f(T)$, $C_r \neq f(T)$, and $d\phi_1/dR = 0$.

RESULTS FOR LIQUIDS AND AMORPHOUS POLYMERS

Experimental Surface Energy

Fowkes proposed that the surface energy is the sum of the components due to each type of interaction,

$$\gamma = \gamma^m + \gamma^h + \gamma^d + \cdots, \tag{33}$$

where the terms with superscripts m, h, and d are the components due to metallic bonding, hydrogen bonding, and dispersion bonding, respectively, and are included as applicable for a given material. He derived an expression,

$$\cos\theta = -1 + 2\left|(\gamma_S^d)(\gamma_L^d)\right|^{1/2}(\gamma_L)^{-1} \tag{34}$$

in terms of the contact angle, θ, of a sessile liquid drop, the dispersion surface energies, γ_L^d and γ_S^d, of the liquid and the solid, respectively, and the total surface energy, γ_L, of the liquid.[22] Measured contact angle values and "experimental" dispersion surface energies obtained using Equation (34) are given in Table II.

The Surface Tension - Tensile Strength Ratio

The theoretical dispersion surface tension, γ^d, and tensile strength, TS, are given by

$$\gamma^d = 216.2\ C_d\ A^{-8}\ \text{dyn/cm} \tag{35}$$

and

$$TS = 209.3\ C_d\ A^{-9}\ \text{dyn/cm}^2, \tag{36}$$

TABLE II

Contact Angles and Surface Tensions of
Liquids and Solids

Liquid	Solid	θ(Deg.)	γ_L	γ_L^d	γ_S^d (dyn/cm)
Water	PTFE	110.0	72.8 [22]	21.8 [22]	19.5 [22]
HAc	PTFE	36.6	27.2 [23]	30.77	19.5 [22]
CCl_4	---	---	26.15 [23]	---	---
Chloroform	---	---	27.14 [24]	---	---
---	PCTFE	---	---	---	30.8 [22]
---	Polystyrene	---	---	---	44.0 [22]
Glycerol	BAPC	66.6	63.4 [22]	37.0 [22]	53.1
Mercury	BAPC	126.1	484 [22]	200 [22]	49.4
Glycerol	PMMA	59.8	63.4 [22]	37.0 [22]	61.4
Mercury	PMMA	121.9	484 [22]	200 [22]	65.0
Glycerol	828-Z	55.1	63.4 [22]	37.0 [22]	67.2
Mercury	828-Z	121.8	484 [22]	200 [22]	65.4
Glycerol	Nylon 6/6	54.4	63.4 [22]	37.0 [22]	68.0
Mercury	Nylon 6/6	121.7	484 [22]	200 [22]	65.6

PTFE = Polytetrafluoroethylene
HAc = Glacial Acetic Acid
CCl_4 = Carbon Tetrachloride
PCTFE = Polychlorotrifluoroethylene
BAPC = Bisphenol A-Polycarbonate
828-Z = Aromatic Amine Cured Epoxy Polymer[25]

where $A = R_o\sqrt{2}$. Then,

$$\gamma^d/TS = 1.033\,A \approx A, \tag{37}$$

and the near-neighbor spacing, R_o, can be determined by two independent calculations using Equation (18) or Equation (37). The linear relation shown in Figure 5 between the experimental tensile strengths[14, 26-28] and dispersion surface tensions indicates that these materials have similar near-neighbor spacings. The formula weight, M, for each molecule, group of molecules, or molecular segment that is separated from another by R_o in the fcc model can be calculated by

$$M = (\gamma^d/TS)^3\,\rho N/4, \tag{38}$$

where ρ is the density and N is the Avogadro number. Computed formula weights are given in Table III along with the calculated number of repeat units per lattice site and the expectations based on molecular size, conformation and binding. Table IV gives values of

A, the unit cell edge length, computed from experimental surface tensions and tensile strengths and, for comparison, A-values calculated from entries in the last column of Table III. Average near-neighbor separations are tab ulated in column three (Table IV).

Physical Properties for Liquids and Amorphous Polymers

Values for R_o and C_d were chosen by an iterative computer calculation to give the best fit to the experimental tensile strengths, surface tensions, thermal expansion coefficients, work of vaporization, and compression curves. The theoretical and experimental properties are compared in Table V and Figures 6 through 13. Brydson reported experimental tensile strengths and thermal expansion coefficients for the polymers.[26] Values for the commercial grades were chosen when the properties of more than one grade of material were reported. Both static compression[17] and adiabatic shock[29] data are shown in the figures.

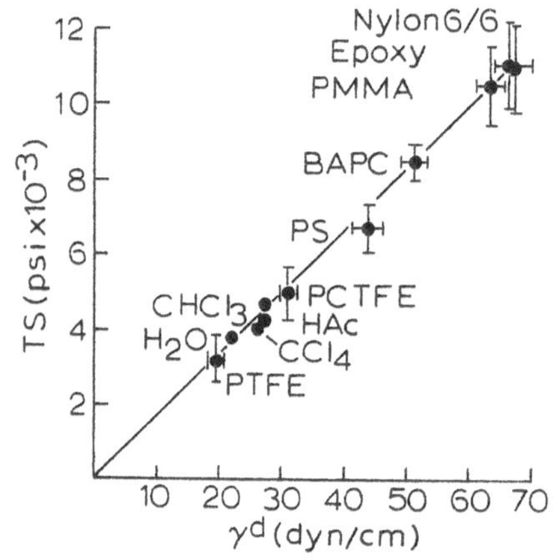

Figure 5. Tensile strength-dispersion surface tension relations for liquids and amorphous polymers.

The experimental tensile strengths of liquids were reported by Briggs who employed a "Z-shaped capillary tube, open at both ends, rotating in the Z plane about an axis passing through the center of the Z and perpendicular to the plane. The liquid menisci are located in the bent-back short arms of the Z. The speed of rotation is increased gradually until the liquid in the capillary breaks."[14,28]

The force constants listed in Table VI were calculated using experimental surface tensions, Equation (30), and the near-neighbor spacings in Table IV.

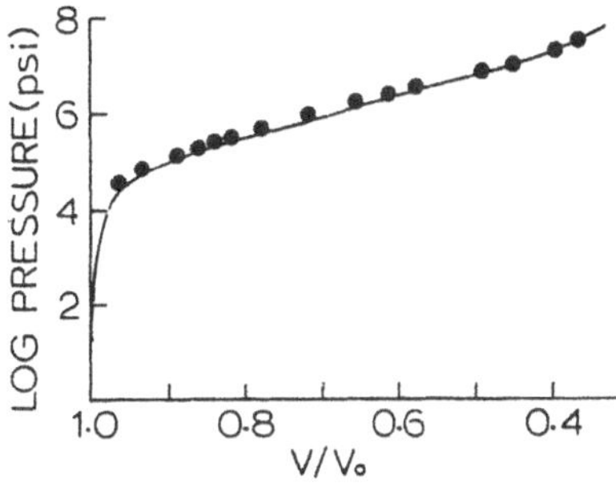

Figure 6. Compression curve for polymethylmethacrylate.

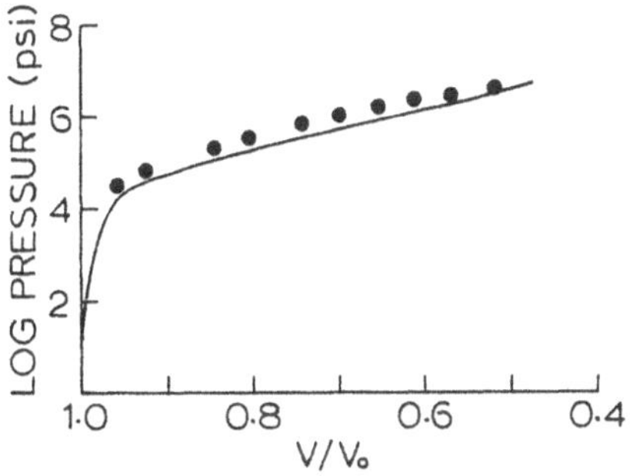

Figure 7. Compression curve for polystyrene.

TABLE III

Theoretical Formula Weights for FCC Lattice Sites

Material	ρ (g/cm^3)	M ($g\ mole^{-1}\ site^{-1}$)	Repeat Units per Site Calculated	Repeat Units per Site Expected
PTFE	2.156	226	2.25	2
Water	0.997	104	5.76	5
CCl_4	1.594	167	1.08	1
Chloroform	1.492	156	1.31	1
HAc	1.0491	110	1.83	2
PCTFE	2.12	222	1.91	2
Polystyrene	1.054	117	1.06	1
BAPC	1.20	126	0.495	0.5
PMMA	1.189	125	1.25	1
Nylon 6/6	1.1457	120	1.06	1

TABLE IV

Theoretical Unit Cell Edge Lengths and Near-Neighbor Spacings

A = Values, (Å)

Material	γ^d/TS	$(4M/\rho N)^{1/3}$	$\bar{R}_o$(Å)
PTFE	8.98	8.52	5.87
Water	7.77	8.46	5.74
CCl_4	9.64	8.63	6.46
Chloroform	8.45	8.12	5.86
Acetic Acid	9.32	9.13	6.52
PCTFE	8.94	9.02	6.35
Polystyrene	9.46	8.70	6.42
BAPC	8.74	8.90	6.24
PMMA	8.73	8.24	6.00
Nylon 6/6	8.81	8.71	6.20

TABLE V

Comparison of Theoretical and Experimental Properties of Liquids and Polymers*

Theoretical/Experimental

Material	γ^d (dyn/cm)	TS (psi)	β (°C^{-1}) x 10^{-4}	ΔF (KCal/Mole)
PTFE	19.3/19.5	3184/3150	19.5/2.286	- - - -
Water	23.1/21.3	3850/3814	16.5/2.07	0.82/2.05
Chloroform	27.0/27.14	4682/4660	15.4/12.73	0.89/1.10
Acetic Acid	27.4/27.2	4209/4234	11.9/10.71	- - - -
PCTFE	31.4/30.8	4901/5000	10.7/- - -	- - - -
Polystyrene	42.9/44.0	6929/6750	8.4/2.1	- - - -
BAPC	52.5/51.25	8295/8500	6.6/2.1	- - - -
PMMA	62.5/63.2	10,632/10,500	6.3/- - -	- - - -
Nylon 6/6	67.4/66.8	10,896/11,000	5.4/2.85	- - - -

* The R_o and C_d values used in these computations were obtained from the best fit of the experimental data:

γ^d = Dispersion Surface Tension
TS = Tensile Strength
β = Cubical Coefficient of Thermal Expansion
ΔF = Work of Vaporization

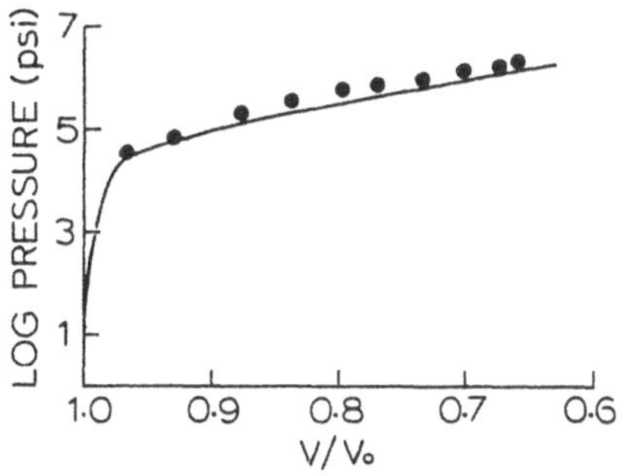

Figure 8. Compression curve for Nylon 6/6.

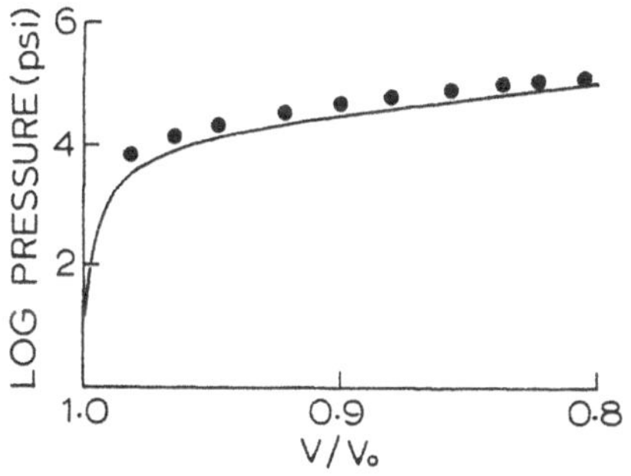

Figure 9. Compression curve for water at 25°C.

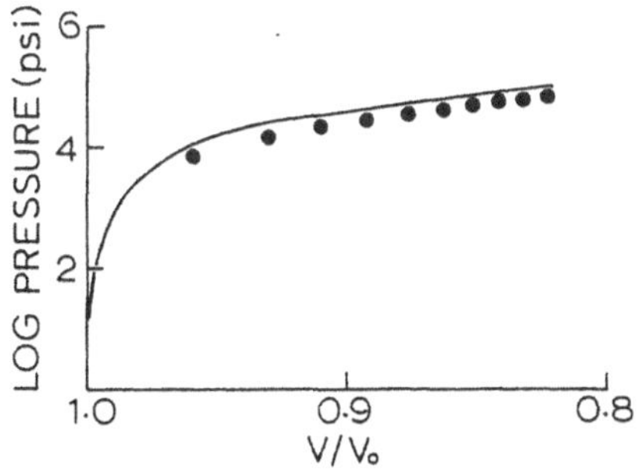

Figure 10. Compression curve for chloroform.

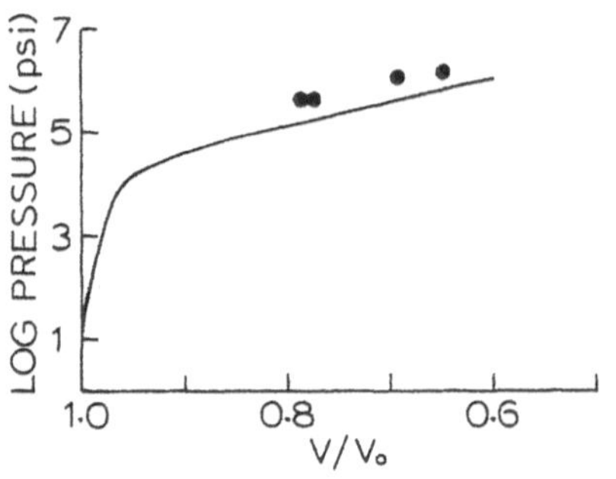

Figure 11. Compression curve for polychlorotrifluoroethylene.

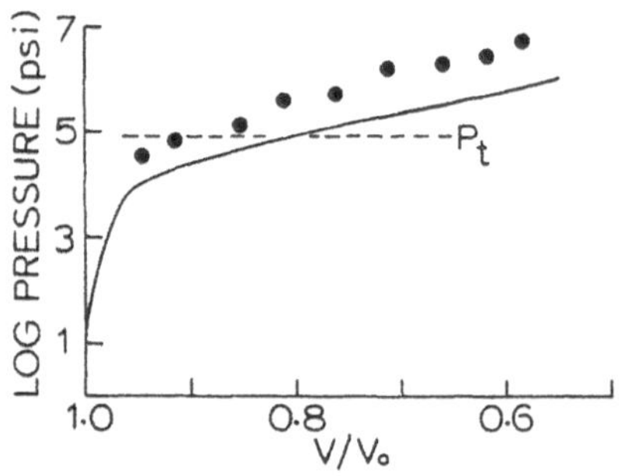

Figure 12. Compression curve for polytetrafluoroethylene. P_t = Pressure induced phase transition reported.

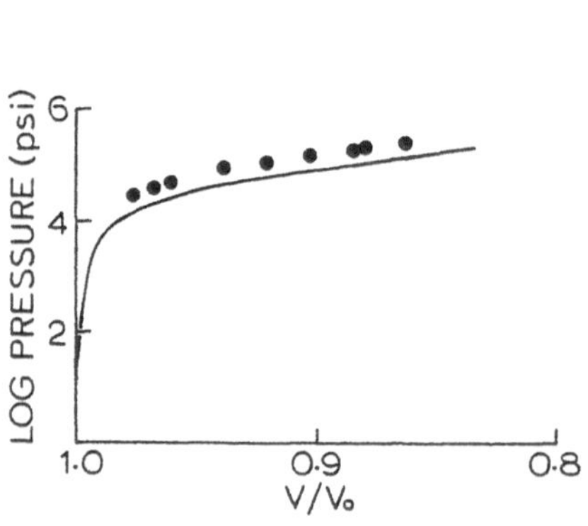

Figure 13. Compression curve for epoxy 828-Z.

TABLE VI

Dispersion Attraction Constants Computed from Experimental Dispersion Surface Tensions

Material	$C_d \times 10^{58}$
Polytetrafluoroethylene	2.035
Water	1.858
Carbon Tetrachloride	5.915
Chloroform	2.793
Acetic Acid	6.560
Polychlorotrifluoroethylene	6.027
Polystyrene	9.399
Bisphenol-A-Polycarbonate	8.720
Polymethylmethacrylate	7.857
Nylon 6/6	10.796

DISCUSSION AND CONCLUSIONS

Theoretical tensile strengths of polymers usually exceed experimental by a factor of about 1,000 when chemical binding is assumed. This difference has been attributed to the effects of stress concentrations and material flaws. While the influence of material defects may be quite important, the observed strengths of most amorphous polymers are comparable to those of simple liquids in which there are no chemical bonds between molecules. We conclude, as others have for hydrocarbon polymers,[2,4] that van der Waals forces determine the strengths of amorphous materials including highly crosslinked epoxy polymers. A simple geometrical model and London's theory were used to calculate a variety of theoretical physical properties. There are indications that these properties should be divided into two classes, those depending strongly upon van der Waals forces (tensile strength and dispersion surface tension), and those that may depend upon other types of bonding in addition (compressibility, thermal expansion, and work of vaporization). This may be clarified by considering the acetic acid dimer,[30]

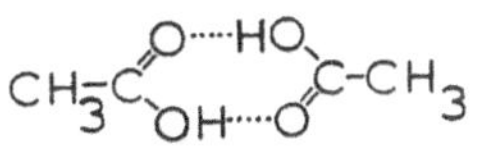

(39)

The dimer is approximately neutral and can interact with neighboring dimers primarily by van der Waals bonding. Tensile failure could occur between dimers unaffected by hydrogen bond strengths. By definition, the dispersion surface tension is the van der Waals component of the surface energy. The other properties (compressibility, thermal expansion, and work of vaporization) should depend on all types of bonding present. Van der Waals bonds, hydrogen bonds, and chemical bonds could be affected in compression and in thermal expansion. The work of vaporization depends upon the van der Waals lattice energy and upon hydrogen bond strengths when monomers are formed.

Experimental properties, used with the theory, indicate states of aggregation for simple liquids which are consistent with other chemical evidence. Van der Waals bonding units in the space model comprise one molecule for carbon tetrachloride and chloroform, two molecules for acetic acid, and five molecules for water. The group sizes calculated for the polymers are consistent with molecular configurations, bond angles, and the range of effective van der Waals forces.

The effects of different molecular weights and degree of crystallinity for a given polymer, though not considered specifically here, may be handled tentatively in the density dependent terms and the geometry relations. This elastic model does not include the viscous flow behavior necessary to account for the rate dependence of observed cohesive strengths. Born and Green derived expressions for hydrodynamic flow which can be used to calculate the viscosity when the energy equations for intermolecular interactions are known.[31] The inclusion of such expressions could result in more generally applicable theories.

Failing in obtaining completely crosslinked polymers, an increase in van der Waals strengths may be effected by substituting more polarizable atoms periodically along the molecular chain. Progressively greater strengths would be predicted for polymers containing the substituents fluorine, chlorine, bromine, or iodine. The electronic polarizability is related to the refractive index by the Mosotti-Clausius equation,[32] so that greater strength would be predicted for a polymer of higher refractive index. This trend was observed for the materials considered here.

ACKNOWLEDGEMENTS

The author is indebted to Dr. R. L. Schwoebel of Sandia Laboratories for his interest in, and discussion of, this paper, for his help in making many of the calculations, and for his detailed criticism of the final manuscript.

REFERENCES

1. J. H. de Boer, Trans. Faraday Soc. 32, 11 (1936).
2. D. J. Pastine, J. Chem. Phys. 49, 3012 (1968).
3. R. Ulbrich, Z. Naturforsch., a21, 763 (1966).
4. A. Muller, Proc. Roy. Soc. (London) A154, 624 (1936); A178, 227 (1941).
5. J. O. Hirschfelder, C. F. Curtiss, and R. B. Bird, "Molecular Theory of Gases and Liquids," John Wiley & Sons, Inc., New York, 1967, p. 1040.
6. C. Kittel, "Introduction to Solid State Physics," John Wiley & Sons, Inc., New York, 1957, p. 78.
7. S. Glasstone, "Thermodynamics for Chemists," D. Van Nostrand Co., Inc., New York, 1960, p. 205.
8. C. Kittel, op. cit., p. 152.
9. R. J. Good, Intermolecular and Interatomic Forces, R. L. Patrick, ed., "Treatise on Adhesion and Adhesives," Vol. 1: Theory, Marcel Dekker, Inc., New York, 1967, pp. 48-55.
10. F. London, Trans. Faraday Soc. 33, 8 (1937).
11. H. Margenau, Rev. Mod. Phys. 11, 1 (1939).
12. K. S. Pitzer, Adv. Chem. Physics 2, 59 (1959).
13. J. C. D. Brand and J. C. Speakman, "Molecular Structure," Edward Arnold (Publishers), Ltd., London, 1960, p. 178.
14. D. E. Gray, et. al., ed., "American Institute of Physics Handbook," McGraw-Hill Book Co., Inc., New York, 2nd Edition, 1963, p. 2:186.
15. Interpolated from data in N. A. Lange, ed., "Handbook of Chemistry," Handbook Publishers, Inc., Sandusky, Ohio, 9th Edition, 1956, p. 1649.
16. Calculated from data in R. C. Weast, et. al., "Handbook of Chemistry and Physics," The Chemical Rubber Co., Cleveland, Ohio, 45th Edition, 1964, p. D:49.
17. Calculated from data in P. W. Bridgman, "Collected Experimental Papers," Harvard University Press, Cambridge, Mass., 1964.
18. D. E. Gray, op. cit., p. 4:75.
19. From radial electron density distribution curve for carbon tetrachloride at 298°K, T. J. Hughel, ed., "Liquids: Structure, Properties, Solid Interactions," Elsevier Publishing Corp., New York, 1965, p. 191.
20. N. A. Lange, op. cit., pp. 1441-1442.
21. B. Linder, Disc. Faraday Soc. 40, 164 (1965).
22. F. M. Fowkes, Ind. Eng. Chem. 56, 40 (1964); "Surfaces and Interfaces I," Burke, et. al., ed., Syracuse University Press, 1967, pp. 197-224.
23. R. C. Weast, et. al., op. cit., p. F:19.
24. D. E. Gray, et. al., op. cit., p. 2:188.
25. T. R. Guess, "Some Dynamic Mechanical Properties of an Epoxy," SC-DR-343, June 1968, Sandia Corporation, Albuquerque, New Mexico.

26. J. A. Brydson, "Plastics Materials," D. Van Nostrand Co., Inc., New Jersey, 1966.
27. The spall threshold reported by T. R. Guess, loc. cit., for the 828-Z epoxy polymer and the tensile strength are assumed equivalent.
28. L. J. Briggs, J. App. Phys. 21, 721 (1950); J. Chem. Phys. 19, 970 (1951).
29. M. Van Thiel, et. al., ed., "Compendium of Shock Wave Data," Lawrence Radiation Laboratory, University of California, Livermore, California, UCRL-50108, 1966.
30. S. Glasstone, "Textbook of Physical Chemistry," D. Van Nostrand Co., Inc., New York, 1959, p. 510.
31. M. Born and H. S. Green, Proc. Roy. Soc. (London) A190, 455 (1947).
32. S. Glasstone, "Textbook of Physical Chemistry," D. Van Nostrand Co., Inc., New York, 1959, p. 542.

DISCUSSION

J. J. Bikerman (Cleveland, Ohio):

The hypothesis accepted by the author, namely that "the interfacial interaction between many liquids and solids can be described by dispersion forces alone," even if these substances have permanent dipoles etc., cannot be correct. It is equivalent to saying that an electron in a symmetric molecule is free to refuse being affected by a part of the field emanating from an asymmetrical molecule. This assumption is contradicted by all we know about electric fields. Consequently, whenever we see γ with superscript d, it is very likely that the theory is not valid.

R. E. Cuthrell:

When two materials interact across an interface there is an attractive force which is universally applicable to polar and non-polar molecules and atoms. This interaction is the result of induced electronic polarization and is dependent upon R^{-6} where R is the distance between interacting volume elements. When one of the materials is polar and the other is non-polar, the interaction is of the dipole-induced dipole type and is, in practice, indistinguishable from the above interaction since the polar molecule has no preferential orientation. In the interaction between two polar molecules the R-dependence is a function of the absolute temperature. When the interaction energy is less than kT, the dipoles are free to rotate about their mass centers and integration over the total solid angle gives an R^{-6} dependence. The dipoles are fixed in orientation with respect to each other only when the interaction energy is significantly larger than kT. In this case the potential energy is dependent upon R^{-3}. In this paper, first order calculations are presented which are based upon dispersion interactions alone. There are, of course, other possible interactions in addition.

A mathematical model representing amorphous organic polymers could be derived, in principle, by considering the structure of diamond, the most highly crosslinked organic material, and incrementally decreasing the number of chemical bonds between volume elements until calculations based on the model give the physical properties of a given polymer. Alternatively, one could begin with a model for carbon tetrachloride or methane, for which van der Waals forces are the only effective intermolecular interactions, and incrementally increase the number of crosslinks considered. This paper describes the first step in the second method. Physical properties were calculated based on a van der Waals model and compared with experimental values for a variety of liquids and polymers. The favorable comparison thus obtained leads to the conclusion that amorphous polymers are not as highly crosslinked as previously assumed.

A. Peterlin (Research Triangle Institute, N. C.):

The calculations presented in this lecture based on a face centered cubic cell arrangement of monomers seem to me extremely important for an estimate of the upper limit of values for mechanical strength and elastic modulus which can be derived from van der Waals cohesive forces between macromolecules. The cohesive forces in the amorphous regions are certainly smaller than in a crystal as a consequence of larger intermolecular distances. This was already considered in the calculations presented. One may deduce that any experimental value of tensile strength and elastic modulus which is larger than the values calculated by this model can be explained only by invoking the strong covalent bonding of monomers in the polymer chain. The linear chain structure of macromolecules is the main characteristics of polymeric material determining the high strength in the glassy and crystalline state, the anisotropy of mechanical properties in the oriented polymers, and the high elasticity in the rubbery state. This feature is explicitly disregarded in the approach which considers merely the cohesive forces between monomers. Therefore one cannot expect reasonable values for the elastic modulus, the stress to break and still less for the strain to break because neither the plastic deformation in the glassy and crystalline state nor the high elastic deformation in the rubbery state were included in the model and the contribution of polymer chains is completely neglected. Hence any agreement of calculated and experimental data is at least misleading and must not be considered as a supporting evidence for the calculation. The value of the data presented is in establishing the upper limit of the contribution of van der Waals cohesive forces to the mechanical properties of polymer solids and liquids.

The agreement between the experimental and calculated tensile strength of polymers in Table V is to a large extent based on nominal stress (load) data. If one considers the change of sample cross-section as a consequence of the large deformation λ_b = length at break/ original length one obtains in the case of nylon 6.6 values which are between 4 and 5 times higher than the quoted "experimental" value 10,896 and the theoretical value 11,000 psi. That means that the van der Waals forces cannot explain more than 25% of the observed value. The rest is supplied by the covalent bond strength of nylon chains. One is certainly surprised that these "experimental" values fit so well Eq. 37 (Fig. 5).

R. E. Cuthrell:

The volume elements placed at each lattice site in the face centered cubic model were assumed to be bound to each other by van der Waals forces alone. We realize that this is not an accurate representation for polymer chains and should result in theoretical estimates for thermal expansion coefficients and compressibilities larger than experimental, and theoretical elastic moduli smaller than experimental. The treatment did, in general, give these results. Therefore, one must consider a somewhat stiffer representation than a model based solely on van der Waals forces. Furthermore, when a crosslinked polymer is characterized by grains of high crosslink density separated by regions of low crosslink density, some properties would be more strongly dependent upon van der Waals forces between grains (tensile strength and dispersion surface tension), while other properties would depend on grain stiffness as well as intergranular stiffness (compressibility, and thermal expansion).

Dr. Peterlin is correct in his assertion that the anisotropy of mechanical properties in the oriented polymers and the high elasticity in the rubbery state are determined by the linear chain structure of macromolecules. When chains of chemically bound atoms are aligned parallel to each other, obviously there are anisotropic mechanical properties determined by chemical binding parallel to the chains and van der Waals binding perpendicular to the chains. The strength perpendicular to the chains is determined by van der Waals forces, however, except when other forms of binding, such as hydrogen bonds, dipole bonds, and chemical crosslinks, are present in significant numbers.

The theoretical yield strain of 25% was calculated for a comletely elastic body deformed by tensile stress in the idealized manner described in the text. We found an average elongation of 24.7 ± 0.5% for ten Nylon 6/6 dumbell tensile samples equilibrated at 50% relative humidity for three months. When a sample behaves in a completely elastic manner, the load at tensile failure is a measure of the force required to break a specific number of bonds. This number of bonds can be estimated from the original crossectional area and the density. Therefore, the tensile stress, based on the load and this area, has physical meaning only when completely elastic behavior is observed. In these cases experimental values should fit Equation 37 and Figure 5.

GLASS POINTS OF POLYMER NETWORKS

A. J. Chompff

Scientific Research Staff, Ford Motor Company

P. O. Box 2053, Dearborn, Michigan 48121

A model for a polymer network is developed to establish analytical relations between the glass point, T_g, and structural features of amorphous networks and to allow prediction of T_g with changes in chemistry. In this model the network is treated as a ternary copolymer of branch points, chain segments and chain ends, each of which contributes a different amount of free volume to the total system according to rules similar to those used in ternary copolymers or mixtures. It is shown that changes in T_g due to plasticization, copolymerization, molecular weight, branching, crosslinking, chain scission, etc. do occur as logical consequences of a well known free volume theory. Where possible, these calculated results have been corroborated with literature data.

The present equations contain only one adjustable parameter, ξ_1, the "effective volume" of a chain end. The effective volume of a j-functional branch point is $2(j-2)\xi_1$. It is shown how the glass point of a network then depends on the volume fractions of ends, segments and branch points, and the glass points and expansion coefficients of the ends and segments only. The best fit between theory and experiments is obtained if ξ_1 equals the volume of a chain segment containing six backbone atoms.

An analytically tractable approximation of the general equation is used to exemplify this method for a network of chains with trifunctional branch points. In addition, it is shown how actually obtained conversions of the reactive groups in non-stoichiometric compositions may be estimated, and that T_g reaches a maximum at the stoichiometric composition provided no side reactions occur.

INTRODUCTION

The applications of a polymer network are frequently limited by the magnitude of its glass transition temperature, or "glass point," T_g. In structural applications, for instance, the useful temperature range lies often below T_g. The glass point, therefore, is an important material parameter. Not only does it prescribe an upper limit to the use temperature, but it also determines roughly the viscoelastic properties above T_g.

Upon crosslinking, T_g changes, sometimes in an unpredictable way. An attempt to solve this problem has been made before by Fox and Loshaek[1,2] who recognized two separate influences on T_g during crosslinking: a copolymer effect due to modification of some chemical groups on the chains, and a crosslink effect due to immobilization of chain segments at the branch points. The same line of thought regarding these two effects has been preserved in the present paper; the approach, however is different. Here, these two effects are considered a direct consequence of the mixing rules for free volumes in homogeneous binary or ternary polymer mixtures. Inhomogeneity of the polymer systems has not been taken into account since T_g is an average property of the material.

The purpose of the present paper is to establish analytical relations between the glass point, T_g, and structural features of amorphous networks, so as ultimately to predict how T_g changes with the chemistry of the system. To perform this task, one theory of the glass transition has been chosen and the consequences of this choice will be discussed. It will be shown that changes in T_g due to plasticization, copolymerization, molecular weight, branching, crosslinking, etc. do occur as logical consequences of a well known free volume theory.

In the first following section some of the principles of the glass transition will be outlined, free volumes will be related to T_g in accordance with established viewpoints, and mixing rules for binary and ternary mixtures will be reformulated and discussed in detail. In the second section this treatment will be extended to polymer networks, which will be represented in a novel manner as a ternary copolymer. In the third section some of the predictions of this model will be verified with experimental data available in the literature. In the fourth section changes in T_g due to some variations in network chemistry will be calculated, and the utility of the method to characterize partially cured networks will be indicated.

FREE VOLUMES IN BINARY MIXTURES

Early in this century[3] it has been suggested that the rate of molecular transport in a liquid is controlled by its free volume. On this basis, many theories have evolved which define the free volume in just as many different ways.

In the field of viscoelasticity of high polymers one of these theories has been particularly successful.[4] This theory, initiated by Doolittle[5] and formulated quantitatively by Cohen and Turnbull,[6] will be used throughout the present paper. The details and consequences of this free volume theory can be found in recent reviews[7-10] and most of it will not be repeated here. It turns out that this free volume concept is a very useful way to describe various phenomena above the glass point, T_g, and in some cases below T_g.

When an amorphous system, either as a low molecular weight liquid or as a macromolecular viscoelastic solid, is cooled at a prescribed rate the specific volume, v, will depend on the temperature as shown schematically by line I in Fig. 1. The thermal expansion coefficient shows a rapid change near T_g. In the immediate neighborhood of this glassy transition ($T_g \pm 100°C$) the specific volume may be approximated by two straight lines: a liquid line at temperatures above T_g and a glass line below T_g. This approximation may be written as

$$v = v_g[1 + \alpha_L(T-T_g)] , \qquad \text{for } T > T_g , \tag{1}$$

$$v = v_g[1 - \alpha_G(T_g-T)] , \qquad \text{for } T < T_g , \tag{2}$$

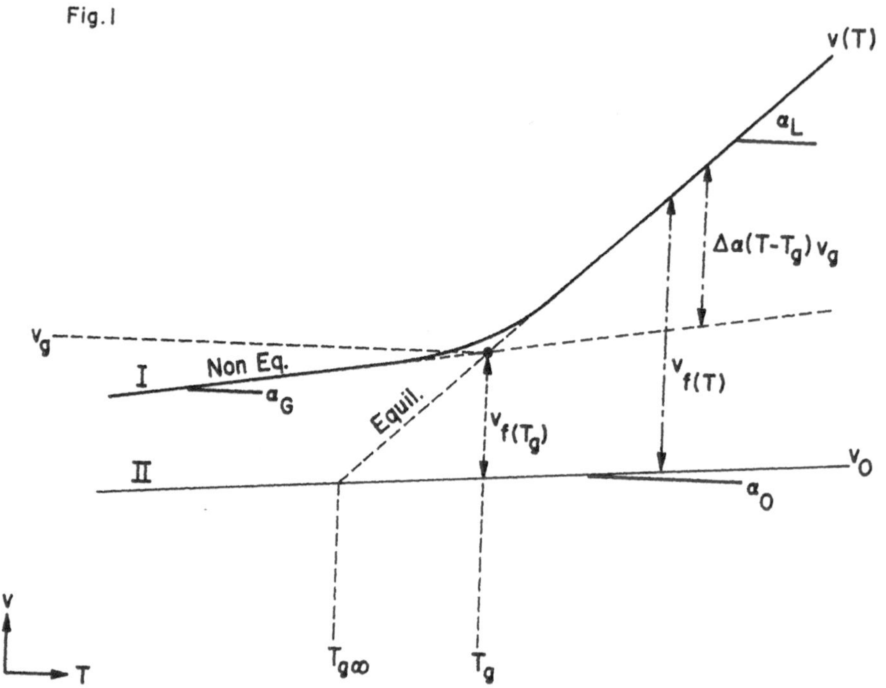

Fig. 1. The temperature dependence of the specific volume, v(T), near T_g and the expansion coefficients, α_0, α_G and α_L, of the various phases.

$$\Delta\alpha = \alpha_L - \alpha_G \quad , \tag{3}$$

where α_L and α_G are the volumetric expansion coefficients of the liquid and of the glass respectively, and v_g is the specific volume at the interpolated intersection of the two straight lines.

At slower cooling rates, the corresponding glass points are found to be lower and are located on an extension of the equilibrium liquid line to lower temperatures, presumably until a point $T_{g\infty}$ is reached, which is the glass point of the system after cooling at an infinitesimally slow rate. The arguments leading to the existence of $T_{g\infty}$ have been discussed in the literature.[11-16] For polymer systems not yet investigated a useful approximation for $T_{g\infty}$ appears to be

$$T_g - T_{g\infty} \approx 50 \text{ to } 100°C \tag{4}$$

At any temperature the specific volume, v, consists of an occupied volume, v_o, and a free volume, v_f:

$$v_f = v - v_o, \tag{5}$$

whereby the free volume is defined as the excess volume available for molecular transport which is generated over and above the thermal expansion of the amorphous solid. The fractional free volume, f, is defined as v_f/v. Since the fractional free volume f can in principle be determined from the temperature dependence of the relaxation behavior of the material, the semiempirical parameter v_o is implicitly defined and need not be specified separately.

In the specified temperature range ($T_g \pm 100°C$) the temperature dependence of v_o may be assumed to be linear, as indicated by line II in Fig. 1. Its thermal expansion coefficient, α_o, should be smaller than α_G but larger than α_C, the cubic thermal expansion coefficient of the same material in crystalline form (if that exists).[12] Thus, the free volume v_f at any temperature is the vertical distance between lines I and II and can be described approximately by

$$v_f = v_{f(T_g)} + v_g\alpha_f(T - T_g) \quad , \qquad \text{for } T > T_g \quad , \tag{6}$$

$$\alpha_f = \alpha_L - \alpha_o \ . \tag{7}$$

It should be noted, that in the present model description v_f has a singularity at T_g, or at $T_{g\infty}$, depending on the conditions of equilibrium. Complementary to Eqs. (6) and (7) are the following equations

$$v_f = v_{f(T_g)} - v_g\alpha_r(T_g - T) \quad , \qquad \text{for } T < T_g, \tag{8}$$

where $\alpha_r = \alpha_G - \alpha_o$. (9)

A more practical quantity to use is the fractional free volume f, here defined as v_f/v_g. Then Eqs. (6) and (8) yield

A: under non-equilibrium conditions

$$f = f_g + \alpha_f(T - T_g) \quad , \qquad \text{for } T > T_g \; , \tag{10}$$
$$f = f_g - \alpha_r(T_g - T) \quad , \qquad \text{for } T < T_g \; , \tag{11}$$

B: at equilibrium

$$f = f_g + \alpha_f(T - T_g) = \alpha_f(T - T_{g\infty}) \quad , \quad \text{for } T > T_{g\infty}, \tag{12}$$
$$f = 0 \quad , \quad \text{for } T < T_{g\infty}. \tag{13}$$

Since this concept is a theory of transport properties, the two empirical parameters f_g and α_f, or α_f and $T_{g\infty}$, can be evaluated when the transport properties of the sample are measured. In many cases sufficient data are lacking. Then, f_g may be approximated by 0.025 and α_f may be estimated to be either proportional to or equal to $\Delta\alpha$. Frequently,[7] Eqs. (10), (11) and (12) contain another factor 1/B, which is omitted from the present equations.*

Since the value of T_g depends on the patience of the experimenter, it seems irrelevant to argue about the exact value or constancy of f_g. Both T_g and f_g are practical numbers to describe the transition from equilibrium to non-equilibrium while cooling under practical conditions, and are not constants but material parameters. If one accepts the present scheme, given by Eqs. (10) to (13), a number of consequences can be drawn which qualitatively concur with experimental data.

When an amorphous polymer is homogeneously mixed with a low molecular weight solvent or a plasticizer,[17] or when two monomers are copolymerized in a random manner,[18] the specific volumes of the components are generally not additive. The specific volume, v, of the mixture can usually be described by

$$v = v_1 g_1 + v_2 g_2 + v_e \; , \tag{14}$$

where v_1 and v_2 are the specific volumes of the pure components, g_i are their weight fractions, and $-v_e$ is the thermodynamic "excess

*The empirical factor 1/B arises from inadequacies of the model; its physical meaning is not clear. Experimentally, one usually finds B = 0.9 ± 0.3.

volume" which is proportional to the product $\phi_1 \phi_2$. The index 1 will be assigned to the component with the lowest T_g. Thus, ϕ_2 is the volume fraction of polymer. Recently a theory has been proposed for the calculation of v_e for mixtures.[19]

Since there is no reason to assume that the free volumes are additive, the fractional free volume of the mixture, f_m, at equilibrium can be described in a similar way

$$f_m = f_1\phi_1 + f_2\phi_2 + k_{12}\phi_1\phi_2 \quad , \tag{15}$$

where k_{12} is the interaction constant, and

$$f_i = f_{gi} + \alpha_{fi}(T - T_{gi}) \quad , \qquad T > T_{g\infty i} , \tag{16}$$

and $i = 1,2$ or m. It should be noted, that k_{12} may still depend on temperature. Although the interaction constant will be neglected in the following sections, some consequences of these mixing rules (with k_{12}) will be discussed here; first, because they have not been described adequately in the literature and second, to elucidate the approximations made in the following sections.

In a limited temperature range ($T_g \pm 50°C$) one can approximate the temperature dependence of k_{12} as

$$k_{12} = \ell_{12} + m_{12}\, T \quad , \tag{17}$$

where ℓ_{12} and m_{12} are independent of temperature and concentration. Since ℓ_{12} and m_{12} can be of opposite sign, a negligibly small k_{12} does not imply that the values of ℓ_{12} and m_{12} are also small.

It should be noted, that a negative v_e does not imply a negative k_{12}. It has been shown that v_e is related to the thermodynamic interaction between polymer and solvent,[19] whereas k_{12} is related to the average mobility increase due to mixing. A definite correlation between these two phenomena has not yet been established. It is better, therefore, to regard k_{12} as additional information about solvation and mobilities, rather than as complementary information about v_e. Similarly, to assume that the occupied volumes are additive[11] is the same as assuming that k_{12} and v_e are proportional, and cannot be justified. Obviously, a

Fig.2

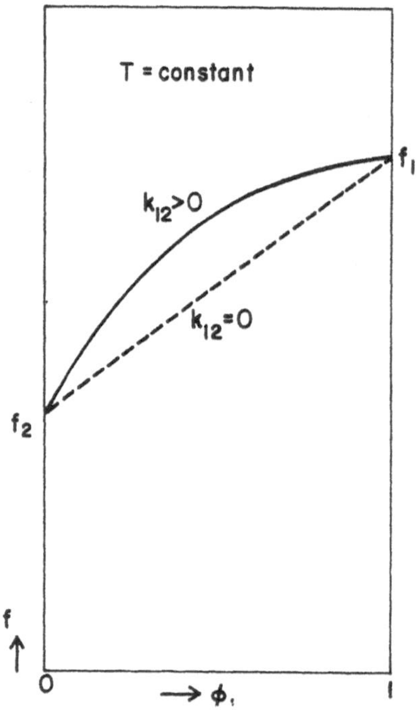

Fig. 3

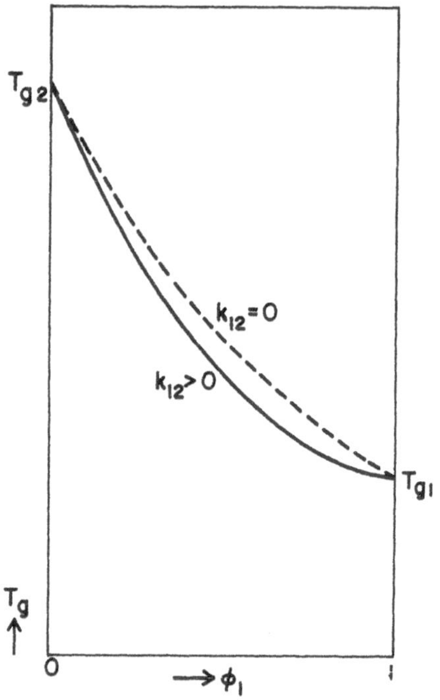

Fig. 2. The fractional free volume f in a polymer/solvent mixture, or in a copolymer, for two values of k_{12} and $T > T_{g\infty 2}$.

Fig. 3. The corresponding glass points of the binary systems of Fig. 2, for the same values of k_{12} and $T > T_{g\infty 2}$.

negative k_{12} leads to a smaller f (or higher T_g) than for the case in which $k_{12} = 0$. This is illustrated in Figs. 2 and 3. Most polymer/diluent mixtures undergo a contraction during mixing of the components, i.e. v_e is negative. The glass points of the mixtures, however, are usually lower[17] (or f is larger) than for the case in which $k_{12} = 0$.

For most polymer systems f_g is almost a constant (0.025 ± 0.005). A useful boundary condition, therefore, is to assume[11] that f_g varies only linearly with composition

$$f_g = \phi_1 f_{g1} + \phi_2 f_{g2} \quad . \tag{18}$$

In this equation and those to follow the index m has been omitted. Equations (15) to (18) form the basis for the calculations in this section.

a. For $T > T_{g\infty_2}$ substitution of Eqs. (16), (17) and (18) in Eq. (15) yields

$$\alpha_f(T-T_g) = \alpha_{f_1}\phi_1(T-T_{g_1}) + \alpha_{f_2}\phi_2(T-T_{g_2}) + \phi_1\phi_2(\ell_{12} + m_{12}T) \quad , \quad (19)$$

where α_f and T_g are respectively the expansion coefficient and the glass point of the mixture. Differentiation of both sides of Eq. (19) with respect to T yields

$$\alpha_f = \alpha_{f1}\phi_1 + \alpha_{f2}\phi_2 + m_{12}\varphi_1\varphi_2 \quad . \qquad (20)$$

Substitution of Eq. (20) in (19) yields

$$T_g = \frac{\alpha_{f1}T_{g1}\phi_1 + \alpha_{f2}T_{g2}\phi_2 - \ell_{12}\phi_1\phi_2}{\alpha_{f1}\phi_1 + \alpha_{f2}\phi_2 + m_{12}\phi_1\phi_2} \qquad (21)$$

A similar equation has been derived before by Jenckel and Heusch.[17] For the case that both ℓ_{12} and m_{12} are equal to zero, Eq. (21) yields the $T_g(\phi_1,\phi_2)$ relationship proposed by Gordon and Taylor[20] for copolymers or by Kelley and Bueche[21] for polymer/diluent mixtures. Although, in Eq. (21), the crossterms containing ℓ_{12} and m_{12} are small compared with the other terms, their effect on T_g can be significant since T_g is expressed in °K. When ℓ_{12} and m_{12} are of opposite sign (i.e. probably $|k_{12}|$ is small) their effect on T_g is small according to Eq. (21). When ℓ_{12} and m_{12} are of the same sign (probably in polymer/diluent systems) their effect on T_g can be large.

b. For $T < T_{g\infty2}$ one may set f_2 equal to zero. Clearly, when a glassy polymer is mixed with a low molecular weight liquid the amount of free volume the polymer contributes to the mixture is identical to that amount if the polymer had been cooled under equilibrium conditions. This, of course, is not limited to temperatures below $T_{g\infty2}$. The state of a polymer solution at any temperature above its T_g (or at equilibrium) is independent of the path chosen for the mixing process. Thus, substitution of Eqs. (16), (17), (18) and $f_2 = 0$ in Eq. (15) yields

$$\alpha_f(T - T_g) + f_{g2}\phi_2 = \alpha_{f1}\varphi_1(T - T_{g1}) + \phi_1\phi_2(\ell_{12}+m_{12}T). \tag{22}$$

Differentiation of both sides of Eq. (22) with respect to T yields

$$\alpha_f = \alpha_{f1}\phi_1 + m_{12}\phi_1\phi_2 \quad . \tag{23}$$

Substitution of Eq. (23) in (22) yields

$$T_g = \frac{\alpha_{f1}T_{g1}\phi_1 + f_{g2}\phi_2 - \ell_{12}\phi_1\phi_2}{\alpha_{f1}\phi_1 + m_{12}\phi_1\phi_2} \; . \tag{24}$$

It should be noted, that Eq. (24) is independent of T_{g2} or α_{f2}; an important consequence in the following section.

Equations (21) and (24) describe the dependence of T_g on ϕ_1 in the two discrete regions $T_{g2} > T > T_{g\infty 2}$ and $T_{g\infty 2} > T > T_{g1}$. Consequently, the function $T_g(\phi_1)$ should possess a discontinuity at $T = T_{g\infty 2}$. It has already been pointed out by Kovacs[8] that a singularity in $T_g(\phi_1)$ should exist at $T_{g\infty 2}$. This conclusion does not depend on the free volume theory, but results from the theorem that the specific volume of a glass cannot be smaller than that of the same material in crystalline form (presumably the state of the densest packing). Any other derivation of mixing rules, e.g. that of Gordon and Taylor,[20] will lead to a similar singularity located at the same temperature.

From the data published in the literature,[17,18,22] one can conclude that the effect of ℓ_{12} and m_{12} on T_g in polymer/diluent systems is frequently significant, whereas their influence in copolymers is often small. Some copolymers, e.g. those of methylmethacrylate, seem to behave anomalously. It should be noted, how- that the tacticity of the MMA component in the copolymer may not be comparable to that in the "pure" amorphous PMMA, which is largely syndiotactic[23] and, therefore, has a T_g which is higher than that of a random copolymer of 50% isotactic and 50% syndiotactic PMMA.

In the sections to follow the calculations are of a semiquantitative character, and the magnitude of the corrections imposed by the coupling terms are frequently unimportant. The effects of mixing and of crosslinking, however, will appear in two separate terms, so that the treatment can be extended to include the coupling constants, if necessary.

NETWORKS AS TERNARY COPOLYMERS

The models used in the calculations in this section have been directed to densely crosslinked networks, as for instance the crosslinked epoxides described in another paper of this symposium.[24] In those networks a linear prepolymer containing active groups is reacted with a crosslinking agent of arbitrary functionality. Contrary to most vinyl/divinyl systems studied, the size of the crosslinking agent in these highly crosslinked networks frequently may not be neglected when compared with the size of the average backbone chain between crosslinks.

When a linear polymer containing a number of active A groups is crosslinked with a bifunctional crosslinking agent containing B groups and A reacts with B, a network with trifunctional branch points is formed as shown schematically in Fig. 4.

In the preparation of this network one can distinguish two separate processes: the act of homogeneously blending the components, which is considered here as a combined mixing and copolymerizing process, and the crosslinking process itself whereby branch points are formed and chain ends are eliminated. In the terminology of Fox and Loshaek,[1] the ultimate change, ΔT_g, of the glass point of the prepolymer is the algebraic sum of the changes in T_g caused by these two processes.

$$\Delta T_g = \Delta_{cop} T_g + \Delta_{x\ell} T_g \quad , \qquad (25)$$

where $\Delta_{cop}\, T_g$ is the change in T_g due to "mixing and copolymerization" and $\Delta_{x\ell}\, T_g$ is the change in T_g due to the crosslinking reaction. When T_{gP} is the glass point of the prepolymer, T_{gcop} is the glass point of the mixture just before the reaction and T_g is the glass point of the network, the terms in Eq. (25) can be expressed as $\Delta T_g = T_g - T_{gP}$, $\Delta_{cop} T_g = T_{gcop} - T_{gP}$, and $\Delta_{x\ell} T_g = T_g - T_{gcop}$.

The first process can be performed along two different routes:

1. In the first route the prepolymer is physically mixed with

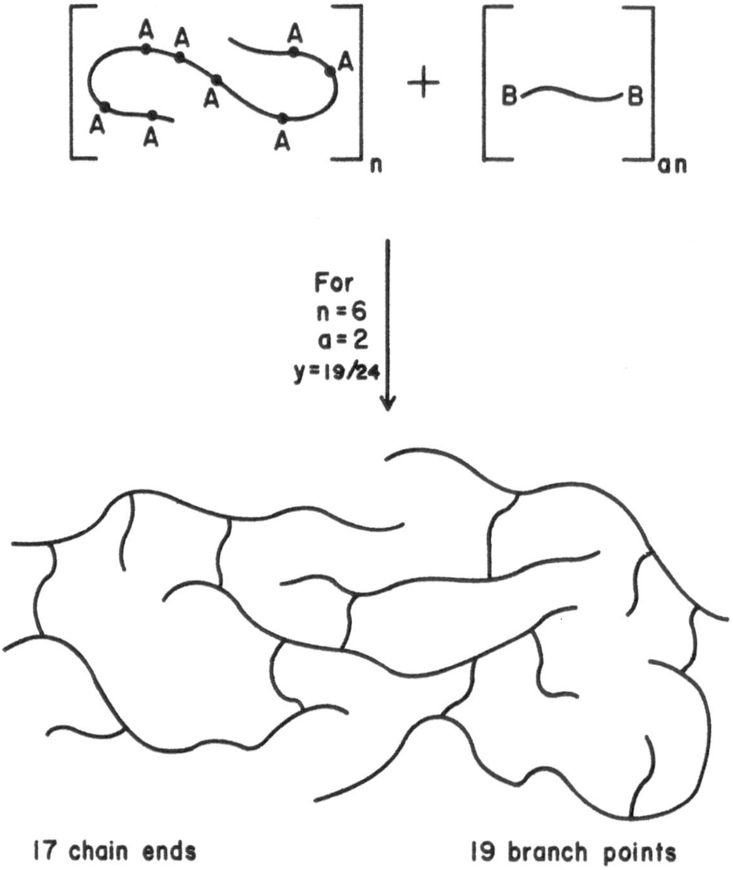

Fig. 4. A model of a network containing trifunctional branch points. The conversion y is the ratio of the branch points present over the highest number of branch points obtainable.

the lower molecular weight crosslinking agent. Thus, the T_g of this mixture is given by Eqs. (21) or (24), presumably with a small correction by k_{12}. Then the mixture is thought to undergo a randomizing segmental rearrangement, resulting in a mixture of long and short copolymer molecules but both of the same kind. This causes a small adjustment in T_g, probably eliminating the effect of k_{12}. This adjustment can be calculated by following the second route:

2. Before mixing, the prepolymer and the crosslinking agent are thought to undergo a randomizing segmental exchange so that their average compositions become identical but their molecular weights remain unchanged. Thus their glass points are given by Eqs. (21) or (24), presumably with a negligibly small effect of k_{12} for this copolymerizing process. Then the high molecular weight copolymer is mixed with the low molecular weight copolymer; here again k_{12} may be neglected since both components are of the same kind. This yields an explicit form for T_{gcop}. The resulting expression for T_g can be simplified if one applies the empirical result of Simha and Boyer[25] that $\alpha_f T_g$ is constant. This constant may not be the same for the compositions of the prepolymer and crosslinker, but if $\alpha_f T_g$ may be considered independent of molecular weight, it follows that

$$T_{gcop} = \frac{\alpha_{fx} T_{gx} \phi_x + \alpha_{fP} T_{gP} \phi_P}{\alpha_{fx} \phi_x + \alpha_{fP} \phi_P}, \quad T_{gcop} > T_{g\infty P}, \tag{26}$$

where the indices x and P stand for the initial components, the "pure" crosslinker and the prepolymer respectively. The case that $T_{gcop} < T_{g\infty P}$ seldom occurs. Although the form of Eq. (26) seems identical to the result obtained by Kelley and Bueche[21] for plasticized systems, it should be noted that Eq. (26) applies to a stage beyond that of physical mixing. This result takes care of the $\Delta_{cop} T_g$ term, which can be either positive or negative.

The calculation of the last term, $\Delta_{x\ell} T_g$, is the main theme of this paper. This term is always positive, due to the immobilization of segments at branch points. Any mobility change due to transfer of atoms or ring openings during the reaction is not neglected but implied by the material constants contained in the $\Delta_{x\ell} T_g$ term.

In a homogeneous network (homogenized by the preceding copolymerization process) one can distinguish three distinct structural elements: free chain ends (with index 1) chain segments (with index 2) and branch points (with index 3 and functionality j). Aside from differences in chemical composition of these structural elements one may regard the chain ends as a plasticizing component and the branch points as local mobility reducers,[26] similar to copolymerization with a low T_g and a high T_g monomer respectively. If one could assign an "effective volume" to each highly mobile chain end and another "effective volume" to each branch point, the free volume of a network could be described conveniently as that of a (three dimensional) three component copolymer:

$$f = f_1\varphi_1 + f_2\varphi_2 + f_3\varphi_3 \quad , \tag{27}$$

where

$$\varphi_1 + \varphi_2 + \varphi_3 = 1 \quad , \tag{28}$$

$$f_i = f_{gi} + \alpha_{fi}(T - T^*_{gi}) \quad , \qquad i = 1, 2 \text{ or } 3 \quad , \tag{29}$$

$$f = f_g + \alpha_f(T - T^*_g) \quad , \tag{30}$$

where φ_1, φ_2 and φ_3 are the effective volume fractions of "ends," segments and branch points respectively, T^*_{gi} are the glass points of the "pure" components and T^*_g is the glass point of the network as obtained by this method. Thus f_2 is the fractional free volume of a linear polymer of infinite molecular weight. Since the properties of the "pure" components still have to be specified (or adjusted) cross terms may be neglected in Eq. (27).

It should be noted, that the present approach is not a new theory but only a formal description of the effect of chain ends and branch points on the fractional free volume. The volume fractions φ_1 and φ_3 are not calculable a priori. The use of this formal description, however, is to show how many phenomena can be predicted without introducing additional theoretical assumptions, and to obtain the functional dependence of T_g on φ_1 and φ_3. A similar approach, but only for the effects of chain ends, has been used before e.g. by Bueche,[27] Ferry et al,[28] and Eisenberg et al.[29] Their results will be discussed in the following section.

A solution of Eq. (27) can be found along routes similar to

those for binary mixtures. Analogous to the treatment in the previous section, one may assume that

$$f_g = f_{g1}\varphi_1 + f_{g2}\varphi_2 + f_{g3}\varphi_3 \ . \tag{31}$$

Again two separate cases can be formulated:

a. At very high crosslink densities, when $T_g^* > T_{g\infty 3}$, substitution of Eqs. (29), (30) and (31) in (27) yields

$$\alpha_f(T-T_g^*) = \alpha_{f1}\varphi_1(T-T_{g1}^*) + \alpha_{f2}\varphi_2(T-T_{g2}^*) + \alpha_{f3}\varphi_3(T-T_{g3}^*) \ . \tag{32}$$

Differentiation of both sides with respect to T yields

$$\alpha_f = \alpha_{f1}\varphi_1 + \alpha_{f2}\varphi_2 + \alpha_{f3}\varphi_3 \ . \tag{33}$$

Substitution of Eq. (33) in (32) yields

$$T_g^* = \frac{\alpha_{f1}\varphi_1 T_{g1}^* + \alpha_{f2}\varphi_2 T_{g2}^* + \alpha_{f3}\varphi_3 T_{g3}^*}{\alpha_{f1}\varphi_1 + \alpha_{f2}\varphi_2 + \alpha_{f3}\varphi_3} \ . \tag{34}$$

b. It should be noted that T_{g3}^* is very high. Published values of $\Delta_{x\ell}T_g$ are frequently 100°C or more.[2,30] Thus, for moderately crosslinked networks, one can expect that

$$T_g^* \ll T_{g\infty 3} \ . \tag{35}$$

In those networks φ_3 is usually small, or $\Delta_{x\ell}T_g < 100°C$. The contribution by f_3 can now be neglected, and Eq. (27) reduces to

$$f = f_1\varphi_1 + f_2\varphi_2 \ , \tag{36}$$

indicating that f_3 is negligibly small. It should be noted, however, that φ_3 is not negligible and that f_g is still given by Eq. (31). Substitution of Eqs. (29), (30) and (31) in (36) yields

$$\alpha_f(T-T_g^*) = \alpha_{f1}\varphi_1(T-T_{g1}^*) + \alpha_{f2}\varphi_2(T-T_{g2}^*) - f_{g3}\varphi_3 \ . \tag{37}$$

Differentiation of both sides with respect to T yields

$$\alpha_f = \alpha_{f1}\varphi_1 + \alpha_{f2}\varphi_2 \ . \tag{38}$$

Substitution of Eq. (38) in (37) yields

$$T_g^* = \frac{\alpha_{f1}\varphi_1 T_{g1}^* + \alpha_{f2}\varphi_2 T_{g2}^* + f_{g3}\varphi_3}{\alpha_{f1}\varphi_1 + \alpha_{f2}\varphi_2} \ . \tag{39}$$

Fortunately, Eq. (39) is independent of α_{f3} and T_{g3}^*, which are inaccessible experimentally. It should be noted, that Eq. (39) gives a non-linear dependence of T_g^* on the numbers of chain ends or branch points. Due to the semi-quantitative character of this paper, only moderately crosslinked networks will be discussed in the following paragraphs. Eqs. (38) and (39) will be accepted as the correct expressions for the effects of chain ends and branch points. The use of Eq. (34) would imply additional complications, which lie beyond the scope of the present paper.

Eq. (39) can be applied in two different ways. In the first method T_{g2}^* is identified as T_{gcop}, and φ_1 and φ_3 are only the volume fractions of chain ends and branch points introduced by the crosslinking process. Thus, before reaction $\varphi_2 = 1$; after reaction $\varphi_1 + \varphi_2 + \varphi_3 = 1$ and φ_1 can be either positive (scission) or negative. Then, T_g^* is identical to T_g, the glass point of the final network

$$T_g = \frac{\alpha_{f1}\varphi_1 T_{g1}^* + \varphi_2(\alpha_{fx}\phi_x T_{gx} + \alpha_{fp}\phi_p T_{gp}) + f_{g3}\varphi_3}{\alpha_{f1}\varphi_1 + \varphi_2(\alpha_{fx}\phi_x + \alpha_{fp}\phi_p)} \quad . \tag{40}$$

A parallel calculation where, prior to crosslinking, the copolymer is first decomposed into three volume fractions containing ends, segments and branch points yields, after crosslinking, a result which can be transformed into Eq. (40). This equivalence, of course, had to be expected.

In principle, the basic question underlying this paper is now answered: within the framework of the present model Eq. (40) gives the glass point of the network rigorously. It is difficult, however, to see the implications of this equation without many numerical examples. Frequently, the material constants α_{fi} and T_{gi} are not available and the temperature range of interest is so narrow (or φ_1 and φ_3 so small) that the nonlinearity in the T_g vs. φ_1 relation is not very important. Therefore, a second (linearized) approach has been taken, more lucid in its physical implications but applicable only to small changes in φ_1 and φ_3:

The T_g increase due to the crosslinking process can be written as

$$\Delta_{x\ell}T_g = T_g - T_{gcop} \approx \left(\frac{\partial T_g^*}{\partial \varphi_1}\right)_{\varphi_3} \Delta\varphi_1 + \left(\frac{\partial T_g^*}{\partial \varphi_3}\right)_{\varphi_1} \Delta\varphi_3 \;. \tag{41}$$

The derivatives in Eq. (41) can be obtained from Eq. (39). By using the approximation

$$f_g = f_{g1} = f_{g2} = f_{g3} \;, \tag{42}$$

the derivatives can be simplified further to

$$\left(\frac{\partial T_g^*}{\partial \varphi_1}\right)_{\varphi_3} = - \frac{\alpha_{f1}\alpha_{f2}}{(\alpha_{f1}\varphi_1 + \alpha_{f2}\varphi_2)^2} \left(T_{g2}^* - T_{g1}^* - \varphi_3 (T_{g\infty 2}^* - T_{g\infty 1}^*)\right), \tag{43}$$

$$\left(\frac{\partial T_g^*}{\partial \varphi_3}\right)_{\varphi_1} = \frac{\alpha_{f1}\alpha_{f2}}{(\alpha_{f1}\varphi_1 + \alpha_{f2}\varphi_2)^2} \left(\frac{f_g}{\alpha_{f1}} - \varphi_1 (T_{g\infty 2}^* - T_{g\infty 1}^*)\right) . \tag{44}$$

Equation (43) predicts that the decrease in T_g caused by chain scission will be larger for branched molecules (i.e. $\varphi_3 \neq 0$) than for linear molecules. Equation (44) predicts that the increase in T_g caused by chain coupling will be smaller for low molecular weight polymers (i.e. $\varphi_1 \neq 0$) than for high molecular weights.

The effects of φ_1 and φ_3 on the common denominator $(\alpha_{f1}\varphi_1 + \alpha_{f2}\varphi_2)^2$ is schematically illustrated in Figs. 5 and 6. In Fig. 5 an increase in φ_1 implies a decrease in φ_2; since α_{f1} is greater than α_{f2}, however, the denominator increases with φ_1 and the slope decreases. A similar effect is noted in Fig. 6. An increase in φ_3 implies a decrease in φ_2; therefore, the denominator decreases with increasing φ_3 and the slope increases. The extrapolated line has only mathematical significance and indicates that T_{g3}^* is out of sight.

If one could estimate φ_1, α_{f1} and T_{g1}^*, the quantities in equations (43) and (44) may be calculated numerically. Then Eq. (25) yields for the glass point of the network

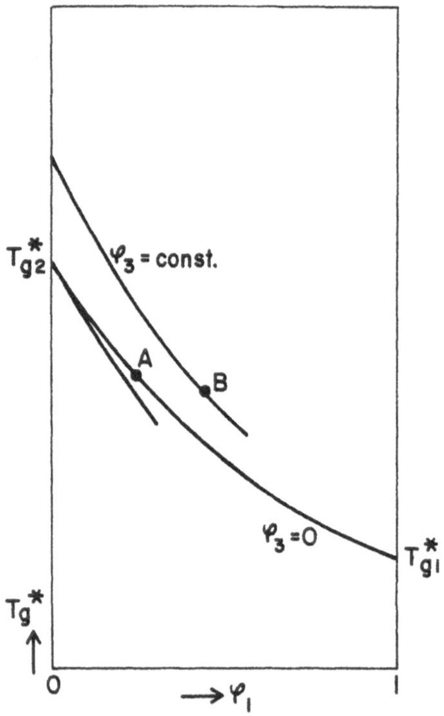

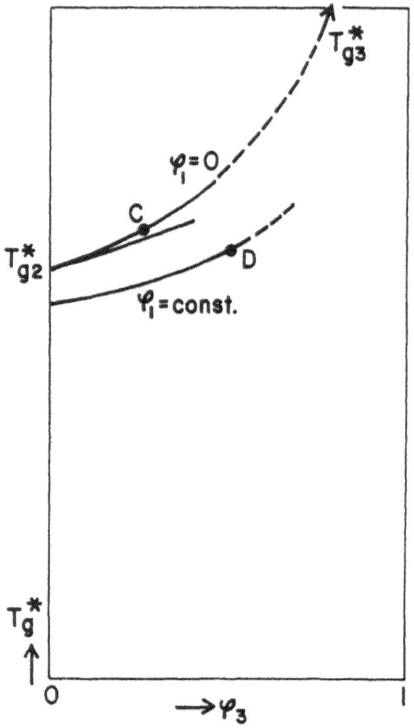

Fig. 5. The effect of chain ends on T_g of a linear ($\varphi_3 = 0$) polymer or a branched (φ_3 = const.) macromolecule.

Fig. 6. The effect of chain coupling on a very high molecular weight ($\varphi_1 = 0$) macromolecule or a low molecular weight (φ_1 = const.) prepolymer.

$$T_g = T_{gcop} + \Delta_{x\ell} T_g =$$

$$\frac{\alpha_{fx}\phi_x T_{gx} + \alpha_{fP}\phi_P T_{gP}}{\alpha_{fx}\phi_x + \alpha_{fP}\phi_P} + \left(\frac{\partial T_g}{\partial \varphi_1}\right)_{\varphi_3} \Delta\varphi_1 + \left(\frac{\partial T_g}{\partial \varphi_3}\right)_{\varphi_1} \Delta\varphi_3 \ . \qquad (45)$$

Unfortunately, experimental data in the literature[1,2,30] are usually limited to $\left(\frac{\partial T_g}{\partial \varphi_1}\right)_{\varphi_1=\varphi_3=0}$ or $\left(\frac{\partial T_g}{\partial \varphi_3}\right)_{\varphi_1=\varphi_3=0}$.

These quantities are indicated by the straight lines in Figs. 5 and

6. The experimental values of these differential quotients will be compared with the theoretical ones in the following section. In a later section Eq. (45) will be applied to practical networks and the effect of their chemical composition on T_g will be discussed.

EXPERIMENTAL VERIFICATION OF THE THEORY

a. The plasticizing function of chain ends has been recognized before by other authors.[27-29,31-33] In those studies only linear polymers were considered.

Some authors[28,34] write for the fractional free volume of a polymer of finite molecular weight

$$f = f_2 + \frac{A}{\bar{M}_n} \quad , \tag{46}$$

where f_2 is the fractional free volume of the polymer of infinite molecular weight as given by Eq. (29), $\bar{M}_n$ is the number average molecular weight, and A is a material constant.

Other authors[1,21,29] write for the glass point of a polymer of finite molecular weight

$$T_g = T^*_{g2} - \frac{K}{\bar{M}_n} \quad , \tag{47}$$

where T^*_{g2} is the glass point of the polymer of infinite molecular weight, and K is a material constant.

These two intuitive approaches are closely related. Substitution of Eqs. (29) and (30) in (46) and using the (very close) approximation $f_g = f_{g2}$ yields

$$T_g = T^*_{g2} - \frac{A/\alpha_f}{\bar{M}_n} \, . \tag{48}$$

As expected, Eqs. (46) and (47) are equivalent expressions.

A check for consistency can be obtained by differentiating both sides of Eq. (46) with respect to T:

$$df/dT = \alpha_f = \alpha_{f2} \; . \tag{49}$$

This result is wrong, of course. Data from Fox and Loshaek[1] show that α_L depends very much on $\bar{M}_n$, therefore, α_f must also depend on $\bar{M}_n$. Thus, the present form of Eq. (46) is incomplete.

The error in the argument, leading to Eq. (49), can be found by using the formal description of the end effect as given in the previous section. It is true that a plot of f versus $1/\bar{M}_n$ (which is proportional to φ_1) should be linear, but the reason for this phenomenon is not Eq. (46) but rather

$$f = f_1\varphi_1 + f_2\varphi_2 = (f_1-f_2)\varphi_1 + f_2 \; . \tag{50}$$

Thus, the constant A, in Eq. (46), should be proportional to $f_1 - f_2$, rather than to f_1 only as suggested by Bueche.[27] Differentiation of both sides of Eq. (50) with respect to T yields correctly

$$\alpha_f = \alpha_{f1}\varphi_1 + \alpha_{f2}\varphi_2 = (\alpha_{f1}-\alpha_{f2})\varphi_1 + \alpha_{f2} \; . \tag{51}$$

Indeed, experimental data of $\Delta\alpha$ (proportional to α_f) vary linearly[1] with $1/\bar{M}_n$ (proportional to φ_1). Therefore, the resulting expression for T_g, obtained from Eq. (50), does not vary linearly with $1/\bar{M}_n$ as in Eq. (47) but should be given by Eqs. (34) or (39) for $\varphi_3 = 0$.

At this point, it is appropriate to make an estimate of φ_1 and compare the values of $df/d(1/\bar{M}_n)$ or $dT_g/d(1/\bar{M}_n)$ obtained from the present theory with the published A and K values. If the "effective volume" of one chain end is denoted by ξ_1, then

$$\varphi_1 = 2\rho N_A \xi_1/\bar{M}_n \; , \tag{52}$$

where ρ is the density of the sample and N_A is Avogadro's number. (In case a diluent is present ρ should be replaced by c, the polymer concentration in grams per cubic centimeter). The volume ξ_1 can be chosen equal to the volume of e.g. one, or two, or three monomer units. The properties of "pure" chain ends are then obtained from those of dimers, or tetramers, or hexamers respectively. Consequently, the value of $\rho N_A \xi_1$ is equal to the molecular weight of a monomer, or dimer, or trimer respectively. A value for ξ_1 smaller than the volume of one monomer unit would not be acceptable within the framework of the present approach. On the other hand, a value for ξ_1 larger than the volume of five or six monomer units (about one statistical element, according to Kuhn)[35] would also be unacceptable, since only short range cooperative motions are possible near or in the glassy state.

Substitution of Eqs. (43) and (52) in

$$\frac{dT_g}{d(1/\bar{M}_n)} = \frac{dT_g}{d\varphi_1} \cdot \frac{d\varphi_1}{d(1/M_n)} , \tag{53}$$

and taking $\varphi_3 = 0$ yields

$$\frac{dT_g}{d(1/\bar{M}_n)} = - \frac{\alpha_{f1}\alpha_{f2}(T^*_{g2} - T^*_{g1})}{(\alpha_{f1}\varphi_1 + \alpha_{f2}\varphi_2)^2} \cdot (2\rho N_A \xi_1) . \tag{54}$$

Equation (54) depends on φ_1 but not on T. A limiting value is

$$K = - \left[\frac{dT_g}{d(1/\bar{M}_n)}\right]_{\varphi_1=0} = \frac{\alpha_{f1}}{\alpha_{f2}} (T^*_{g2} - T^*_{g1}) (2\rho N_A \xi_1)$$

$$\approx \frac{\Delta\alpha_1}{\Delta\alpha_2} (T^*_{g2} - T^*_{g1}) (2\rho N_A \xi_1) . \tag{55}$$

Values of these parameters have been estimated and are listed in Table I together with the experimental values of K both for polystyrene and for polymethylmethacrylate. The values of T^*_{g1} have been interpolated from data by Ueberreiter and Kanig,[32] or estimated e.g. for PMMA. The values of α_{L1} have been reported by Fox and Loshaek,[1] and are given by the following empirical relations.

for PSty: $\alpha_L = (5.5 + 643/M) \times 10^{-4}\ \ {}^\circ C^{-1}$, (56)

for PMMA: $\alpha_L = (5.0 + 714/M) \times 10^{-4}\ \ {}^\circ C^{-1}$. (56b)

The value of α_G is reported to be independent of molecular weight,[27,31] and was determined by Rusch.[13,36] The values of T^*_{g2} have been taken from Lee and Knight.[37]

Table I

The effect of chain ends on T_g of polystyrene and polymethylmethacrylate calculated for various choices of ξ_1, the effective volume of a chain end.

	PSt	PSt	PSt	PMMA	PMMA
$\rho N_A \xi_1$	104	208	312	200	300
T^*_{g1}	195	242	268	220	240
α_{L1}	8.6×10^{-4}	7.1×10^{-4}	6.5×10^{-4}	6.8×10^{-4}	6.2×10^{-4}
α_G	2.1×10^{-4}	2.1×10^{-4}	2.1×10^{-4}	1.8×10^{-4}	1.8×10^{-4}
$\Delta\alpha_1$	6.5×10^{-4}	5.0×10^{-4}	4.4×10^{-4}	5.0×10^{-4}	4.4×10^{-4}
T^*_{g2}	373	373	373	378	378
$\Delta\alpha_2$	3.4×10^{-4}	3.4×10^{-4}	3.4×10^{-4}	3.2×10^{-4}	3.2×10^{-4}
K(Calc.)	0.71×10^5	0.80×10^5	0.85×10^5	0.99×10^5	1.14×10^5
K(Exp.)	1.0×10^5	1.0×10^5	1.0×10^5	---	---
α_{f2}(Exp.)	6.3×10^{-4}	6.3×10^{-4}	6.3×10^{-4}	3.4×10^{-4}	3.4×10^{-4}
α_{f1}(Calc.)	12.0×10^{-4}	9.27×10^{-4}	8.15×10^{-4}	5.3×10^{-4}	4.7×10^{-4}
A(Calc.)	44.6	50.5	53.4	34	39
A(Exp.)	54	54	54	60	60
φ_1(c)	0.131	0.228	0.320	0.358	0.454
$1/\bar{M}_n$(c)	6.28×10^{-4}	5.48×10^{-4}	5.13×10^{-4}	8.95×10^{-4}	7.57×10^{-4}

The values of K, calculated according to Eq. (55) for volumes of ξ_1, of two or three monomer units, are within experimental error of K(Exp), determined by Fox and Flory.[38] Since K(Calc) depends very little on the choice of ξ_1, the value of ξ_1 cannot be evaluated from these data.

Quite analogously, substitution of Eqs. (50) and (52) in

$$\frac{df}{d(1/\bar{M}_n)} = \frac{df}{d\varphi_1} \cdot \frac{d\varphi_1}{d(1/\bar{M}_n)} , \tag{57}$$

yields

$$\frac{df}{d(1/\bar{M}_n)} = (f_1 - f_2)\ (2\rho N_A \xi_1) \ . \tag{58}$$

Equation (58) depends on T but not on φ_1. A limiting value is

$$A = \left[\frac{df}{d(1/\bar{M}_n)}\right]_{T=T^*_{g2}} = \{f_{g1}+\alpha_{f1}(T^*_{g2}-T^*_{g1})-f_{g2}\}\ (2\rho N_A\xi_1)$$

$$\approx \alpha_{f1}(T^*_{g2}-T^*_{g1})\ (2\rho N_A\xi_1) = \alpha_{f2}K \ . \tag{59}$$

Values of α_{f2} have been determined by Pierson[39] for polystyrene and by Onogi et al[40] for polymethylmethacrylate, and are listed in Table I together with the calculated values of A. The experimentally observed values, A(Exp), for polystyrene have been obtained by Pierson and Kovacs,[34] and have been calculated for PMMA from the data of Onogi et al.[40]

The calculated and experimental values of A for polystyrene are virtually identical. The differences in A values for PMMA are not surprising if one reconsiders Onogi's data; the data points are all within a range of $1/\bar{M}_n$ from 0.27×10^{-4} to 1.1×10^{-4}, (see Fig. 7) and the relative error in f is quite large. The polystyrene data, however, were obtained over a much wider range of molecular weights. The agreement between theoretical and experimental K and A values would be even better if the actual T^*_{g1} values of the styrene oligomers would be somewhat lower than those reported by Ueberreiter and Kanig.[32] Fox and Loshaek[1] give reasons why this systematic error in Ueberreiter's data may be expected.

If one assumes the validity of

$$\frac{\alpha_{f1}}{\alpha_{f2}} = \frac{\Delta\alpha_1}{\Delta\alpha_2} , \qquad (60)$$

the values of α_{f1} can be calculated. These are also listed in Table I.

Experimental values of T_g, for polystyrene, from Fox and Flory[31] and from Ueberreiter and Kanig[32] have been plotted in Fig. 7. In analogy to the treatment described in the previous section one can derive from Eq. (50)

$$T_g = \frac{\alpha_{f1}\varphi_1 T^*_{g1} + \alpha_{f2}\varphi_2 T^*_{g2}}{\alpha_{f1}\varphi_1 + \alpha_{f2}\varphi_2} , \qquad T_g > T^*_{g\infty 2}, \qquad (61)$$

or

$$T_g = \frac{\alpha_{f1}\varphi_1 T^*_{g1} + f_{g2}\varphi_2}{\alpha_{f1}\varphi_1} , \qquad T_g < T^*_{g\infty 2} . \qquad (62)$$

Substitution of Eq. (52), and the values for α_{fi} and T^*_{gi} from Table I in Equations (61) and (62), together with the approximation $f_{g2} = 0.025$, yields the theoretical lines a, b and c drawn in Fig. 7.

At $T_g = T^*_{g\infty 2}$ a singularity should exist and the values of T_g obtained either from Eq. (61) or (62) should be identical. The volume fraction of chain ends, $\varphi_{1(c)}$, at which this occurs is given by

$$\frac{1}{\varphi_{1(c)}} = 1 - \frac{\alpha_{f1}}{\alpha_{f2}} + \frac{(T^*_{g2} - T^*_{g1})\alpha_{f1}}{f_{g2}} . \qquad (63)$$

These values of $\varphi_{1(c)}$ and their corresponding $1/\bar{M}_{n(c)}$ values are shown in Table I.

Obviously, the location of $T^*_{g\infty 2}$ shown in Fig. 7 is fixed at 40°C below T^*_{g2} because

$$T^*_{g2} - T^*_{g\infty 2} = f_{g2}/\alpha_{f2} , \qquad (64)$$

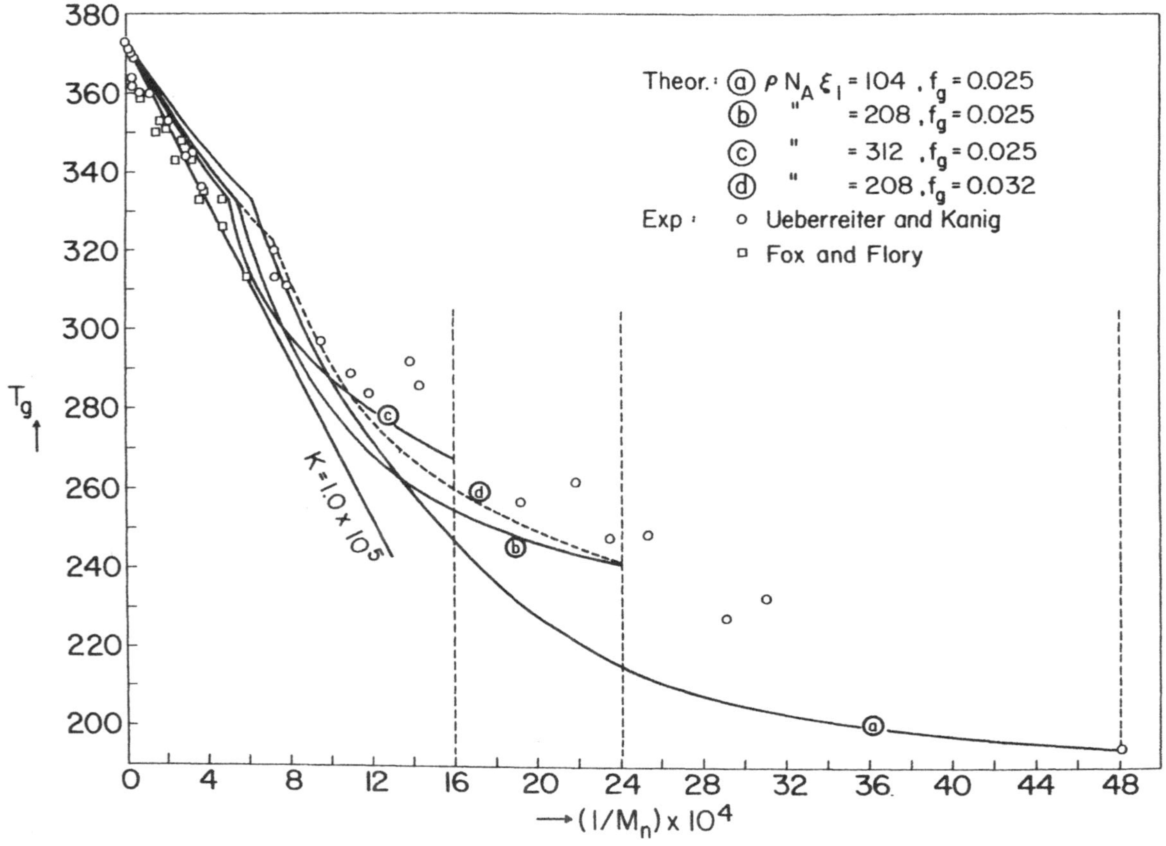

Fig. 7. Molecular weight dependence of T_g for linear polystyrene. Curves a, b, c and d are calculated according to Eqs. (61), (62) and (63).

where f_{g2} and α_{f2} were taken to be 0.025 and 6.3×10^{-4} respectively. A more appropriate set of values for f_{g2} and α_{f2} would be to take the set from the same author and the same sample; Pierson[39,7] reports 0.032 and 6.3×10^{-4} respectively. The corresponding theoretical curve, with $T^*_{g\infty 2} = 323$, is indicated in Fig. 7 as curve d. The scatter in the experimental data still does not permit an unambiguous estimate of the effective volume of a chain end from this curve fitting procedure. It seems that the volume of either a dimer or a trimer would be a satisfactory choice. The values of K and A have been calculated for several other polymers. These will be published in a forthcoming paper.

b. The treatment of the immobilizing effect of branch points on T_g can be given similarly. Fox and Loshaek[1] wrote for the glass-point of a network

$$T_g = T^*_{g2} + \frac{K_x \Gamma_4}{N_A} , \tag{65}$$

where Γ_4 is the number of four functional branch points per unit volume, and K_x is a material constant. The last term in Eq. (65) represents the term $\Delta_{x\ell}$ in Eq. (25).

Analogous to relations (47) and (48) it is possible to rewrite Eq. (65) in the equivalent form

$$f = f_2 - \frac{A_x \Gamma_j}{N_A} , \tag{66}$$

where

$$A_x = \alpha_f K_x , \tag{67}$$

and Γ_j is the number of j functional branch points per unit volume.

It should be noted, however, that Eqs. (65) and (66) are limiting cases of the more general form

$$f = f_1\varphi_1 + f_2\varphi_2 + f_3\varphi_3 \tag{68}$$

In most practical cases $f_3 = 0$. Therefore, Eq. (68) reduces to

$$f = f_1\varphi_1 + f_2\varphi_2 \,. \tag{69}$$

Hence,

$$\alpha_f = \alpha_{f1}\varphi_1 + \alpha_{f2}\varphi_2 \tag{70}$$

Thus, the glass points of networks are given by Eq. (39) and do not vary linearly with Γ_j as suggested by Eq. (65).

A situation which frequently occurs, is that crosslinking takes place while φ_1 stays constant. Then Eqs. (69) and (70) yield

$$f = f_2 + (f_1 - f_2)\varphi_1 - f_2\varphi_3 \,, \tag{71}$$

$$\alpha_f = \alpha_{f2} + (\alpha_{f1} - \alpha_{f2})\varphi_1 - \alpha_{f2}\varphi_3 . \tag{72}$$

Eq. (71) replaces (66). Thus, both f and α_f should decrease linearly with Γ_j (which is proportional to φ_3).

Analogous to Eq. (52), one can write formally

$$\varphi_3 = \xi_{3j}\Gamma_j \tag{73}$$

where ξ_{3j} is the "effective volume" of a j functional branch point. In estimating ξ_{3j} it seems logical to assume that ξ_{3j} is about j times larger than the effective volume of a chain end:

$$\xi_{3j} \approx j\,\xi_1 \tag{74}$$

Since Eq. (71) should apply to temperatures from T_g up to T_g + 100°C, it is probably more accurate to make φ_3 proportional to an effective crosslink density as used in theories of viscoelasticity.[26)] An analogous treatment of branched, uncrosslinked polymers has been published recently.[41)] The principal result of this theory is that the mechanical response of j polymer chains ending in a j functional branch point can be given by the response of an equivalent configuration whereby j - 2 chain ends are attached to fixed points in space. If this decoupling process is assumed to apply at temperatures down to T_g, then the immobilizing effect of a branch point is accounted for by the j - 2 fixed chain ends. The decoupled configuration would also explain why T^*_{g3} is out of sight. These arguments suggest that ξ_{3j} should be proportional to j - 2.

Another requirement for ξ_{3j} is unambiguity with respect to the absolute "size" of the crosslinking agent. This is illustrated in Fig. 8 for an infinite series of networks made with bifunctional crosslinking agents of various "lengths." Compare a

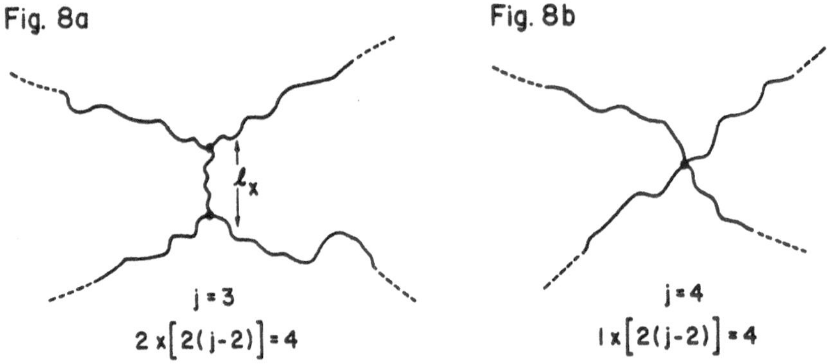

Fig. 8. Continuous variation in "size" of a bifunctional crosslinker transforming a network containing trifunctional branch points to a network containing tetrafunctional branch points.

section of the network prepared with a "long" crosslinking agent as schematically indicated in Fig. 8a, with an analogous part prepared with a "short" crosslinking agent, indicated in Fig. 8b, where the "tie chain" ℓ_x has been reduced to infinitesimal length. The transformation from two trifunctional branch points to one four functional branch point has not changed the number of elastically effective network chains. Furthermore, this transformation may not generate a discontinuity in ξ_{3j} at some arbitrary value of ℓ_x. A possible choice for ξ_{3j}, therefore, would be

$$\xi_{3j} = 2(j - 2)\xi_1 \quad , \quad \text{for } j > 1, \tag{75}$$

and $$\varphi_3 = 2(j - 2)\xi_1\Gamma_j \tag{76}$$

The front factor 2 in Eqs. (75) and (76) has been accommodated to the following mechanism of forming a trifunctional branch point: a chain end (of volume ξ_1) reacts with a chain segment (of volume ξ_1 also) to form a trifunctional branch point of volume $d\varphi_3$, equal to $-(d\varphi_1+d\varphi_2)$ or $2\xi_1$.

Equation (75) yields $4\xi_1$ for either case shown in Fig. 8. Similarly, the effective volume of two j functional branch points is

$$2 \times 2\,(j-2)\xi_1 \quad ,$$

and the effective volume of one (2j - 2) functional branch point is

$$2[2j-2)-2]\xi_1 = 4(j-2)\xi_1.$$

Equation (76) will be used in the following derivations.

Substitution of Eqs. (44) and (76) in

$$\frac{dT_g}{d(\Gamma_j/N_A)} = \frac{dT_g}{d\varphi_3}\cdot\frac{d\varphi_3}{d(\Gamma_j/N_A)}, \tag{77}$$

and taking $\varphi_1 = 0$ yields

$$\frac{dT_g}{d(\Gamma_j/N_A)} = \frac{2(j-2)f_g}{\rho\alpha_{f2}(1-\varphi_3)^2}\cdot(\rho N_A\xi_1). \tag{78}$$

Equation (78) still depends on φ_3 but not on T. A limiting value is

$$K_x = \left[\frac{dT_g}{d(\Gamma_j/N_A)}\right]_{\varphi_3=0} = \frac{2(j-2)f_g}{\rho\alpha_{f2}}(\rho N_A\xi_1). \tag{79}$$

Values of these parameters have been estimated and are listed in Table II together with the experimental values of K_x both for polystyrene and polymethyl methacrylate. The values of f_g and α_{f2} have been determined by Pierson[39] for PSt and by Onogi et al.[40] for PMMA. The values of K_x have been calculated for j = 4, in accord with the structures used for the experimental determination of K_x.[2] A slightly different value of K_x has been reported by Nielsen.[42] Values of A_x have not yet been reported in the literature.

Table II

The effect of crosslinking on T_g of polystyrene and polymethyl methacrylate calculated for various choices of ξ_1 and $j = 4$.

	PSt	PSt	PMMA	PMMA
$\rho N_A \xi_1$	208	312	200	300
f_g	0.032	0.032	0.026	0.026
α_{f2}	6.3×10^{-4}	6.3×10^{-4}	3.4×10^{-4}	3.4×10^{-4}
ρ	1.07	1.07	1.19	1.19
K_x(Calc.)	0.40×10^5	0.60×10^5	0.52×10^5	0.78×10^5
K_x(Exp.)	0.65×10^5	0.65×10^5	0.78×10^5	0.78×10^5
A_x(Calc.)	25.2	37.8	17.7	26.5
Q(Calc.)	2.0	1.4	1.9	1.5

The agreement between theoretical and experimental values of K_x is very good if ξ_1 is taken to be the volume of three monomer units, i.e. six backbone chain atoms in length. That this length is identical to Schatzki's crankshaft rotation model,[43] however, must be considered accidental.

It can be seen that experimental evaluation of K and K_x present complementary information about the free volume contribution by chain ends, even without knowing T^*_{g1} or α_{f1}. From Eqs. (55) and (79) the ratio K/K_x can be obtained as

$$Q = \frac{K}{K_x} = \frac{\rho}{j - 2}\left(\frac{f_1(T^*_{g2})}{f_g} - 1\right) , \tag{80}$$

from which $f_1(T^*_{g2})$ may be calculated. The value of Q is usually found larger than unity.

When all the data are known, as for instance given in Tables I and II, either Eq. (40) or (45) may be used to predict experimental results. Most frequently, however, the free volume parameters are lacking. Even though the calculations then have to be restricted to crude linearized equations, the results are still useful approximations in the molecular engineering of practical thermosets. The following example illustrates this.

It is well known that an increase in either $\bar{M}_n$ or Γ_j will increase T_g. It is not certain, however, how T_g will change when a linear polymer is converted into a branched polymer of the same number average molecular weight. The linearized version of Eq. (41) can be written as

$$T_g = T^*_{g2} - \frac{K}{\bar{M}_n} + K'_x \frac{\Gamma_3}{N_A} =$$

$$= T^*_{g2} - K\frac{\Omega}{2\rho N_A} + K'_x \frac{\Gamma_3}{N_A} , \qquad (81)$$

where the prime in K'_x refers to trifunctional branch points, Γ_3 is the number of trifunctional branch points per unit volume and Ω is the number of chain ends per unit volume. Eq. (79) yields $K'_x = \frac{1}{2}K_x$, where K_x refers to four functional branch points. The T_g increment due to the branching process is obtained from Eq. (81) as

$$\Delta T_g = - \frac{K}{2\rho N_A} \Delta\Omega + \frac{K_x}{2N_A} \Delta\Gamma_3 \; . \qquad (82)$$

It should be noted that formation of each trifunctional branch point generates also one chain end. Thus

$$\Delta\Omega = \Delta\Gamma_3 \; .$$

Therefore, when $K > \rho K_x$ as in PSt, or PMMA, or poly vinyl acetate the change in T_g due to branching (at constant $\bar{M}_n$) is negative. This fact has been observed experimentally by Meares[44)] for branched poly vinyl acetate. This decrease in T_g is schematically illustrated in Fig. 5 by points A and B or, analogously, in Fig. 6 by points C and D.

For some polymers, e.g. poly ethylene oxide[45)] or poly butylene oxide,[46)] K can be quite small because $T^*_{g2} - T^*_{g1}$ is small; K_x, however, is always large because f_g is always much larger than α_{f2}.

In thermosets which are glassy at room temperature, like those to be discussed in the following section, K is always larger than K_x. In many cases the value of K_x may be estimated from data of α_{f2} in the literature; the evaluation of K, however, is more cumbersome.

The physical interpretation of ξ_3 as given here can be a useful guide in the choice of network components. For instance, it has been found frequently that crosslinkers composed of epoxides derived from cyclohexene derivatives yield higher T_g increases than glycidyl derivatives. This may be attributed to a larger ξ_3 value of the bulky cyclohexyl groups.

ANALYSIS OF VARIABLES.IN NETWORK CHEMISTRY

Before the preceding equations can be applied to networks, the accounting procedures for branch points and chain ends have to be specified. The following example has been calculated for a network formed from a linear prepolymer which reacts with a bifunctional crosslinking agent to form a network of trifunctional branch points, as shown schematically in Fig. 4. Although the free volume parameters are lacking, it will be shown in the following paragraphs that the results obtained from the approximate equations are still useful.

Consider n prepolymer molecules containing an arbitrary number of active groups along the chains and having a number average molecular weight of $\bar{M}_{n(P)}$. This prepolymer has a glass point of T_{gP}, a free volume expansion coefficient of α_{fP} and a density of ρ_P. The prepolymer will be mixed with an crosslinker molecules, each containing two reactive groups. Thus the effective functionality of the prepolymer is 2a, provided the number of active groups of the crosslinking agent does not exceed those of the prepolymer. The crosslinking agent has a glass point of T_{gx}, a free volume expansion coefficient of α_{fx} and a density of ρ_x. Generally, T_{gx} will be smaller than T_{gP} and α_{fx} will be larger than α_{fP}.

The molecular volume of a crosslinker molecule is indicated by ξ_x, and the number average volume of a prepolymer chain between two branch points at 100% reaction is indicated by ξ_C. Thus the volume of crosslinking agent, V_x, is

$$V_x = \xi_x \, an \, , \tag{83}$$

and the volume of prepolymer, V_P, and its molecular weight are respectively

$$V_P = 2a\xi_C n \, , \tag{84}$$

$$\bar{M}_{n(P)} = 2a\xi_C \rho_P N_A \, . \tag{85}$$

The total volume after mixing, or reaction will be

$$V = (2\xi_C + \xi_x) \, an \, , \tag{86}$$

where volume changes due to mixing, or reaction are neglected. Thus the volume fractions of polymer and crosslinker are respectively

$$\phi_P = 2\xi_C/(2\xi_C + \xi_x), \tag{87}$$

$$\phi_x = \xi_x/(2\xi_C + \xi_x) \, . \tag{88}$$

If a free volume expansion ratio is defined as

$$q = \alpha_{fx}/\alpha_{fP} \, , \tag{89}$$

the glass point of the homogenized copolymer before crosslinking can be written according to Eq. (26) as

$$T_{gcop} = \frac{qT_{gx}\phi_x + T_{gP}\phi_P}{q\phi_x + \phi_P} =$$

$$= \frac{qT_{gx}\xi_x + 2T_{gp}\xi_C}{q\xi_x + 2\,\xi_C} \, , \tag{90}$$

where T_{gP} may be approximated by

$$T_{gp} = T_{gP}(M=\infty) - \frac{K}{2a\xi_C \rho_P N_A} . \qquad (91)$$

Initially, the number of branch points, C_o, in the mixture is zero and the number of chain ends, ω_o, is

$$\omega_o = 2(an + n). \qquad (92)$$

After an infinite reaction time, or presumably 100% reaction, the number of branch points and chain ends is respectively

$$C_\infty = 2an , \qquad (93)$$

$$\omega_\infty = 2n . \qquad (94)$$

Of course, due to steric limitations a 100% conversion can never be achieved. A conversion y is defined as

$$y = C/C_\infty , \qquad (95)$$

and the number of chain ends at this particular conversion is

$$\omega = \omega_o - C \qquad (96)$$

The number of trifunctional branch points produced per unit volume at a particular conversion is

$$\Delta\Gamma_3 = \Delta C/V = C/V , \qquad (97)$$

and the number of chain ends produced per unit volume is according to Eq. (96)

$$\Delta\Omega = \Delta\omega/V = - C/V \qquad (98)$$

The effect of crosslinking on T_g can be approximated according to Eq. (82) by

$$T_g - T_{gcop} = - \frac{K}{2\rho N_A} \Delta\Omega + \frac{K_x}{2N_A} \Delta\Gamma_3 =$$

$$= \frac{C}{2VN_A}\left(\frac{K}{\rho} + K_x \right) = \frac{CA}{2VN_A} \qquad (99)$$

where A stands for $K/\rho + K_x$ and is of the order of 10^5. Substitution of Eqs. (86), (90), (93) and (95) in (99) yields for the glass point of the network

$$T_g = \frac{qT_{gx}\xi_x + 2T_{gP}\xi_C}{q\xi_x + 2\xi_C} + \frac{Ay}{N_A(2\xi_C + \xi_x)} . \qquad (100)$$

Equation (100), finally, is the equation which connects the macroscopic glass point with the microstructure of the network. Variations in network chemistry can be readily read from it.

A frequently occurring experiment in thermosets is to vary the amount of crosslinking agent. This implies that ξ_C is varied while $\bar{M}_{n(P)}$ is kept constant. To predict this experiment $(\partial T_g/\partial \xi_C)_{\bar{M}_{n(P)},y}$ should be calculated. It should be noted, however, that two different ways to change ξ_C exist; the other differential quotient is $(\partial T_g/\partial \xi_C)_{a,y}$. If the molecular weight of the prepolymer is sufficiently large, variations in q may be neglected. Equations (100) and (91) yield

$$\left(\frac{\partial T_g}{\partial \xi_C}\right)_{\bar{M}_{n(P)},y} = \frac{2(T_{gP} - T_{gcop})}{q\xi_x + 2\xi_P} - \frac{2Ay}{N_A(2\xi_P + \xi_x)^2} . \qquad (101)$$

$$\left(\frac{\partial T_g}{\partial \xi_C}\right)_{a,y} = \frac{2[T_{gP}(M{=}\infty) - T_{gcop}]}{q\xi_x + 2\xi_P} - \frac{2Ay}{N_A(2\xi_P + \xi_x)^2} . \qquad (102)$$

The experiment implied in Eq. (102) is one whereby the molecular weight of the prepolymer is varied but its effective functionality is kept constant. Another appropriate differential quotient in that experiment would be $(\partial T_g/\partial \bar{M}_{n(P)})_{a,y}$ which can be obtained from Eqs. (102) and (85) as

$$\left(\frac{\partial T_g}{\partial \bar{M}_{n(P)}}\right)_{a,y} = \frac{1}{2a\rho_P N_A}\left(\frac{\partial T_g}{\partial \xi_C}\right)_{a,y} . \qquad (103)$$

It should be noted, however, that a more common way to vary $\bar{M}_{n(P)}$ is to keep ξ_C constant instead of a. In that case Eq. (100) yields

$$\left(\frac{\partial T_g}{\partial \bar{M}_{n(P)}}\right)_{\xi_C,y} = \frac{2\xi_C}{q\xi_x + 2\xi_C}\left(\frac{K}{\bar{M}^2_{n(P)}}\right) . \qquad (104)$$

In Eqs. (101) through (104) both ξ_x and y were kept constant. Similar relations can be derived, however, for the cases that the molecular weight of the crosslinker is varied. These will be discussed in a forthcoming paper. The increase in T_g with increasing conversion is expressed by

$$\left(\frac{\partial T_g}{\partial y}\right)_{a,\xi_C} = \frac{A}{N_A(2\xi_C + \xi_x)} \; . \tag{105}$$

The linear dependence of T_g on y and on $1/\bar{M}_{n(P)}$ is not a coincidence. This has been introduced through Eq. (99), which is only a linearized form of Eq. (40) or (41).

Equations (104) and (105) show that an increase of the conversion or the molecular weight will always increase T_g. Equations (101), (102) and (103), however, indicate that an increase in spacing between branch points may not always cause a decrease in T_g, as might be expected. In fact, at conversions y of approximately

$$y < y_C \approx \xi_x N_A(T_{gP} - T_{gx})/A \tag{106}$$

in the case of Eq. (101), and of

$$y < y_C \approx \xi_x N_A[T_{gP}(M{=}\infty) - T_{gx}]/A \tag{107}$$

in the case of Eq. (102), one should observe an increase in T_g when the volume fraction of crosslinking agent is decreased. A calculated example on the application of these equations is given below.

Consider a linear random copolymer consisting of 9 moles styrene and 1 mole acrylic acid, polymerized to a relatively large molecular weight. The T_g of this copolymer is 110°C, according to Illers.[18] Let this prepolymer be crosslinked in the presence of some tetrabutyl ammonium bromide with a diepoxide of the structure (ϕ stands for phenylene):

$$CH_2-CH-CH_2-O-\phi-\overset{\overset{\displaystyle CH_3}{|}}{\underset{\underset{\displaystyle CH_3}{|}}{C}}-\phi-O-CH_2-\overset{\overset{\displaystyle OH}{|}}{CH}-CH_2-O-\phi-\overset{\overset{\displaystyle CH_3}{|}}{\underset{\underset{\displaystyle CH_3}{|}}{C}}-\phi-O-CH_2-CH-CH_2$$

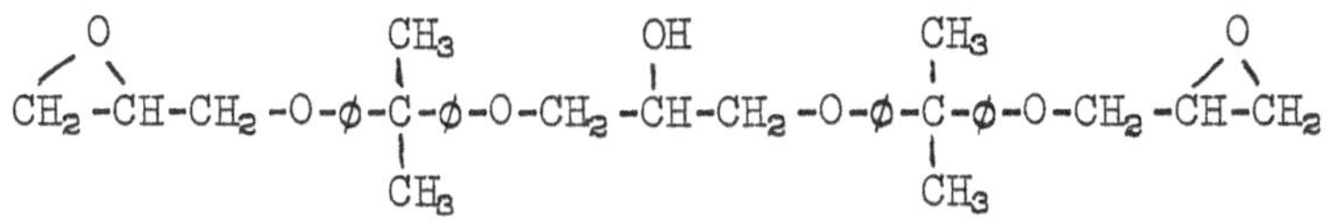

which approximately corresponds to the composition of a commercial[47] resin Epon 836, which has a glass point of about 30°C and a density of 1.2 grams per cm^3. The molecular weight of a diepoxide with a structure as given above is 624. Thus 936 grams prepolymer are stoichiometrically equivalent with 312 grams crosslinker, or 875 cm^3 prepolymer are stoichiometrically equivalent with 260 cm^3 crosslinker. If q is taken to be about 1.3 and A is 1.4×10^5 cm^3 °K/mole (as it would be for "pure" polystyrene) the glass points of the networks, obtained from various component ratios, can be calculated according to Eq. (100) and are listed in Table III or plotted in Fig. 9.

The first two columns in Table III, V_x and V_p, were obtained from Eqs. (83) and (84) where the value of an was chosen equal to N_A. The third column represents the glass points of the homogenized mixtures at zero conversion. In the fourth column the values of the ultimate conversion, y_∞, are given, which is the idealized conversion obtained when all the epoxy and acid groups had the chance to find each other. The first six samples do not cause any complication; as indicated in the beginning of this section the functionality of the prepolymer has been defined as 2a and does not depend on the arbitrary number of carboxyl groups present, and thus $y_\infty = 1$. In sample 6) carboxyl and epoxy groups are present in stoichiometric amounts. In samples 7) through 10) excess epoxy groups are present which would invalidate Eq. (93). To maintain coherence in the calculations, however, samples 7) through 10) will be treated as if enough active groups were present, but the conversion is terminated when all carboxyl groups are consumed. This explains the y_∞ values smaller than unity. The last column in Table III represents the ultimate T_g increase due to crosslinking. The numbers from the last three columns of Table III were used to

TABLE III

Glass points of networks obtained by mixing an amount, V_X, of crosslinker with various amounts, V_P, of prepolymer, at conversions of $y = 0$ and $y = y_\infty$

$T_{gX} = 303°K$, $T_{gP} = 383°K$, $q = 1.3$, $A = 1.4 \times 10^5\ cm^3\ °K/mole$.

Sample number	V_X, or $N_A \xi_X$ in cm^3.	V_P, or $2N_A \xi_C$ in cm^3.	T_{gcop} in °K	y_∞	$T_g - T_{gcop}$ at $y = y_\infty$ in°K
1)	520	∞	383	1	1
2)	520	3500	370	1	35
3)	520	2625	367	1	45
4)	520	2100	364	1	53
5)	520	1925	362	1	57
6)	520	1750	361	1	62
7)	520	1575	359	0.9	60
8)	520	1400	357	0.8	58
9)	520	1225	355	0.7	56
10)	520	875	348	0.5	50

construct Fig. 9. The critical conversion, y_C, calculated according to Eqs. (106) or (107) is

$$y_C \approx 0.3,$$

which, in Fig. 9, turns out to be somewhat higher.

It should be noted, however, that the T_g increase as given in the last column of Fig. 9 cannot be achieved in reality. Due to steric limitations, the ultimate conversion, y_∞, will never be reached. If instead another boundary condition can be found, e.g. if the product of the epoxy and acid concentrations would be limited to a lowest limiting value unequal to zero, the actually realizable conversions for each of the compositions can be calculated in the following way.

The concentration of carboxylic acid groups (in moles per cm^3)

is indicated by [Z], and the concentration of epoxy groups is indicated by [E]. The initial concentrations of acid and epoxy groups in the mixture is indicated respectively by $[Z]_o$ and $[E]_o$, and are given in the first two columns of Table IV. At any moment during the reaction one may write

$$[Z] = [Z]_o - y\,\frac{C_\infty}{VN_A}\,, \tag{108}$$

$$[E] = [E]_o - y\,\frac{C_\infty}{VN_A} \tag{109}$$

Three initial concentration ratios r, R_E and R_Z are defined as

$$r = [Z]_o/[E]_o, \tag{110}$$

$$R_E = [S]_o/[E]_o, \tag{111}$$

$$R_Z = [S]_o/[Z]_o, \tag{112}$$

where $[S]_o$ is the stoichiometric concentration of 8.811×10^{-4} of either acid or epoxy groups in sample 6). The lowest realizable concentrations are indicated by $[Z]_m$ and $[E]_m$, or $[S]_m$, and the maximum attainable conversion is indicated by y_m, where $y_m < y_\infty$.

Let it be assumed that the minimum concentrations attainable in each composition can be approximated by

$$[Z]_m\,[E]_m = L = [S]_m^2\,, \tag{113}$$

where L is a constant which only depends on temperature, and has to be evaluated experimentally. If it is found, for instance, that

$$[S]_m = 0.1\,[S]_o\,, \tag{114}$$

indicating a 90% conversion in the stoichiometric case, it follows that the value of L is

$$L = [S]_m^2 = 10^{-2}\,[S]_o^2\,. \tag{115}$$

This boundary condition makes y_m solvable. The calculation now has to be performed separately for a series with excess carboxylic acid and a series with excess epoxy.

a. In cases 1) to 5) excess acid is present. Therefore

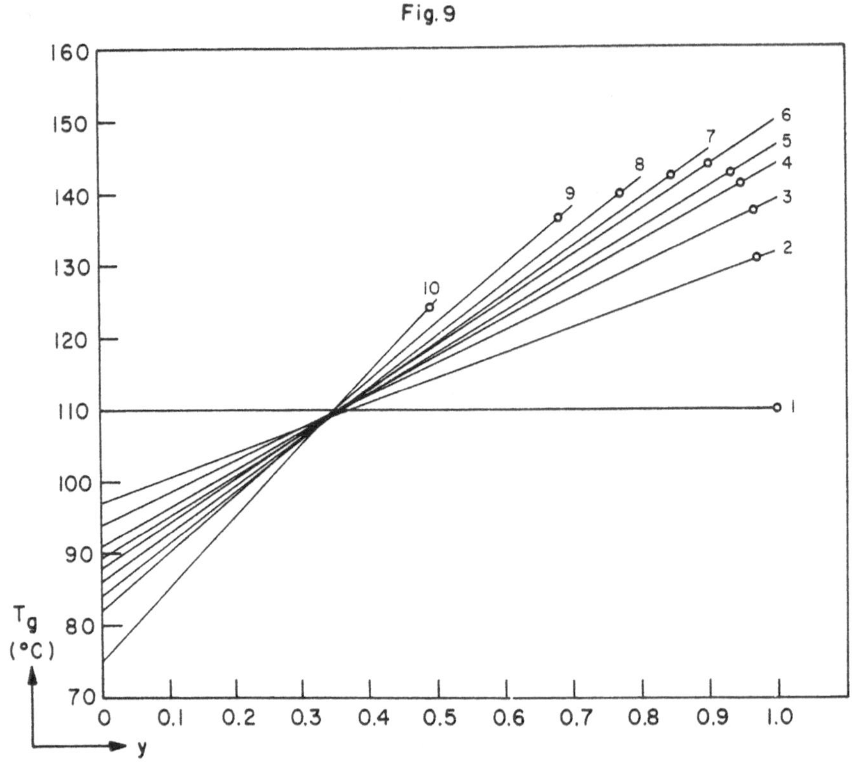

Fig. 9. Dependence of T_g on conversion y for networks containing excess amounts of polycarboxylic acid (lines 1 to 5) or excess amounts of diepoxide (lines 7 to 10). The data points are calculated according to Eq. (118) or Eq. (121).

$$\frac{C_{\infty}}{VN_A} = [E]_o \ . \tag{116}$$

Substitution of Eqs. (108), (109) and (116) in (113) yields rigorously

$$L = [E]_o^2 \ [y_m^2 - (1 + r) \ y_m + r] \ . \tag{117}$$

Substitution of condition (115) in Eq. (117) yields y_m

$$2y_m = 1 + r - [(1 - r)^2 + 4 \times 10^{-2} \ R_E^2]^{\frac{1}{2}} \ . \tag{118}$$

b. In cases 7) to 10) excess epoxy is present. Therefore

Table IV

Initial concentrations, concentration ratios, and "actual" conversions attainable for the same compositions as indicated in Table III.

Sample number	$[E]_o$ in moles/cm^3 x 10^4	$[Z]_o$ in moles/cm^3 x 10^4	r	R_E or R_Z	y_m
1)	0	11.43	∞	$R_E = \infty$	1.000
2)	4.975	9.950	2	$R_E = 1.771$	0.970
3)	6.359	9.539	1.5	$R_E = 1.386$	0.964
4)	7.634	9.160	1.2	$R_E = 1.154$	0.947
5)	8.180	8.998	1.1	$R_E = 1.077$	0.931
6)	8.811	8.811	1.0	$R_E = 1 = R_Z$	0.900
7)	9.547	8.592	0.9	$1.026 = R_Z$	0.845
8)	10.417	8.333	0.8	$1.057 = R_Z$	0.769
9)	11.461	8.023	0.7	$1.098 = R_Z$	0.681
10)	14.337	7.168	0.5	$1.229 = R_Z$	0.493

$$\frac{y_\infty C_\infty}{VN_A} = [Z]_o. \tag{119}$$

Substitution of Eqs. (108), (109) and (119) in (113) yields

$$L = [Z]_o^2 \left[\left(\frac{y_m}{y_\infty}\right) - \left(1 + \frac{1}{r}\right)\frac{y_m}{y_\infty} + \frac{1}{r}\right]. \tag{120}$$

Substitution of Eq. (115) in (120) yields y_m/y_∞

$$\frac{2y_m}{y_\infty} = 1 + \frac{1}{r} - \left[\left(1 - \frac{1}{r}\right)^2 + 4 \times 10^{-2}\ R_Z^2\right]^{\frac{1}{2}}. \tag{121}$$

Values of y_m obtained from Eqs. (118) or (121) are listed in Table IV and are plotted in Fig. 9. It appears that the maximum attainable T_g coincides with the stoichiometric composition. Experimentally, this maximum may be shifted somewhat to the right or to the left; this shift must then be attributed to side reactions not yet taken into account in the present calculations.

Of course, the description of the mechanical properties of the networks given in this section has not been exhausted by the present calculations. There are many more variations on this theme possible, e.g. the functionality and the type of crosslinker, or the evaluation of side reactions. The critical conversion y_C is also a point of consideration. If a long aliphatic diepoxide would have been chosen as crosslinking agent, y_C would appear to be greater than unity. On the other hand, when fatty polyamides, or aliphatic polyamines are crosslinked with "high" molecular weight Epon resins, as done frequently in coatings and adhesives, y_C will appear to be negative. When prepolymer compositions are varied, it should be noted that not only ξ_C changes but T_{gP}, q and frequently y are also influenced. A proper interpretation of the variables, e.g. with the use of Eq. (100), will provide the correct series of experiments.

CONCLUSIONS

It is shown that the present model for a polymer network as a copolymer of chain ends, chain segments and branch points provides a workable hypothesis, from which useful conclusions can be derived. To arrive at a set of consistent expressions, the mixing rules for free volumes of copolymers and polymer mixtures had to be reformulated. Where possible, these results have been checked with data from the literature on the effects of chain ends and branch points on T_g. Thus it was shown that a unified free volume theory for polymer solutions, copolymers, or polymer networks exists which is in agreement with most experimental data.

The present treatment does not provide a new theory of free volumes, but merely assumes there exists one. In the present approach some consequences of the free volume concept have been derived, and it is shown how the dependence of glass points of polymers or polymer networks on structural entities can be formulated analytically by Eqs. (39) or (40). Approximate solutions for the individual effects of chain ends or branch points on T_g have also been derived, and are given by Eqs. (55), (79) and (82).

The adjustable parameter in these equations is the "effective volume" of a chain end, ξ_1. The best fit between theory and experimental results is obtained if ξ_1 equals the volume of three vinyl monomer units, or equivalently, a segment volume containing six backbone atoms. Once ξ_1 has been chosen, changes in T_g due to changes in concentration of free ends or branch points are predicted without the use of other adjustable parameters. The equations also guide the chemist in "simple" series of experiments, where e.g. the molecular weight or the composition of the prepolymer is varied, but actually several variables (e.g. number of free ends, spacing between branch points, conversion, etc.) are changed simultaneously.

In addition, it is shown how "actual" conversions of the active groups in non-stoichiometric compositions may be estimated, and that T_g reaches a maximum at the stoichiometric composition. This maximum may serve as an indicator for side reactions, but it also explains the S shape of curves frequently found when data of T_g versus crosslinker concentration are plotted.

At present only one type of network has been calculated in detail. The present approach may form the basis for extensions to many other possible network structures. In the synthesis of polymer networks for specific purposes it is now possible to design a strategy to obtain the best results in the shortest time.

ACKNOWLEDGEMENT

It is a pleasure to acknowledge the stimulating discussions this author had with Drs. S. Newman and H. VanOene on the contents of this paper.

REFERENCES

1. T. G. Fox and S. Loshaek, J. Polymer Sci. 15, 371 (1955).
2. S. Loshaek, J. Polymer Sci. 15, 391 (1955).
3. A. J. Batschinski, Z. Physik. Chem. 84, 644 (1913).
4. A. K. Doolittle and D. B. Doolittle, J. Appl. Phys. 28, 901 (1957),
   A. Kovacs, J. Polymer Sci. 30, 131 (1958),
   J. D. Ferry and R. A. Stratton, Kolloid Z., 171, 107 (1960),
   H. Fujita and A. Kishimoto, J. Chem. Phys. 34, 393 (1961),
   H. Fujita, Fortschr. Hochpolym. Forsch. 3, 1 (1961),
   A. Teramoto, R. Okada and H. Fujita, J. Phys. Chem. 67, 1228 (1963),
   D. J. Plazek, and J. H. Magill, J. Chem. Phys. 45, 3038 (1966),
   A. J. Matheson, J. Chem. Phys. 44, 695 (1966),
   K. H. Hellwege, W. Knappe, F. Paul and V. Semjonow, Rheol. Acta 6, 165 (1967),
   S. P. Chen and J. D. Ferry, Macromolecules 1, 270 (1968).
5. A. K. Doolittle, J. Appl. Phys. 22, 1471 (1951) and 23, 236 (1952).
6. M. H. Cohen and D. Turnbull, J. Chem. Phys. 31, 1164 (1959).
7. J. D. Ferry, "Viscoelastic Properties of Polymers," 2nd Edition (John Wiley and Sons, Inc., N. Y., 1970), Chs. 11, 12 and 17A.
8. A. J. Kovacs, Fortschr. Hochpolym. Forsch. 3, 394 (1963).
9. M. C. Shen and A. Eisenberg, Rubber Chem. and Tech. 43, 95(1970).
10. R. F. Boyer, Rubber Chem. and Tech. 36, 1303 (1963).
11. G. Braun and A. J. Kovacs, in "Physics of Non-Crystalline Solids," J. A. Prins, Ed. (North Holland Publishing Comp., Amsterdam, 1965), pp. 303-319.
12. A. J. Kovacs, Rheologica Acta 5, 262 (1966).
13. K. C. Rusch, J. Macromol. Sci. B2, 179 (1968).
14. K. C. Rusch and R. H. Beck, Jr., J. Macromol. Sci. B3, 365 (1969).

15. J. H. Gibbs and E. A. DiMarzio, J. Chem. Phys. 28, 373 (1958).
16. G. Adam and J. H. Gibbs, J. Chem. Phys. 43, 139 (1965).
17. E. Jenckel and R. Heusch, Kolloid Z. 130, 89 (1953).
18. K. H. Illers, Kolloid Z. 190, 16 (1963).
19. P. J. Flory, J. L. Ellenson and B. E. Eichinger, Macromolecules 1, 279 (1968).
20. M. Gordon and J. S. Taylor, J. Appl. Chem. 2, 493 (1952).
21. F. N. Kelley and F. Bueche, J. Polymer Sci. 50, 549 (1961).
22. L. J. Garfield and S. E. Petrie, J. Phys. Chem. 68, 1750 (1964).
23. F. A. Bovey and G. V. D. Tiers, J. Polymer Sci. 44, 173 (1960), F. E. Karasz and W. J. MacKnight, Macromolecules 1, 537 (1968).
24. S. S. Labana, S. Newman and A. J. Chompff, Paper number 22 of these "Proceedings."
25. R. Simha and R. F. Boyer, J. Chem. Phys. 37, 1003 (1962).
26. A. J. Chompff and J. A. Duiser, J. Chem. Phys. 45, 1505 (1966).
27. F. Bueche, "Physical Properties of Polymers," (Interscience Publ., New York, 1962) pp. 113-116.
28. K. Ninomiya, J. D. Ferry and Y. Ōyanagi, J. Phys. Chem. 67, 2297 (1963).
29. A. Eisenberg and T. Sasada, in "Physics of Non-Crystalline Solids," J. A. Prins, Ed. (North Holland Publishing Comp., Amsterdam, 1965) pp. 99-116.
30. K. Ueberreiter and G. Kanig, J. Chem. Phys. 18, 399 (1950).
31. T. G. Fox and P. J. Flory, J. Appl. Phys. 21, 581 (1950).
32. K. Ueberreiter and G. Kanig, J. Colloid Sci. 7, 569 (1952).
33. M. C. Shen and A. V. Tobolsky, Adv. in Chem. Series, 48, 27 (1965).
34. J. F. Pierson and A. J. Kovacs, (in preparation).
35. W. Kuhn and H. Kuhn, Helv. Chim. Acta 26, 1394 (1943).
36. K. C. Rusch, J. Macromol. Sci. B2, 421 (1968).
37. W. A. Lee and G. J. Knight, in "Polymer Handbook," J. Brandrup and E. H. Immergut, Eds., (Interscience Publishers, New York, 1966) Ch. III, pp. 61-91.

38. T. G. Fox and P. J. Flory, J. Polymer Sci. 14, 315 (1954).
39. J. F. Pierson, Thesis, University of Strasbourg, 1968.
40. T. Masuda, K. Kitagawa and S. Onogi, Polymer J. 1, 418 (1970).
41. A. J. Chompff, J. Chem. Phys. 53, 1566 (1970).
42. L. E. Nielsen, J. Macromol. Sci.-Revs. Macromol. Chem. C3, 69 (1969).
43. T. F. Schatzki, Polym. Prepr., Am. Chem. Soc., Div. Polym. Chem., 6, 646 (1965).
44. P. Meares, Trans. Faraday Soc. 53, 31 (1957).
45. J. A. Faucher, J. V. Koleske, E. R. Santee Jr., J. J. Stratta, and C. W. Wilson, J. Appl. Phys. 37, 3962 (1966).
46. J. A. Faucher, Polymer Letters, 3, 143 (1965).
47. Shell Chemical Company.
48. H. Lee and K. Neville, "Handbook of Epoxy Resins," (McGraw-Hill, Inc. New York, 1967) pp. 10-5 to 10-10.

DISCUSSION

M. Gordon (University of Essex): You said your copolymer effect on T_g agreed with the Gordon-Taylor equation. That equation was derived in 1952 without using the notion of free volume; it follows from considering ideal mixing of (overall) molar volumes. As the notional way of splitting overall volumes into free and occupied volumes is controversial, I wonder if for your theory it is necessary at all.

A. J. Chompff: It may be possible to derive the preceding equations in a different way. It should be noted, however, that if the method of Gordon and Taylor were used for mixtures the resulting glass points would depend directly on the nonideality of the mixing process. Until it is proven that the mobilities in the mixture depend on the excess volumes, I cannot accept this relationship. Fortunately, however, the free volume theory is a very efficient one and has been proven to work. If one wants to derive simple expressions which relate glass transitions of networks with their chemistry, molecular weights, plasticizer content, etc., one can make life a great deal easier by accepting that free volume exists. That, precisely, is the purpose and scope of the present paper.

The free volume concept is not as controversial as some authors let us believe. It is just another way of describing the phenomena; a semiempirical one. Unfortunately, too much attention is focused on the occupied volume v_o. It should be kept in mind, however, that experiments always yield f(T). Which role the "actual" v_o plays in this game is of secondary importance and cannot (yet) be elucidated, since f(T) is still of a semiempirical nature.

A. Peterlin (Research Triangle Institute, N. C.): Do you think that your method can be applied to the pseudoamorphous surface layers of polymer crystals? They contain regular or irregular loops and free ends, i.e. molecular chains with one or two ends fixed in the boundary between the crystal core and the amorphous layer. In crystals of a bulk sample, either spherulitic or fibrous, one has also tie molecules with the ends fixed in two different crystals. The glass transition of such layers is a very important parameter influencing the mechanical properties of the crystalline solid.

A. J. Chompff: Yes, with some reservations, however. In analogy to Eq. (76) one might write

$$\varphi_3 = 4\xi_1 \nu_e,$$

where ν_e is the number of tie chains and loop chains. It should be noted, however, that the tie chains and loop chains are not actually fixed at both ends because the crystalline lamellae are able to recrystallize e.g. during aging, or under stress. A second complication is introduced by the free chain ends, which are certainly not evenly distributed over the amorphous and crystalline phases; their concentration in the amorphous phase must be higher than the average, but the exact distribution is not known. In addition a third term should be added to the equation of state for f: an additional amount of free volume might be introduced because the tie chains are under tension. A fourth complication is that the interlamellar amorphous phase is actually not representative of the amorphous bulk state. Moreover, it is uncertain whether the present model is valid for these short tie chains; the values of φ_1 and φ_3 are probably quite large.

THE INTERACTION BETWEEN POLYMERIC STRUCTURE, DEFORMATION AND FRACTURE†

M. L. Williams* and F. N. Kelley**

*University of Utah, Salt Lake City, Utah

**Air Force Rocket Propulsion Lab, Edwards, California

SUMMARY

The authors have recently been attempting to establish a connection between the molecular composition and the mechanical characterization of predominantly linear viscoelastic polymers in order to provide a means of directly assessing the impact of chemical structure upon engineering design. Preliminary examples using continuum mechanics and principles of three-dimensional stress analysis showed, for example, the interaction between chain stiffness and deformation and fracture. This morphological approach, utilizing an Interaction Matrix, will be reviewed in order to stimulate discussion. In addition, some possible connections between molecular structure and the specific characteristic fracture energy will be reported.

INTRODUCTION

Over the past three-year period, the authors have been interested in studying the relations which might be developed between the chemical structure of a polymeric material and its engineering manifestations such as mechanical deformation, fracture, flammability, and combustion. Utilizing the morphological concepts advanced by Zwicky [1,2], the authors introduced an *Interaction Matrix* which, in essence, exhibits molecular parameters as row

†Major portions of this research were supported under a Project THEMIS grant, *The Chemistry and Mechanics of Combustion with Applications to Rocket Engine Systems*, administered by the U. S. Air Force Office of Scientific Research.

elements and engineering parameters as column elements in a matrix. By so doing, a scheme is developed whereby *a priori* judgments of major interactions are reduced in favor of a logical development which minimizes sins of omission.

The first investigations were concerned with the association between molecular descriptors and mechanical deformation. As it proves to be a convenient way to describe the Interaction Matrix approach, the essential details of these previous papers will be repeated, and then supplemented by a discussion related to extensions of this technique to fracture.

MECHANICAL DEFORMATION

As a material-dependent expression relating mechanical properties, molecular structure and engineering design, one may consider the tensile relaxation modulus, $E_{rel}(t)$, which characterizes the deformation response of a polymer through a broad time or temperature scale. The relaxation modulus for a given polymeric material may be produced by well-established testing procedures [6] and then be approximated by a five-parameter modified power law [7]. One form of this relationship is

$$E_{rel}(t) = E_e + \frac{E_g - E_e}{[1 + (t_r/\tau_o)]^n} \tag{1}$$

where E_e and E_g are the equilibrium and glassy moduli, respectively; τ_o is the characteristic time at the inflection of the curve on double logarithm paper; n is the slope of the curve in the transition region; and t_r is the temperature-reduced time, $t_r = t/a_T$. Figure 1 illustrates the relaxation curve with its associated parameters.

The relaxation modulus is directly used in viscoelastic time-dependent stress analysis to predict deformations and is therefore a useful relation to illustrate the association between basic molecular characteristics and engineering design parameters. Table I demonstrates the application of the five-parameter relationship in the Interaction Matrix format. After inserting the five modified power-law parameters along the top of the vertical columns, one may then proceed to list molecular and microstructural characteristics along the left side of the horizontal rows. The list of characteristics is fairly representative of measurable features common to polymer science and technology. Individual characteristics may not be mutually independent, and additions or refinements are likely to be used for specific cases. The intersecting columns and rows may then

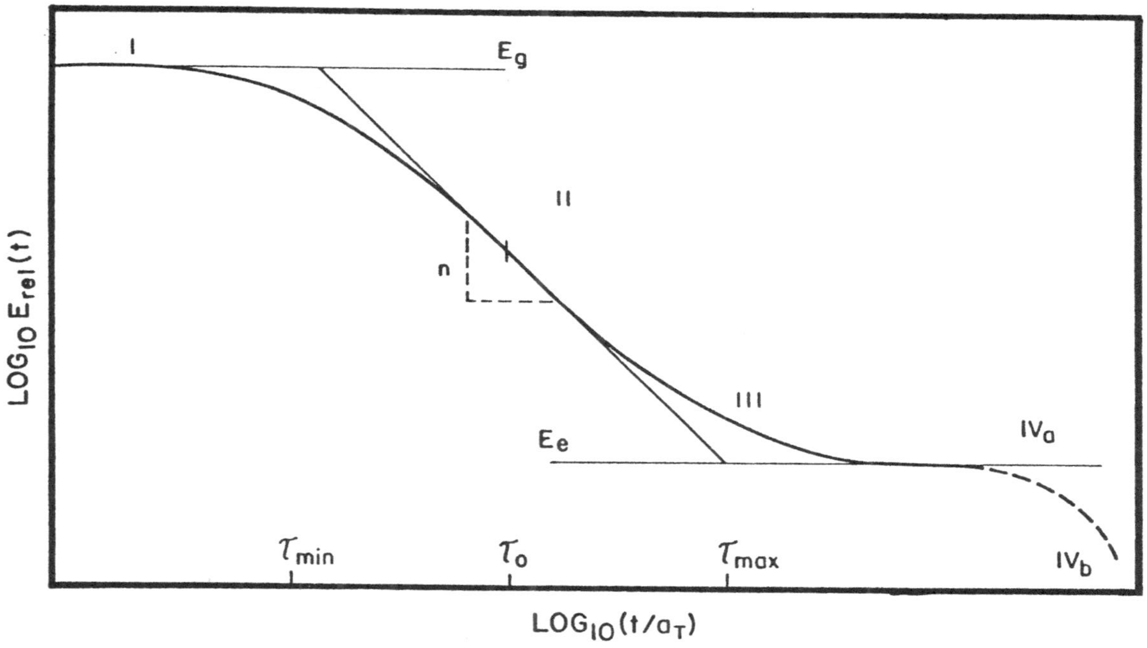

Figure 1. Schematic representation of the relaxation modulus $E_{rel}(t)$ plotted against reduced time, with power law parameters indicated. Various regions of the curve are labeled with Roman numerals.

be examined for known or unknown relationships. Some of the blocks represent well-known interactions. In others, the interaction is weak or strong. The table therefore presents a first attempt to reveal the first order interactions as indicated by the S (strong) symbol.

TABLE I

Molecular Characteristics	Symbol	Modified Power Law Parameters				
		E_g	E_e	τ_o	n	T_g
Cross Link Density	ν_e	N	S(1)	N	M	N(2)
Chain Stiffness	N_s	N	N	U	M(3)	S
Monomeric Friction Coefficient	ζ_0	U	N	S(4)	U	S(5)
Solubility Parameter	δ_p	M(6)	N	U	U	S
Molecular Weight	M	N	S(7)	N(8)	N(9)	S(10)
Heterogeneity Index	M_w/M_n	N	N	M	N	M(11)
Molecular Weight between Entanglements	M_e	N	S(12)	N	N	N
Degree of Crystallinity	Λ	N	S	S	S	N(13)
Volume Fraction of Filler	ϕ	N	S	M(14)	S	M
Volume Fraction of Plasticizer	V_p	N	S	S(15)	N	S(16)

U = Unknown, N = Negligible, M = Moderate, S = Strong

Notes for Table I:

(1) Equation (2)
(2) Except at very high values of ν_e
(3) Figure 2
(4) Reference [9]
(5) Reference [18]
(6) References [3], [10]
(7) Effect of entanglements
(8) At high molecular weights
(9) At high molecular weights
(10) At low molecular weights only
(11) Chain end effects from short chain fractions
(12) At high molecular weights producing a plateau or pseudo-equilibrium modulus
(13) Except at very high Λ
(14) Through WLF; Reference [20]
(15) Through ζ_0
(16) Reference [19]

The value of such an approach may be seen in the communication which results between the design engineer and the formulation chemist. Strong related molecular and engineering parameters become obvious and the relative importance of each may be deduced. In accordance with the aforementioned philosophy, suspected strong interactions suggest areas for concentrated work. Hence, an additional characteristic of the Interaction Matrix is the focus which is placed on unknown interactions requiring further research and quantification. Some relationships may have only qualitative significance and the degree may be indicated as strong, moderate or slight. Certainly the most desirable relationships will be those which are expressible in mathematical form. It will be particularly enlightening to adopt a simple theory, e.g., the Rouse [9] theory, for molecular structure-mechanical property inter-relationships, and then to examine the relative adequacy of these relationships for specific applications.

While the general Interaction Matrix format remains essentially the same, interactions may be continually revised to reflect progress in quantifying the inter-relations. In some cases, a new test or better definition of molecular characteristics will permit the clarification of previously vague relationships. Indeed, the motivation for the development of improved tests or analysis procedures may be supplied by the extended perspective produced by such a chart. It may be true that some relationships are reasonably well-developed for use as *ad hoc* or pseudo-empirical guidelines in engineering design, while others are practically unknown. An equivalent level of development for all of the "compartments" in Table I would be desirable if it could be shown that all interactions are equivalent in importance. One might conclude also that higher returns in overall efficiency would be realized if effort was first applied to important but poorly-established relationships rather than the further refinement of those which are already fairly well-developed.

Molecular and Microstructural Characteristics

The list of items along the left side of Table I are those which strongly influence the mechanical properties of polymeric materials. Most are quite well known and subject to measurement with varying degrees of accuracy and precision. Filler and plasticizer concentration must also be considered in combination with their specific nature, such as filler particle size and surface, or plasticizer type. However, for the purposes of this discussion we may restrict them to non-polar, or non-interacting materials, and concentration will be the measurable quantity.

The subsequent short discussion describes each characteristic and clarifies certain definitions. It should be kept in mind that the choice of relationships has been rather arbitrary for the sake of simplicity of illustration. Several basic texts are available for discussion in depth [6,10,13].

Cross-Link Density. Cross-link density, ν_e, is one of the more significant factors affecting the long-time portion of the response curve, permitting variation of the equilibrium* modulus, E_e, from zero to the glassy limit of approximately 10^{10} dynes/cm^2. The relationship derived from the kinetic-molecular theory of rubberlike elasticity is

$$E_e = 3\nu_e kT \tag{2}$$

where k is the Boltzmann constant and T is the absolute temperature.

Chain Stiffness. The definition of polymer chain stiffness chosen here is derived from the concept of an "equivalent random link" in the statistical theory of rubberlike elasticity [13,14]. Since actual polymer chains do not have freely orienting backbone bonds, in which each atom-to-atom juncture is completely flexible, an accumulation of the limited flexibility of several bonds may produce a larger chain segment with nearly free orientation of the vector joining the ends of the segment. Those polymers in which the freely orienting segment (equivalent random link) consists of fewer backbone bonds are then considered to be more flexible than those which require a larger number of backbone bonds for free orientation [15]. The number of backbone bonds (atoms) per statistical segment, N_s, is the index of chain stiffness and is determined from

$$N_s = \frac{M_c n_b}{N\, M_m} \tag{3}$$

where M_c is the average molecular weight of a network chain, n_b is the number of backbone bonds per monomer unit, N is the number of statistical segments per network chain and M_m is the monomer molecular weight. An approximate value for N may be obtained from the maximum extension ratio of the cross-linked polymer, $\lambda_{max} = N^{\frac{1}{2}}$; based on the inverse Langevin function representation of the stress-strain curve by Treloar [13].

*True equilibrium is perhaps never obtained in these experiments since there is always some perceptible decay in stress at even very long times.

Monomeric Friction Coefficient. The average mobility of chain segments is normally characterized by a friction coefficient, f, which describes the frictional resistance encountered by a segment moving through its surroundings at unit velocity. If the chain length is below the critical entanglement value, the subscript o is attached [11]. A further definition involves the monomeric friction coefficient ζ_o, where $f_o = \zeta_o q$ for q monomer units per segment. In this case chain stiffness (internal viscosity) is not taken into account.

Solubility Parameter. One measure of the energy of interaction between molecules or segments is the cohesive energy density (CED) which is defined as the ratio of the molar vaporization energy to the molar volume E_v/V. The Hildebrand [16] solubility parameter, $\delta_p = (CED)^{1/2}$ may be determined from swelling or solubility measurements in a number of liquids of known CED [17].

Molecular Weight M. The significance of an average molecular weight, or chain length, is most readily apparent if the distinction is made between cross-linked and uncross-linked polymers. In cross-linked samples, the effect of initial molecular weight (before cross-linking) may be seen in the resultant network defects due to dangling chain ends. In uncross-linked polymers, the molecular weight strongly affects the long-time portion of the relaxation curve (Regions IVa and IVb of Figure 1). The length and height of the plateau depends on the tendency to form mechanical entanglements, and this varies from polymer to polymer. At low chain lengths, viscous flow will take place and no plateau may be observed.

Heterogeneity Index. The ratio of weight average to number average molecular weight, M_w/M_n, is one indication of the breadth of the molecular weight distribution.

Entanglement Molecular Weight. If the viscosity of bulk polymer fractions is plotted against molecular weight on a double logarithmic plot, a sharp break appears at a characteristic point for specific polymers. The viscosity increases in proportion to the chain length up to this point and then changes abruptly to a slope of about 3.4. This rapid change in viscosity has been interpreted to be due to the onset of entanglements [11,18] forming a transient network. The plateau region in the relaxation curve for high molecular weight polymers (uncross-linked) will depend upon the characteristic entanglement molecular weight.

Crystallization. The degree of crystallinity expressed in terms of a volume fraction, Λ, may be estimated by density measurements. Its influence on the relaxation curve is quite striking, but quantitative associations are almost completely lacking.

Filler Content. Filler materials may be separated into two types: reinforcing and nonreinforcing. The distinction is usually based on whether or not there is any molecular bonding between the filler and the base material.

Plasticizer. The primary effects of plasticizer may be associated with the dilution of the number of polymer chains per unit volume, and a shift of the transition to shorter times.

A reasonably successful description of the effects of plasticizer concentration has been obtained by assuming that the diluent contributes additional free volume, and reduces entanglement concentration [19].

Interactions

Having introduced some of the important parameters, Table I may now be examined more closely to assess its usefulness to the materials engineer. The three categories of interaction to be supplied for each block may form a guide for further investigation at any stage in the construction of the matrix. Each block which is originally placed in the "unknown" category on a tentative basis may lead to further examination of available published data, or further experimental research. Those interactions which are known qualitatively as "strong" or "moderate" may be upgraded to quantitative associations through systematic study. Finally, where positive interactions are well established, whether they are negligible or strong, a direct link can be made between chemical structure and engineering design.

The fifty blocks in Table I formed by intersection of the engineering parameters and molecular characteristics require much more elaborate treatment than can be provided here. However, consideration of a single example such as that between chain stiffness and slope through the transition region should illustrate an application of the matrix.

Tobolsky [10] observed that the slope of the relaxation curve, n, appears to be related to the relative stiffness of the polymer chain. This would indicate a qualitative interaction between the chain stiffness index N_s and n on the matrix. It became apparent after examining the literature that no quantitative assessment of this interaction has been presented, although the data available from several sources could be combined for such a purpose. Figure 2 contains data on six different unfilled polymers from six sources. One system (polyurethane) contains a variation in basic structure due to systematic changes in catalyst-prepolymer ratio. Accepting for the moment this interaction as being

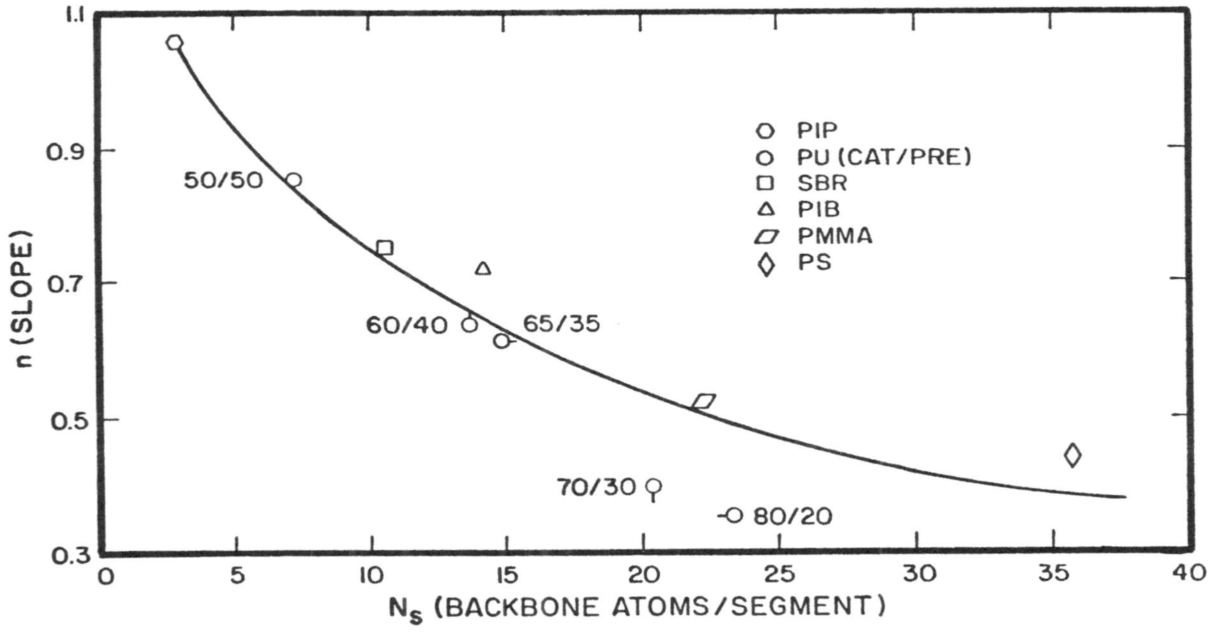

Figure 2. Plot of slope of the relaxation curve, n, against stiffness index, N_s, for several polymers. Circles are polyurethanes of nearly the same cross-link density, but varying catalyst to prepolymer ratios as indicated [23]. Other polymers and associated references are ⬡ - polyisoprene [13,21], □ - SBR [21,22], △ - polyisobutylene [15,21], ▱ - polymethyl methacrylate [15,21] and ◇ - polystyrene [15,24].

generally valid for all polymers, we obtain the following empirical expression for the curve in Figure 2

$$n = 1.50\ [\ell n\ N_s]^{-1} \tag{4}$$

which somewhat resembles the Rouse dependence of slope upon the number of sub-molecules per network chain. See Equation (7), following.

The modified power law may then be expressed in terms of N_s, the chain stiffness index, for purposes of examining the dependence of the design variables upon this particular molecular parameter.

Engineering Applications

Our previous discussions [3,4] calling attention to the use of the Interaction Matrix concept dealt with associations between the chain stiffness in terms of the simple Rouse Theory and (1) the stress at a spherical flaw in a linearly incompressible viscoelastic material loaded in hydrostatic tension, and (2) cyclic fatigue life in essentially the same geometrical configuration. In the first case [3] it was shown that at the flaw radius, a, and at the time, $t = \tau_o$ (the center of the transition region), for example,

$$\frac{\Delta\sigma_\theta\ (a,\tau_o)}{\sigma_\theta\ (a,\tau_o)} = \left[1 + \frac{2/\log_e 10}{2\ \log_{10} P - \log_{10}\sqrt{E_g/E_e}}\right] \frac{\log_{10}\sqrt{E_g/E_e}}{2\ \log_{10} P} \cdot \frac{\Delta P}{P} \tag{5}$$

wherein P is the number of submolecules per network chain as used in the Rouse Theory. Equation (5) therefore permits one to evaluate explicitly the exchange coefficient, or constant of proportionality, between the relative changes in stress, σ_θ, and number of submolecules. This variation is reproduced in Figure 3. Other variations can be calculated in a similar fashion, especially when the remaining interactions from the matrix are quantified.

The second example [4] dealt with a prediction of fatigue life at long times, again based upon the spherical flaw model [25,26]. The strain at failure ε_{cr} is related to the number of cycles to failure, N, by

$$\varepsilon_{cr}^2 = \frac{\gamma_c/\varepsilon}{\pi^2 E_g N} \cdot \frac{1}{n(n+1)} \cdot \frac{1}{(\omega\tau_o)^n} \tag{6}$$

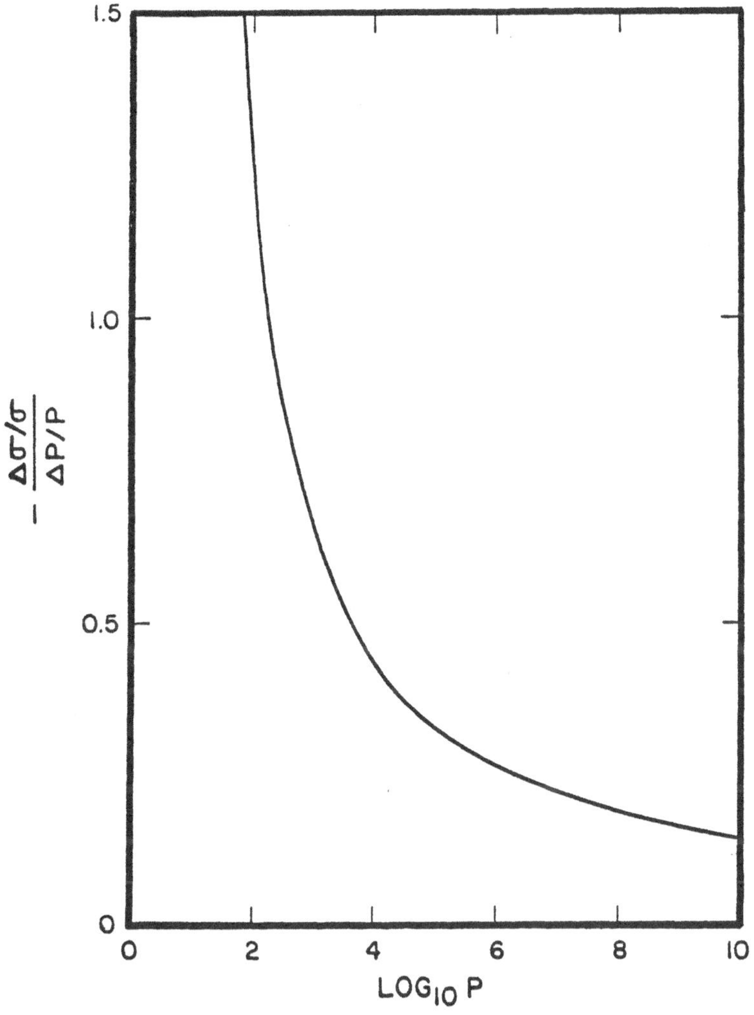

Figure 3. Exchange ratio between stress and segments per molecule for $E_g/E_e = 1000$ and at $t/a_T = \tau_o$

in which γ_c is the cohesive fracture energy.* Using the Rouse Theory for which

$$n = \frac{\log_{10}(E_g/E_e)}{\log_{10}(\tau_{max}/\tau_{min})} = \frac{\log_{10}(100U_o/r_o{}^3)/(3\nu_e kT)}{2\ \log_{10}P} \qquad (7)$$

$$\tau_o = \tau_{min}\tau_{max} = \frac{6\eta_o}{\pi^2\nu kT}\ \frac{1}{P} \qquad (8)$$

in which ω is the applied load frequency, η_o is the steady state viscosity, U_o is the minimum potential energy between isolated atoms at a separation distance, r_o, and ν_e is the number of effective network chains per unit volume.

The exchange coefficient between the critical strain at failure and chain stiffness can be evaluated as

$$\frac{\Delta\varepsilon_{cr}/\varepsilon_{cr}}{\Delta P/P} = \frac{4[\log P + \log(E_g/E_e)]}{\log_e 10[2\ \log P + \log(E_g/E_e)]\ \log\ (E_g/E_e)} +$$

$$\frac{\log(18\eta_o\omega/\pi^2E_e)}{\log P} \qquad \frac{\log E_g/E_e}{4\ \log P} \qquad (9)$$

For the special case of $E_g/E_e = 1000$, and constant flow viscosity, η_o, and rubbery modulus, E_e, equation (9) has also been reproduced for various loading frequencies in Figure 4.

FRACTURE

In the foregoing illustration of the Interaction Matrix the primary emphasis was placed upon the influence of the time-dependent relaxation modulus upon the mechanical behavior, and, in principle, it was indicated how the maximum stress or critical

---

*For this calculation, the cohesive fracture energy γ_c had been assumed constant. Since that time, it appears that γ_c is actually time-dependent as reported by Bennett *et al.* [27] for a cracked sheet of butadiene-acrylonitrile-acrylic acid viscoelastic terpolymer cross-linked with an epoxy curing agent. It has the effect of changing the slope of the fatigue curve continuously with the number of cycles to failure, instead of the constant (-½) value in [6].

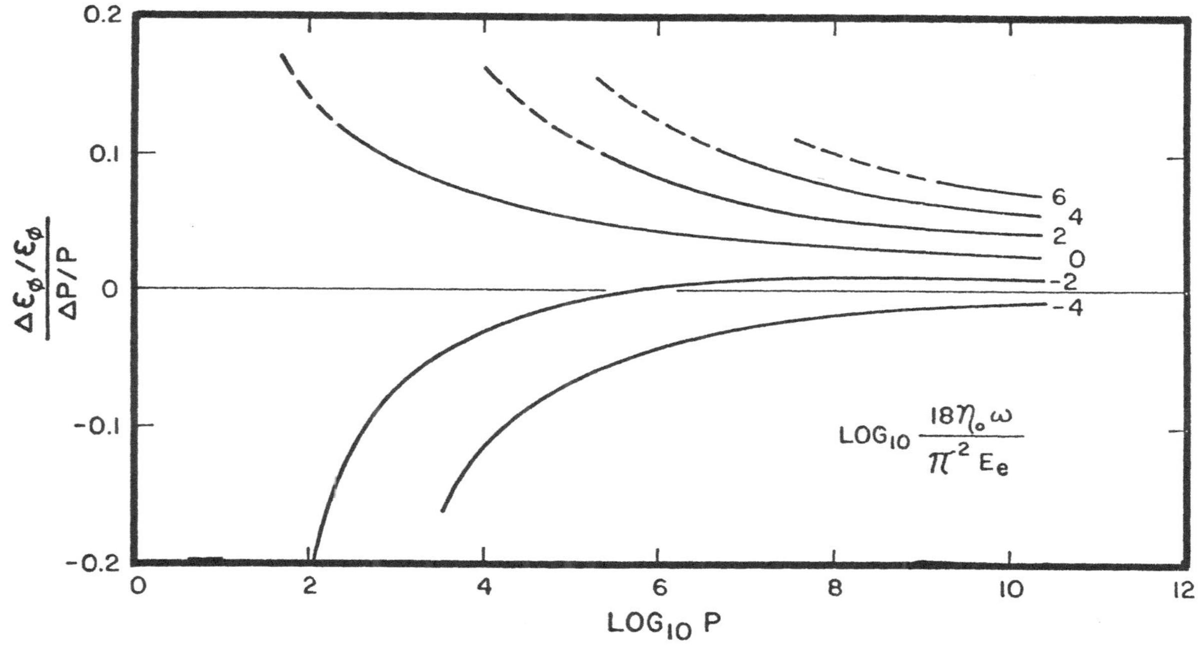

Figure 4. Exchange ratio between fatigue failure strain and Rouse segments for E_e/E_g = 1000. Broken portion indicates the limit of the approximation used, $\omega\tau_o << 1$.

fatigue strain could be associated with a molecular parameter such as (using the Rouse Theory) the number of sub-molecules per network chain or (using the empirical data) chain stiffness. On the other hand, if one accepts an energy criterion of failure, the modulus also influences the critical stress at fracture, σ_{cr}. This behavior is commonly recognized for elastic behavior as the Griffith critical stress criterion, $\sigma_{cr} \sim \sqrt{E\gamma_c/a}$, in which E is the (elastic) modulus and a the initial flaw size. By modeling a flaw as a spherical cavity, Williams [25] showed that the same type of energy consideration can in certain cases be extended to viscoelastic materials yielding, with sufficient accuracy for our present purposes,

$$\sigma_{cr} \sim \sqrt{E_{rel}(t)\gamma_c/a} \tag{10}$$

It is thus apparent that the influence of the molecular structure upon fracture can also be assessed through the associations in the Interaction Matrix, namely

$$\sigma_{cr} \sim E_e \sqrt{\left[\frac{E_g - E_e}{1 + (t_r/\gamma_o)^n}\right] (\gamma_c/a)} \tag{11}$$

Indeed a simple example was already discussed [3].

Recently the experimental data and analysis by Bennett *et al.* [27] have added a new dimension to our discussion by exhibiting a temperature-reduced time dependence of the cohesive fracture energy, $\gamma_c = \gamma_c(t/a_T)$, and consequently a predicted time-dependent critical fracture stress of

$$\sigma_{cr} \sim \sqrt{E_{rel}(t/a_T)\ \gamma_c(t/a_T)/a} \tag{12}$$

Hence, whereas (10) or (11) had already implied a specific influence of the chemical structure upon fracture, the time dependence of the cohesive fracture energy suggests a different sensitivity or expanded dependence of fracture upon molecular parameters.

Before pursuing the implications of this point further, consider first the related case of fracture between two dissimilar materials. In contrast to the former case in which the fracture separated the same material, thus leading to a cohesive type failure, this debonding of different materials may be described as adhesive fracture. From the point of view of continuum mechanics, the problem and energy balance are essentially the same [28,29] except for replacing γ_c by γ_a, leading to a new physico-chemical interpretation of γ_a, the specific energy required to separate the two materials in question--in short, the adhesive fracture energy. Thus, again with sufficient accuracy for our present purpose, we are led to suspect that the time-temperature dependent stress criterion for adhesive fracture would be of the form

$$\sigma_{cr} \sim \sqrt{E_{rel}(t/a_T)\ \gamma_a(t/a_T)/a} \tag{13}$$

Specific Fracture Energy

We turn now to consider the interaction between molecular characteristics and fracture energy, either cohesive (γ_c) or adhesive (γ_a), in terms of the Interaction Matrix.

From the phenomenological standpoint it seems appropriate to inquire first if the fracture energy would be expected to depend upon the same or different molecular properties as the relaxation modulus. Because it has not been commonly appreciated that both the modulus and fracture energy may independently be time dependent, experiments which have measured, for example, time dependent fracture stress have attributed the time dependence to the "tear energy" with a constant modulus [30]. On the other hand, Williams' analysis [25] assumed the entire dependence lay in the modulus, with a constant fracture energy. Thus one must beware of previous intuition which has perhaps been conditioned by a knowledge of only the time-dependent product, i.e., $\sigma_{cr}(t)$, when in fact it can arise from either or both the modulus and fracture energy. This line of reasoning suggests that from the energetic approach the deformation sensitive quantities are reflected in the modulus while the rupture or bond breakage is reflected in the fracture energy, with the overall breaking stress characteristics being proportional to the product of their respective square roots. From the basic molecular structure viewpoint, however, in the quantum mechanics sense, such a complete partitioning of the individual effects seems unlikely. Hence an Interaction Matrix construction as shown in Table II is probably more appropriate.

Interactions occupying only the U, V, and W blocks along the main diagonal represent mutually independent behavior such that there would be no connection between deformation, cohesion, or adhesion as far as molecular parameters were concerned. As this behavior seems unlikely, provision for off-diagonal interaction is provided, although zeroes can always be entered where appropriate. The additional molecular descriptors V_m and W_n are those needed to account specifically for quantities entering only fracture and not deformation. At the present time the only characteristics we choose to place in this category are the chemical bond energies themselves. On the other hand, several sources suggest, for example, that the molecular weight and cross link density have a direct bearing on cohesive fracture energy,* γ_c.

Benbow [31] reports that there is a hundredfold increase in energy in polystyrene as the molecular weight, $\overline{M}_w$, is raised from 60,000 to 260,000. These conclusions have been substantiated by Broutman and Kobayashi [32] who report the following values deduced from splitting a tapered cantilever beam

$\overline{M}_w$	γ_c (ergs/cm^2)
53,211	3.5×10^3
231,000	4.3×10^5
246,148	4.0×10^5

Berry [33] has suggested that the increase in fracture energy is due to the greater degree of involvement of the longer chains (higher molecular weight) in the plastically deformed area surrounding the crack tip. The least amount of energy is dissipated if the chain is short enough to be completely enclosed by the plastic enclave. Intermediate energy is absorbed if the chain is partly inside and partly outside the region, while maximum energy results if the chain passes through the plastically deformed enclave and both chain ends terminate in the elastic region.

Broutman and Kobayashi [32] also examined the effect of cross-linking in polystyrene, accomplished by gamma radiation using a Cobalt 60 source in order not to introduce a second, and possibly complicating, cross-linking chemical species. At molecular weights of approximately 250,000, the fracture energy was reduced by one-half after a dosage of 50 megarads.

The above typical results suggest a S (strong) interaction for molecular weight and cross linking with weaker and indeterminate effects attributed to the remainder of the molecular parameters in the extended Interaction Matrix, except for the inclusion of chemical bond energies as the V_m and W_n.

---

*Note, however, that according to our previous remarks, it may be important to reassess the basic experimental data to ascertain that $\overline{M}_w$ and ν_e influence attributed by the referenced authors to γ_c, could not equally as well have been attributed to the modulus.

TABLE II

Mechanical Molecular Parameters	Deformation Parameters E_e E_g n τ_o T_g	Fracture Parameters γ_c	Fracture Parameters γ_a
U_1 U_2 . . . U_s	U		
V_1 V_2 . . . V_m		V	
W_1 W_2 . . . W_n			W

Ad Hoc Correlation

In the absence of specific data upon which to extend the molecular descriptors, it is possible to better represent the fracture energy than in terms of the simple functions $\gamma_c(t)$ and $\gamma_a(t)$. In this regard, the test data of Bennett *et al.* on cohesive fracture energy shown in Figure 5 is suggestive. (Adhesive fracture energy would probably behave in a similar way.) The functional form of γ_c is seen to be monotonic decreasing with time. Furthermore, the very short time value of γ_c is clearly not unbounded, thus implying a limit.* Reproducing the Bennett data for $\gamma_c(t)$ and $E_{rel}(t)$ on one curve in Figure 5 shows a reasonable similarity in shape, aside from a two-way shift: 3.5 decades to the left in reduced time and 2.0 decades up in magnitude. In any event, it seems that the fracture energy could be approximated by the same type of modified power law used for the relaxation modulus, viz.,

$$\gamma(t/a_T') = \gamma_e + \frac{\gamma_g - \gamma_e}{[1 + t/(a_T'\tau_o)]^{n'}} \tag{14}$$

which could equally well be converted to a "fracture energy" spectrum** if this interpretation was more helpful in deducing a theoretical basis for the behavior. For the specific material used by Bennett, and incorporating the appropriate shifts including $a_T' = 10^{-3.5}a_T$ and noting the slope similarity $n' = n$, one has on an *ad hoc* basis

---

*A personal communication from S. J. Bennett has revealed that some tentative additional short time, low temperature data appears to place a limit value of the order of 100 in-lbs/in^2 on γ_c for his material at t/a_T of 10^{-9}.

**Note the modified power law approximation for the relaxation modulus (1) can be deduced from the relaxation spectrum

$$H(\tau) = \frac{E_g - E_e}{\Gamma(n)} \left(\frac{\tau_o}{\tau}\right)^n \exp(-\tau_o/\tau) \tag{i}$$

In the same way, therefore, then one could write the "fracture energy" spectrum, $H'(\tau)$, as

$$H'(\tau) = \frac{\gamma_g - \gamma_e}{\Gamma(n')} \left(\frac{\tau_o}{\tau}\right)^{n'} \exp(-\tau_o/\tau)$$

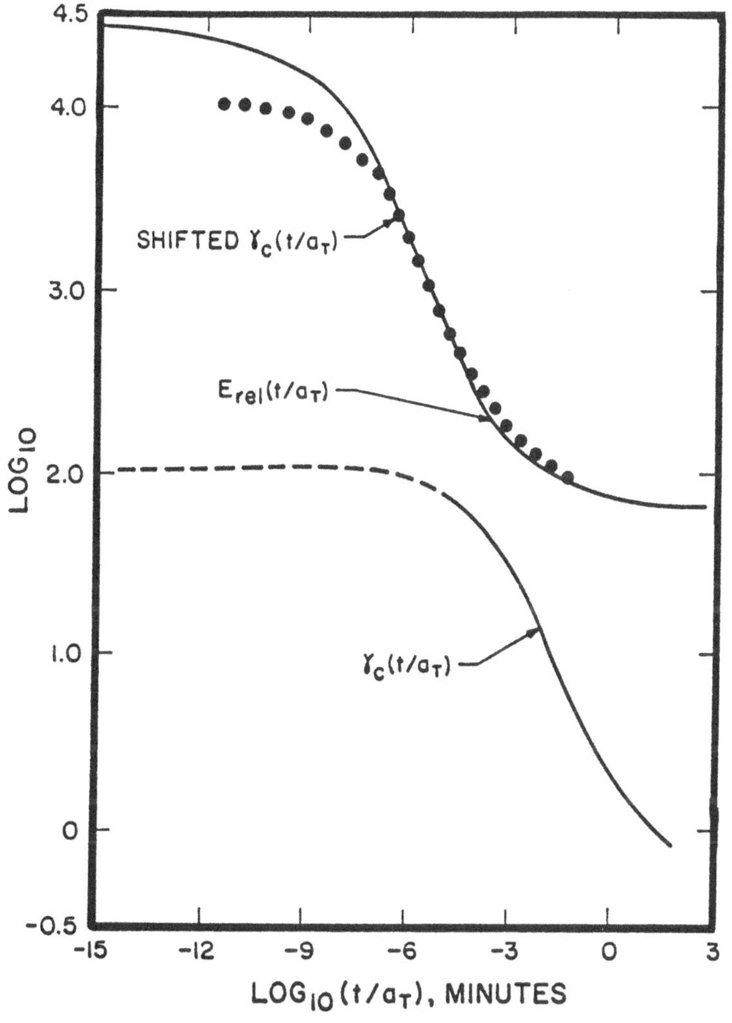

Figure 5. Comparison of relaxation modulus and cohesive fracture energy [27]; ●●●●● – $\gamma_c(t/a_T)$ shifted 3.5 decades left and 2.0 decades up.

$$\gamma_c(t/a_T) = \frac{E_e}{100} + \frac{(E_g/100) - (E_e/100)}{[1 + 10^{3.5}\ t/(a_T'\tau_o)]^n} \tag{15}$$

In a more direct manner, a series of centrally-cracked sheets could be tested in order to determine the parametric dependence of the five mechanical parameters upon the molecular characteristics of the series of materials.

Were the simple case of constant proportionality to apply, i.e.

$$\gamma_c(t) = \mu\ E_{rel}(t) \tag{16}$$

and simultaneous measurement of E and γ_c would only be required at one time. For example in the rubbery or elastic long time region,

$$\gamma_c(\infty) \equiv \gamma_{ce} = \mu E_e = 3\mu\nu_e kT \quad \text{or} \quad \mu = \gamma_{ce}/3\nu_e kT \tag{17}$$

which might be further related to molecular weight or cross-link density through the relationships proposed by Benbow [31] or Brontman and Kobayashi [32], thus eliminating any further need for V_m or W_n quantities in the Interaction Matrix, or our proposed additional dependence of γ upon chemical bond energy. These matters have proven quite interesting and intriguing to us on strictly an *ad hoc* basis--sufficiently so to encourage further study upon a more fundamental basis.

AN INTERACTION MATRIX FOR VISCOELASTIC FRACTURE

Accepting for the moment the analogy suggested in Figure 5 and the above discussion relating the relaxation and fracture energy spectra, we may construct an interaction matrix for fracture similar to that described for the modified power law representation of the relaxation modulus. The purpose of such an exercise, to reiterate our earlier remarks, is to gain perspective on known relationships and to point out possible significant interactions for further study.

Molecular Considerations

Time dependent fracture has been observed for both unfilled and filled polymers [34,35,36]. The remarkable feature of this time dependence has been the good agreement between the temperature dependence of the shifting factor a_T, for both small deformation response and large deformation fracture [27]. Since the molecular interpretation of a_T is based on small-scale motion of molecular segments, which is dominated by the temperature difference above the glass temperature (T-Tg) [8], these same local segmental motions apparently govern the time dependence of fracture. Most current molecular theories presume an ordering of the polymer chains in front of the advancing crack in which several chain segments are oriented perpendicular to the direction of propagation. This process dissipates considerable energy since it is opposed by the ordinary viscous forces which restrict chain motion, i.e., segmental friction--including chain entanglement. Of course, cross-linking would be expected to increase the difficulty of orientation and contributes additional relaxation mechanisms in the long time portion of the spectra.

This view would support the connection between the relaxation and fracture curves at longer times or higher temperatures. Since the maximum dissipation observed in cyclic (small strain) tests corresponds with the transition region of the relaxation curve, one would expect increasing fracture energy when moving to this experimental time scale from the longer times [40]. At very short times the dissipation is greatly reduced since the material exhibits glassy behavior, but the critical stress is so high that the onset of crack growth would be governed by other factors such as Van der Waal's bonding forces [44], or more likely, the greater number of primary chain backbone bonds holding the load in a given cross-section [11, p. 263].

A limiting critical fracture stress or fracture energy in the glassy region of the fracture spectrum seems quite likely, although the data are far too limited for generalization at present. It does appear, however, that the long time, rubbery region has been given sufficient attention for early speculation on the nature of chemical interactions. Lake and Thomas [37] have advanced certain molecular explanations for a limiting tearing energy, T_o, which is largely independent of viscoelastic processes and varies slightly with chemical structure. In this case T_o might be considered roughly equivalent to γ_e, and the reported values for the two quantities for similar polymers agree rather well (Lake and Thomas, 10^5 ergs/cm^2 for butadiene-acrylonitrile copolymer and Bennett *et al.* [27], 1.3 x 10^5 ergs/cm^2 for butadiene-acrylonitrile-acrylic acid terpolymer.) Lake and Thomas derive the following relationship for T_o:

$$T_o = (3/8)^{\frac{1}{2}} g\ell U N \bar{n}^{\,3/2} \tag{18}$$

where g is a stiffness factor related to our notation N_s by $N_s = g^2 n_b$, ℓ is the length of a monomer unit and is proportional to n_b, U is the energy required to rupture a backbone bond, N is the number of network chains per unit volume, and $\overline{n}$ is the average number of monomer units per chain. Although Lake and Thomas indicate that there is experimental evidence that the type of cross-link has some effect on the variation of T_o with cross-linking, Equation (18) may provide an initial basis for some important interactions between molecular structure and γ_e in Table III.

TABLE III

Molecular/Microstructural Characteristics	Symbol	Modified Power Law Parameters				
		γ_g	γ_e	τ_o	n	T_g
Cross Link Density	ν_e	M(1)	S(2)	N	M	M(3)
Chain Stiffness	N_s	N	M(4)	U	M(5)	S
Monomeric Friction Coefficient	ζ_o	U	U	S(6)	U	S(7)
Solubility Parameter	δ_p	S	M	U	U	S
Molecular Weight	M	N	S(8)	N(9)	·N(10)	S(11)
Heterogeneity Index	M_w/M_n	N	N	M	N	M(12)
Molecular Weight between Entanglements	M_e	N	S(13)	N	N	N
Degree of Crystallinity	Λ	U	S	S	S	N
Volume Fraction of Filler	ϕ	S	S	M(14)	S	M
Volume Fraction of Plasticizer	V_p	S	S	S(15)	N	S(16)

U = Unknown, N = Negligible, M = Moderate, S = Strong

Notes for Table III:

(1) Reference [32]
(2) Reference [37]
(3) References [41], [42]
(4) Reference [37]
(5) Figure 2
(6) Reference [9]
(7) Reference [18]
(8) Effect of Entanglements
(9) At high molecular weights
(10) At high molecular weights

Notes (continued):

(11) Reference [43]

(12) Low M fractions may plasticize [11]

(13) Plateau effect similar to cross-linking

(14) Reference [20] related to T_g

(15) Plasticizer effect on T_g and ζ_o

(16) Reference [19]

One of the most interesting features of Figure 5 is the displacement of the fracture transition region along the time scale by three and one-half decades as compared to the relaxation curve. We have no explanation for this as yet and can only speculate about a possible generality of shifts to longer times for fracture of various polymers. The position of τ_o should be determined to a large degree by the glass transition temperature, T_g. Since the fracture process implies very large strains at the tip of the crack at σ_c, therein could be the principal reason for differing time scales between the fracture energy transition and the small strain relaxation transition. Some evidence is available on the effect of strain induced anisotropy on T_g [38, 45] as well as the basic nonlinearity of the response due to finite strains [39]. However, any further examination of this behavior must await the accumulation of experimental fracture data on other polymers. (Smith and Dickie [46] have recently examined a_T-strain effects.)

Table III has been filled in by assuming a correspondence between the relaxation and fracture behavior of polymers. The interactions with various molecular and microstructural features is assumed to be similar with respect to the highly time-dependent phenomena, particularly those associated with the transition regions. Some information on the fracture of glassy polymers clearly indicates a dependence of γ_g on cross-linking [32] and molecular weight [31]. Certainly, many of the interactions assumed by analogy in Table III must be checked out experimentally, and quantitative associations should be developed. In any case, the arrangement of parameters in an array such as that provided by the Interaction Matrix leads to a more direct assessment of the significance of molecular variations with respect to the deformation and fracture requirements of a polymeric material in its ultimate engineering application.

REFERENCES

1. F. Zwicky, *Morphological Astronomy*, Springer-Verlag Berlin, Gottingen, Heidelberg, 1957.

2. F. Zwicky and A. G. Wilson, *New Methods of Thought and Procedure*, Springer-Verlag New York, Inc., 1967.

3. M. L. Williams and F. N. Kelley, "The Relation Between Engineering Stress Analysis and Molecular Parameters in Polymeric Materials," *Proc. 5th Internat'l. Cong. Soc. Rheology*, October 1968, University of Tokyo Press (1970), 185-202.

4. F. N. Kelley and M. L. Williams, *Rubber Chem. & Tech. 42*, 1175 (1969).

5. M. L. Williams and F. N. Kelley, CPIA Publication No. 193, Vol. 1, Johns Hopkins University, Applied Physics Laboratory, March 1970, 89-105.

6. See, for example, J. D. Ferry, *Viscoelastic Properties of Polymers*, John Wiley & Sons, Inc., New York, 1961.

7. M. L. Williams, *AIAA Journal 2*, 5, 785 (1964).

8. M. L. Williams, R. F. Landel and J. D. Ferry, *Journal Am. Chem. Soc. 77*, 3701 (1955).

9. P. E. Rouse, *J. Chem. Phys. 21*, 1272 (1953).

10. A. V. Tobolsky, *Properties and Structure of Polymers*, John Wiley & Sons, Inc., New York (1960).

11. F. Bueche, *Physical Properties of Polymers*, John Wiley & Sons, Inc., New York (1962).

12. P. J. Flory, *Principles of Polymer Chemistry*, Cornell University Press, Ithaca, New York, 1953.

13. L. R. G. Treloar, *The Physics of Rubber Elasticity*, 2nd Ed., Clarendon, Oxford, 1958.

14. W. Kuhn, *Kolloid Z. 76*, 258 (1936); *87*, 3 (1939).

15. F. N. Kelley, PhD. Dissertation, University of Akron (1961).

16. J. H. Hildebrand, *J. Am. Chem. Soc. 51*, 66 (1929).

17. G. Gee, *Trans. IRI 18*, 266 (1943); *Advances in Colloid Sci. II*, Interscience Publishers, New York (1946).

18. F. Bueche, *J. Chem. Phys. 20*, 1959 (1952); *25*, 599 (1956).

19. F. N. Kelley and F. Bueche, *J. Poly. Sci. 50*, 549 (1961).

20. R. F. Landel, *Trans. Soc. Rheology 2*, 53 (1958).

21. A. V. Tobolsky and E. Catsiff, *J. Poly. Sci. 19*, 111 (1956).

22. R. F. Landel and R. F. Fedors, JPL Space Programs Summary 37-36, IV, Jet Propulsion Laboratory, 137 (1965).

23. R. F. Landel, California Institute of Technology Report, CHECIT PL 68-1, June 1968.

24. K. Ninomiya and H. Fujita, *J. Coll. Sci. 12*, 204 (1957); *J. Poly. Sci. 24*, 233 (1957); *J. Phys. Chem. 61*, 814 (1957).

25. M. L. Williams, *Intl. J. Frac. Mech. 1*, 292 (1965).

26. M. L. Williams, *J. Appl. Phys. 38*, 4476 (1967).

27. S. J. Bennett, G. P. Anderson and M. L. Williams, *J. Appl. Poly. Sci. 14*, 735 (1970).

28. M. L. Williams, *J. Appl. Poly. Sci. 13*, 29 (1969).

29. M. L. Williams, *J. Appl. Poly. Sci. 14*, 1121 (1970).

30. H. W. Greensmith and A. G. Thomas, *J. Poly. Sci. 18*, 189 (1955).

31. J. J. Benbow, *Proc. Phys. Soc. (London) 78*, 970 (1961).

32. L. J. Broutman and T. Kobayashi, *ACS Polymer Preprints 10*, September 1969.

33. J. P. Berry, *J. Poly. Sci. 2*, 4069 (1964).

34. For review of the subject see B. Rosen (Ed.), *Fracture Processes in Polymeric Solids*, John Wiley & Sons, Inc. (1964).

35. F. Bueche and J. C. Halpin, *J. Appl. Phys. 35*, 36 (1964).

36. E. H. Andrews, *Fracture in Polymers*, American Elsevier, New York (1968).

37. G. J. Lake and A. G. Thomas, *Proc. Roy. Soc. A, 300*, 1460 (1967).

38. K. E. Polmanteer, J. A. Thorne, and J. D. Helmer, *Rubber Chem. Tech. 39*, 1403 (1966).

39. G. E. Warnaka and H. T. Miller, *Rubber Chem. Tech. 37*, 1421 (1966).

40. L. Mullins, *Trans. Inst. Rubber Ind. 35*, 213 (1959).

41. E. A. DiMarzio, *J. Res. NBS, 68A*, 611 (1964).

42. L. A. Nielsen, *Cross-Linking Effect on Physical Properties of Polymers*, Washington University/ONR/ARPA Report HPC 68-57 (1968).

43. T. G. Fox and P. J. Flory, *J. Appl. Phys. 21*, 581 (1950).

44. J. P. Berry, *J. Poly. Sci. 50*, 107 (1961).

45. G. Gee, P. N. Hartley, J. B. M. Herbert, and H. A. Lanceley, *Polymer 1*, 365 (1960).

46. T. L. Smith and R. A. Dickie, *J. Polymer Sci.*, A2 7, 635 (1969).

PROPERTIES OF A HIGHLY CROSSLINKED ELASTOMER

R. F. Landel

Jet Propulsion Laboratory, California Institute of Technology, Pasadena, California 91109

SUMMARY

The properties of a highly crosslinked polyurethane elastomer, based on ricinoleic acid (castor oil) as the backbone, have been measured. The rubber, termed Galcit I, has been proposed as a standardized, highly birefringent, nearly elastic rubber for use in testing new apparatus on a material with known properties, and in checking or calibrating existing instruments. Various characterizations have been carried out and the results will be presented briefly. These include stress relaxation and dynamic shear modulus, uniaxial and uniform biaxial tests to failure, some additional biaxial tests in combined shear and tension, and swelling in several liquids. The tensile modulus at 25°C is high, about 37 bar (535 psi). $M_c \approx 2300$g/mole as determined from swelling and ultimate properties. This corresponds to 112 carbon atoms between crosslinks, which would be equivalent to $M_c \approx 1500$ in a polybutadiene or 2000 in natural rubber. This is a high enough crosslink density to virtually eliminate the effects of chain entanglements on the time dependence of the modulus and to render the elastomer almost brittle. The stress/strain response is neo-Hookean, i.e. at 25°C and above the material appears to obey the classic, Gaussian-chain theory of rubberlike elasticity for strains up to rupture, though these are only of the order of 60%. At very small strains in biaxial tests, however, there appears to be a departure from the simple kinetic theory response.

INTRODUCTION

An obvious prerequisite to the understanding of the behavior of highly filled elastomers is an understanding of the behavior of the elastomer itself. However, any reasonably complete study encompasses too wide a variety of topics and techniques to permit a successful resolution by any individual research group, hence a cooperative effort is indicated. In the polymer field, such cooperative efforts have been stimulated in the past merely by the availability of a suitable polymer for study.

This paper discusses an attempt to prepare and characterize a standardized crosslinked rubber. By careful control of preparations and cure conditions, it should be possible to obtain specimens whose properties are reproducible from batch to batch. Distribution of such materials to those doing research on the viscoelastic behavior of cross-linked polymeric systems will stimulate a rapid advance in our understanding of such systems. Moreover, even in cases where research studies are not intended, they can serve as useful materials for apparatus calibration or for round robin tests.

The criteria adopted for such a standarized elastomer were that:

1) the material should have as high a stress optical (or strain optical) coefficient as possible and be readily prepared in optically clear specimens. This will permit a study of the origin of the birefringence or the use of the material in experimental stress analysis.

2) the crosslink density should be easily variable over wide ranges because this is one of the more important properties controlling rubberlike behavior.

3) the glass transition temperature should be readily varied by changing the backbone structure rather than by plasticization. This will permit a ready change in the time dependence of the mechanical, electrical, and optical properties.

4) the chemical structure of the backbone and the chemistry of the crosslinking reaction should be well understood so that the physical chemist can assess the effects of changes in the chemical structure of the backbone or the topology of the network.

5) the material should be oxidatively and photolytically stable.

Of these, No. 1 is the most restrictive. Not only must the elastomers be transparent and preferably colorless, the molds must have a mirror finish to prevent distortion of the light beam. In addition, the experimental stress analyst is usually examining a complex shape and his photoelastic models must be cast because rubbers cannot be readily machined to shape.

Prior and concomitant work on this problem at the Jet Propulsion Laboratory[1] indicated that an elastomer based on the Thiokol Chemical Company's Solithane family of polyurethanes most nearly fulfilled these goals. Hence, one particular formulation of Solithane 113 was adopted as an interim standard. This is called GALCIT I. Its major defects are that the chemical structure is not sufficiently well defined, its properties change slowly with time, and its ultimate elongation in a uniaxial tensile test at 25°C is only about 70%. This means that it is an inadequate vehicle for large deformation studies.

Galcit I is a polyurethane obtained by reacting equal volumes of an isocyanate terminated Prepolymer (Thiokol Solithane 113), with a polyhydroxy "Catalyst" (Thiokol Catalyst 300). The latter was thought to be a purified castor oil. Castor oil, which consists mainly of the glyceryl ester of ricinoleic acid, is a triol, the fatty acid chains bearing hydroxyl groups on the twelfth carbon atom. Solithane 113 was originally thought to consist essentially of castor oil capped with tolylene diisocyanate (TDI), i.e. a triisocyanate. Figure 1 shows the presumed structure of the prepolymer and of the network chain which would be produced in the elastomer.

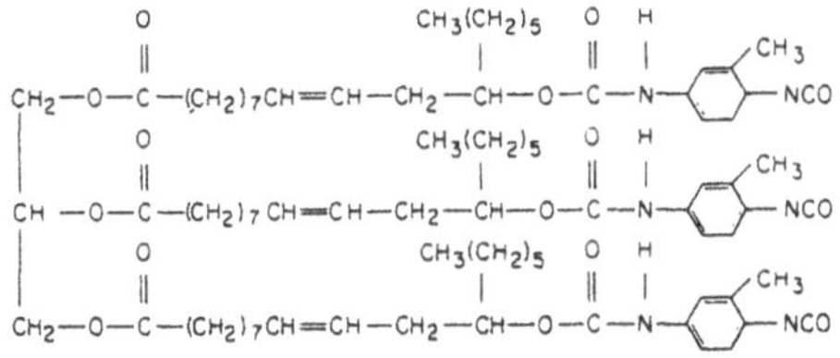

Structure of Prepolymer

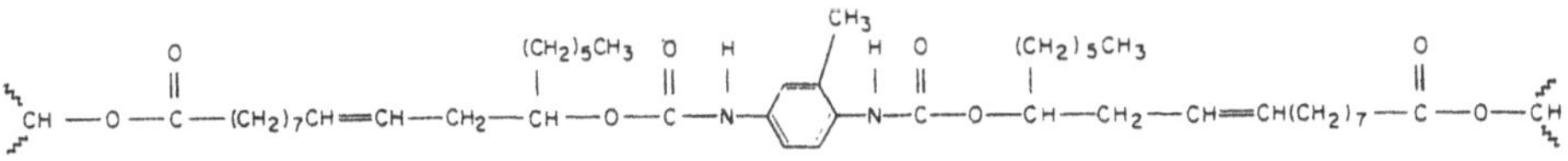

Network Chain

Figure 1. Structure of TDI-capped glyceryl triricinoleate and the Network Chain which would result from its reaction with the ester itself.

The latter is highly crosslinked and as such it is appropriate that a summary of its properties be included in this Symposium. The following sections describe its preparation: the birefringence behavior; the uniaxial response arranged in sequence; small strain, large strain and ultimate properties; swelling behavior; and some aspects of the multiaxial response.

COMPONENTS, PREPARATION AND QUALITY CONTROL

Although gel permeation chromatagraphy was available when this study was initiated, the analysis was performed only recently, when the results of the tensile failure and swelling experiments revealed that the structure of the components could not be that indicated by the manufacturer. Thiokol Solithane 113 prepolymer, Thiokol 300 Catalyst and degassed Baker DB castor oil were all analyzed by both vapor phase osmometry and gel permeation chromatography, Table I.

Table I. Molecular Weight of Solithane Components and Castor Oil

Material	VPO	GPC[b]
Baker DB Castor Oil	1025	1040
Solithane 300 Catalyst	1045	1040
Solithane 113 Prepolymer[a]	2350	<460 1520 2690

a) After reaction with ethanol in tetrahydrofuran and degassing.
b) 0.5 w/v % in THF. Calibrated with poly(propylene oxide).

The results confirm the suspected origin of the Catalyst component, but show that the 113 Prepolymer is not simply TDI-capped castor oil. The three peaks observed in GPC have been tentatively ascribed to TDI dimer, to TDI-capped castor oil and to a dimer of the latter. Consideration of the molecular weights of the components involved indicates that the castor oil used to make the prepolymer has very probably been pre-reacted to give a diol rather than a triol. If so, then all three components are diisocyanates rather than triisocyanates. This is confirmed by the fact that diols will not readily gel the prepolymer (2).

The two high molecular weight peaks in the prepolymer GPC curve could not be readily resolved, but they are present in roughly equal quantities. The third component accounts for 10% of the total area under all peaks but since it may well have a different refractive index than the final two components, it does not necessarily constitute 10% of the prepolymer.

To prepare the elastomer, equal volumes of Thiokol Solithane 113 Prepolymer and Solithane 300 Catalyst were mixed in a special facility which has been previously described (3). Both components are stored under a nitrogen blanket and heated to 60 °C prior to delivery to measuring burets held at the same temperature. Equal volumes of the two components are delivered to a mixing chamber held at 60 °C, mixed for 3 to 5 minutes and delivered to a closed mold which is preheated to about 140°C. Samples are cured at 140°C for 2 hours, followed by a room temperature conditioning for at least one week in a dry box containing silica gel.

The equivoluminal composition was selected to give a near stochiometric composition, according to the manufacturer's stated equivalent weights and densities (397g and 1.075 g/cm^3 for the Prepolymer, 345 and 0.96g/cm^3 for the Catalyst). Earlier work (1) had shown a steady variation in the properties with composition. This was confirmed for the tensile strength, the strain optical coefficient (measured as the number of fringes per 100% strain) and the dynamic shear modulus at ~1 Hz(3) The first two quantities increased linearly with the percent prepolymer as the latter was increased from 42 to 58 volume %. G" also increased linearly for compositions containing more than 45 volume %, while G" was nearly independent of the composition up to 52 volume % and then increased steadily. Since G' was sensitive to composition yet could be determined in a small-strain, nondestructive test, it was selected as a quality control parameter for sheet specimens. With proper care, the reproducibility in modulus is ± 5% for specimens prepared at about the same time.

PROPERTIES

BIREFRINGENCE

The high birefringence of the Solithane family is illustrated in Figure 2 (1). The strain optical constant as defined here is $E\Delta n/2t(\epsilon_1-\epsilon_2)(1+\mu)$ where E is the modulus, Δn is the difference in refractive indices, t is the thickness, ϵ_1 and ϵ_2 are the principle strains, and μ is Poisson's ratio. This is the classical definition for infinitestimal strain; in practice, the strains were limited to less than 10%. The modulus was varied by varying the composition. (The Hysol family is based on Hysol Corporation's Hysol resins, which are polyester polyurethanes. The JPL-H family is based on poly(propylene oxide), trimethylol propane, TDI and, in some cases, a diamine.) According to both this classical definition and the simple Kuhn-Grün molecular theory of birefringence of rubberlike material (4,5) K should be directly proportional to the modulus, while the latter shows that the slope should be a measure of the optical anisotropy of a chain segment. (See Chapter 15). The linearity of the response shows that the birefrigence can be described in conventional terms, in spite of the fact that some of the systems are rather highly crosslinked. (As a practical matter, these results demonstrated to the experimental stress analyst that it was easy for the polymer chemist to modify the strain optical coefficient of a resin if the analyst

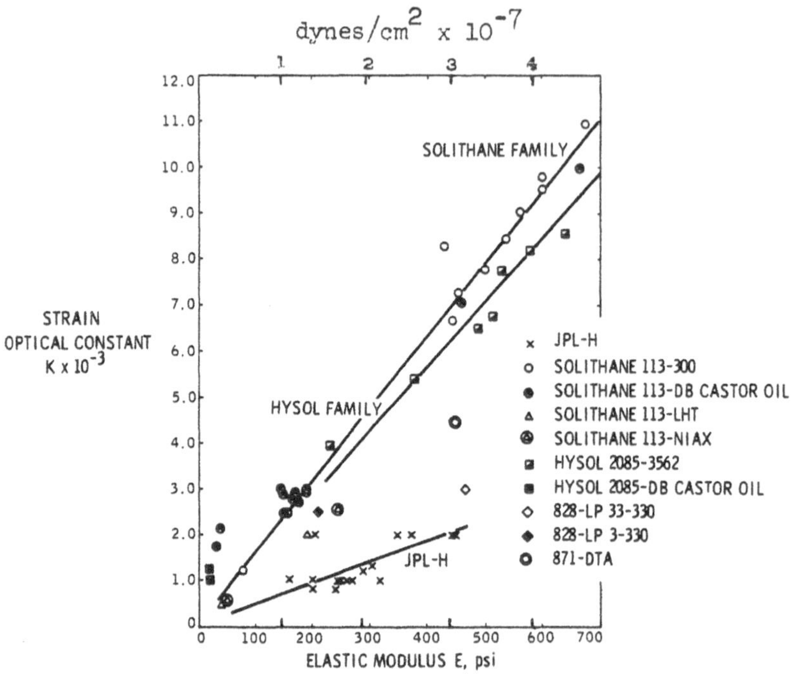

Fig. 2 Variation in strain optical constant with modulus for different formulations of polyurethane materials (after San Miguel and Duran, Ref.(1)

could tolerate a concommitant change in its modulus). The slope of this plot for the Solithane 113 family, and hence the optical anisotropy per segment, is roughly four times that of natural rubber and the highest of any readily available amorphous elastomer we have investigated.

Following the early work at JPL (1), where the emphasis was on low modulus as well as high birefrincence, Arenz, Ferguson and Williams (5) measured the stress-strain and stress-optical properties of a Solithane material with a nominal 50/50 composition, 53% Catalyst and 47% Prepolymer. Both properties were measured on the same specimen and specimens were tested at various rates and temperatures. A master stress-strain plot could be constructed in the usual fashion (7), where it is assumed that the stress-strain curve has the form $\lambda\sigma=E\epsilon$. The results are shown in Fig. 3 and it can be seen that superposition was obeyed very well indeed. They also derived an expression showing that birefringence data can be reduced in the same fashion; Figure 4 gives the results. Here N is the fringe order number. Moreover, just as the slope of Figure 3 is related to the stress relaxation modulus, the slope of Figure 4 is related to a time-dependent strain optical coefficient, defined here as $C=N/(\epsilon_1-\epsilon_2)$, where ϵ_1 and ϵ_2 are the principle

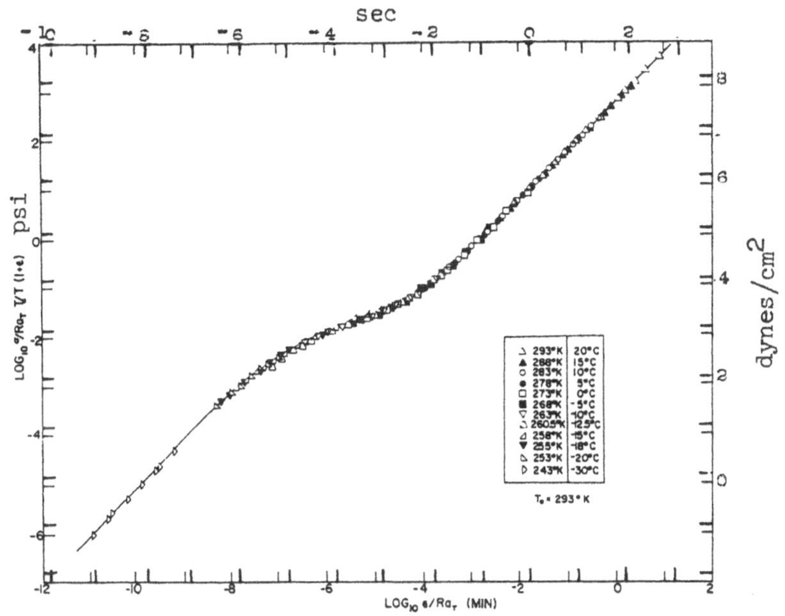

Figure 3. Reduced stress—strain curve for a polyurethane similar to Galcit I. The ordinate refers to stress in psi (from Arenz, Ferguson and Williams, Ref. 5)

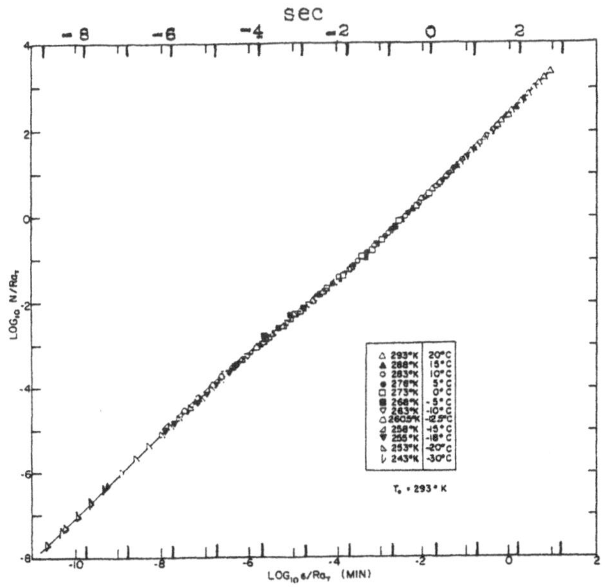

Figure 4. Reduced strain-optical coefficient curve for the same material as Figure 3.(from Arenz, Ferguson and Williams, Ref. 5)

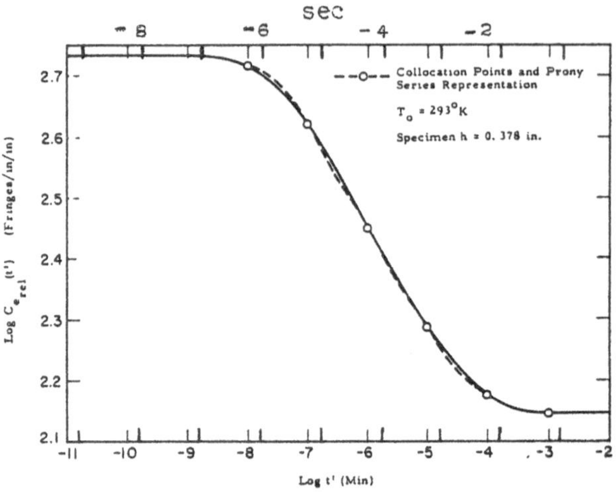

Figure 5. Time dependent strain optical coefficient as determined from Figure 4. The points and interpolated dashed line show that the data can be described very well by a Prony series (from Arenz, Ferguson and Williams, Ref. 5)

strains. Figure 5 illustrates the time dependence of C. The variation is small - slightly over $\frac{1}{2}$ decade. The modulus plot is not shown, but has the typical sigmoidal variation, changing from log E_g=4.9 to log E_e=2.7 (8 x 10^4 to 500 psi or 5.52 x 10^9 to 3.45 x 10^7 dynes/cm^2). The corresponding values for shear would be about 1.8 x 10^9 dyne/cm^2 and 11.5 x 10^6 dynes/cm^2, respectively.

Transient Modulus. A direct measurement of the tensile stress relaxation modulus E(t) of Galcit I was obtained by Mueller and Knauss from experiments carried out on strip specimens strained to 5%. The results are shown in Figure 6. Applications of reduced variables gave the smooth curve shown in Fig. 7. Long a_T values are shown in Fig. 8, along with corresponding values from failure data and crack propagation. The parameters of the WLF equation are indicated in Fig. 8. The creep compliance D(t) was then obtained by inversion, giving the results shown in Fig. 9. The equilibrium modulus, converted to shear, is found to be 8.15 x 10^6 dynes/cm^2; the glassy tensile modulus, ~5.5 x 10^9 dynes/cm^2, corresponding to G_g=1.8 x 10^9 dynes/cm^2. The low value of the latter and the broad transition would seem to indicate that there is an additional transition at still shorter reduced times. The glassy modulus values from the two sets of experiments are in good agreement and the equilibrium values are in reasonable agreement. The transition is much broader in the direct measurements however, probably reflecting the inherent inaccuracies in the value derived from the constant strain rate measurements as the glassy zone is approached.

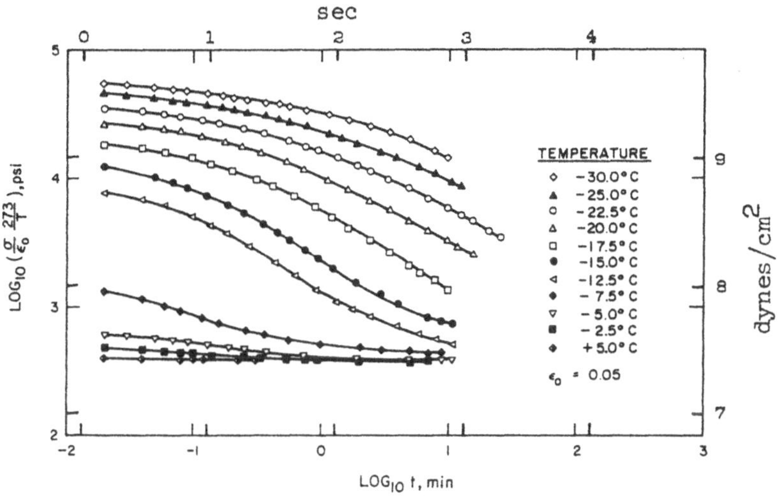

Figure 6. Temperature reduced stress relaxation modulus of Galcit I is a function of time at different temperatures as shown. (From Mueller, H. K. and Knauss, W. G., Ref. 8)

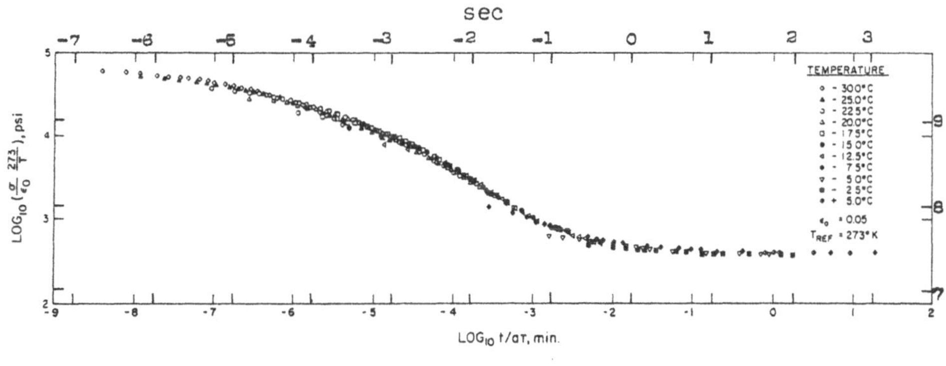

Figure 7. Temperature-reduced stress relaxation modulus of Galcit I as a function of reduced time, $T_o = 0^oC$. (From Mueller, H. K. and Knauss, W. G., Ref. 8)

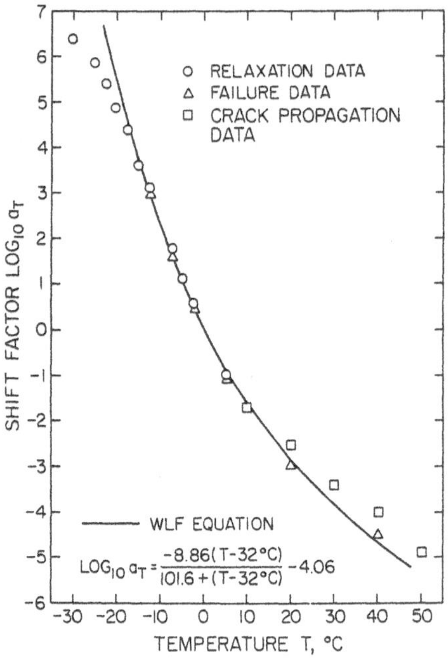

Figure 8. Log a_T plot for Galcit I, showing agreement for data from various types of experiments. (From Mueller, H.K., and Knauss, W.G., Ref. 8)

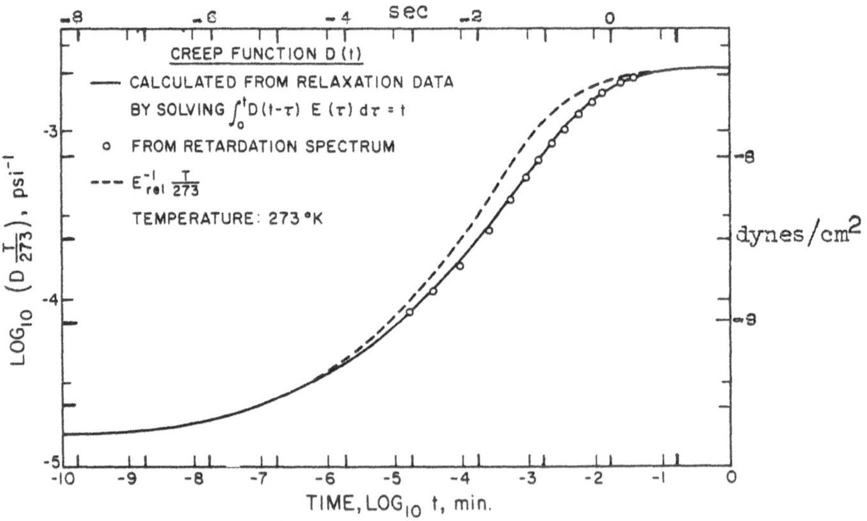

Figure 9. Reduced creep compliance for Galcit I as a function of reduced time, calculated from Fig. 7. (From Mueller, H.K., and Knauss, W.G., Ref. 8)

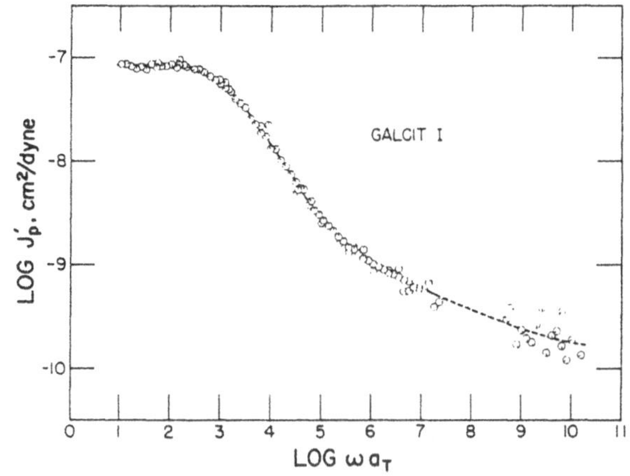

Figure 10. Complex shear storage compliance of Galcit I as a function of reduced frequency. T_0=30°C.

Dynamic Modulus. The complex shear compliance J^+ is being obtained in a Ferry-Fitzgerald transducer apparatus and preliminary results are illustrated here.(9) Figure 10 shows the real part of the compliance, reduced to 25°C; Figure 11, the imaginary part. The indicated equilibrium compliance of $10^{-7.1}$ cm²/dyne corresponds to an equilibrium shear modulus of 12 x 10^6 dynes/cm², close to but considerably higher than that found. This may reflect an unexpected difficulty in preparing reproducible samples or it may reflect a different sample history. The tensile specimens were tested shortly after preparation while the dynamic mechanical properties were measured some 18 months after preparation.

The glassy compliance is not well defined yet in the dynamic experiments, but could well be the value of ~5.5 x 10^{-10}cm²/dyne indicated by the transient data (log J_g=-9.26).

The distribution functions of relaxation and retardation times have not yet been calculated. Earlier, though less accurate stress relaxation data gave a tensile relaxation spectrum with a slope of -3/4, but this value must be viewed with considerable caution, especially since a composition containing 65% Solithane 113 gave a conventional spectrum with a slope of -1/2 over a small portion of the time scale.

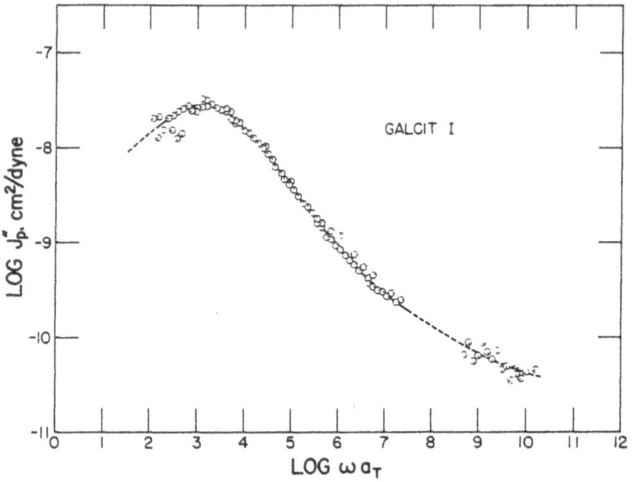

Figure 11. Complex shear loss compliance of Galcit I as a function of reduced frequency. $T_0=30^\circ C$.

Stress-Strain Response. The tensile properties of Galcit I were also measured by Mueller and Knauss (8), using ring shaped specimens. At low temperatures and high strain rates, the response was that of a conventional elastomer. Mooney-Rivlin plots showed a typical linear region, a minimum and a rapid upswing as the strain increased. Figure 12 shows the results at -5°, which is only about 13° above T_g. As the rate was decreased, however, $C_2 \to 0$. At still higher temperatures ($20^\circ C$ or 38° above T_g), the response became strain rate independent, within experimental error, Figure 13. There was still a variation in break properties however, so that at the highest rates the curves showed an upturn below $\lambda=2$ (100% strain). Raising the temperature to $40^\circ C$ removed even this small upturn, since the strain never exceeded 100%, Figure 14. The observed increase in C_1 is just what would be expected if C_1 is proportional to the absolute temperature. The only disturbing feature of the results is that the shear modulus calculated from $2C_1$ at $20^\circ C$ is 6.0×10^6 dynes/cm^2, rather than the values of about 8×10^6 obtained in the stress relaxation experiments.

The important point to note however, is that this material appears to obey the kinetic theory of rubberlike elasticity exactly if the temperature is high enough. In this same high temperature region the modulus is time-independent, supporting the argument that C_2 is related to time effects. Note that from Figures 6 and 10, equilibrium behavior is reached in about 2 milliseconds at $25^\circ C$.

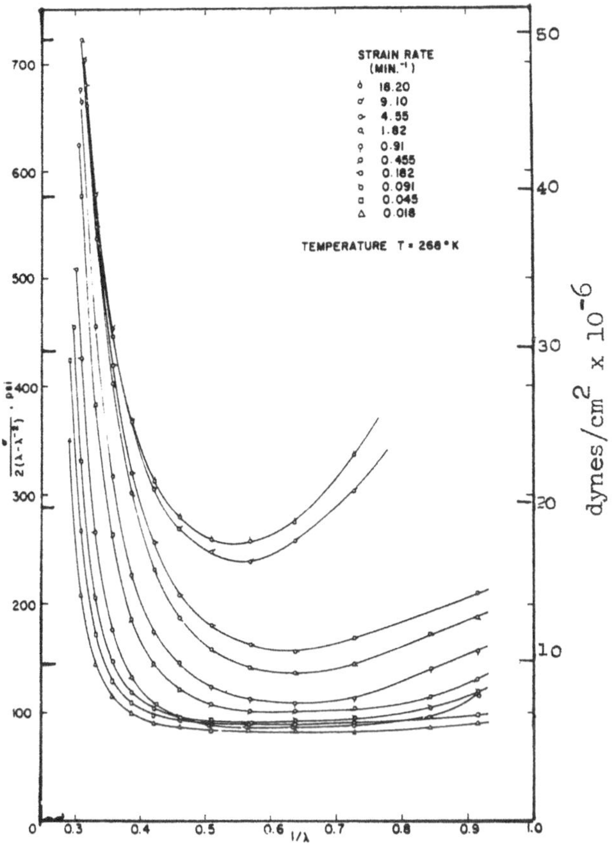

Figure 12. Mooney-Rivlin Plot for Galcit I at -5°C.

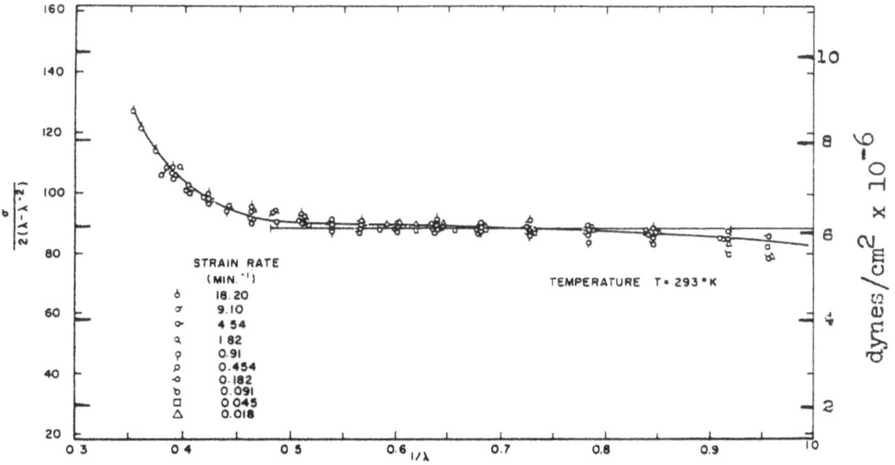

Figure 13. Mooney-Rivlin Plot for Galcit I at 20°C.

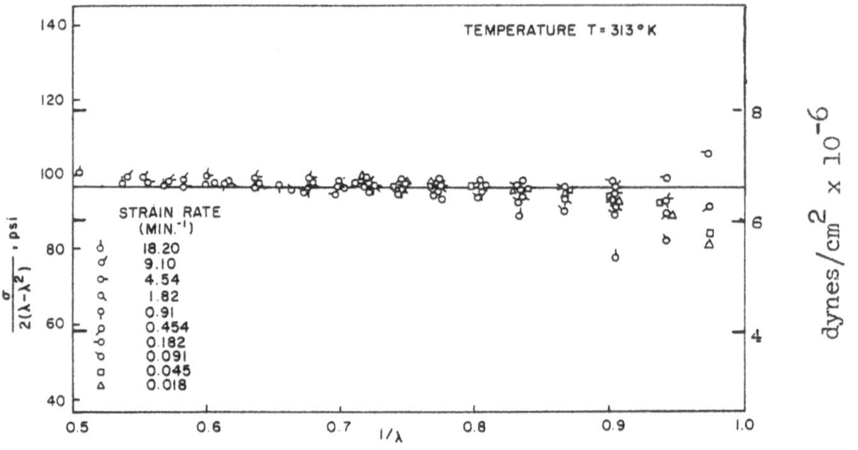

Figure 14. Mooney-Rivlin Plot for Galcit I at 40°C.

The time dependence of the breaking stress and strain measured at different temperatures often follows the time-temperature superposition principle (9, 12), to give a single composite curve. Typically, the rupture stress superposes more smoothly than the rupture strain. When the reduction is carried out for Galcit I[(8)], the breaking strain data superpose smoothly, far more so than usually observed, Figure 15. (Note that the abscissa is the reduced strain rate, not time-to-break). The breaking stresses, on the other hand, do not superpose at the lower temperatures or higher reduced rates, Figure 16.

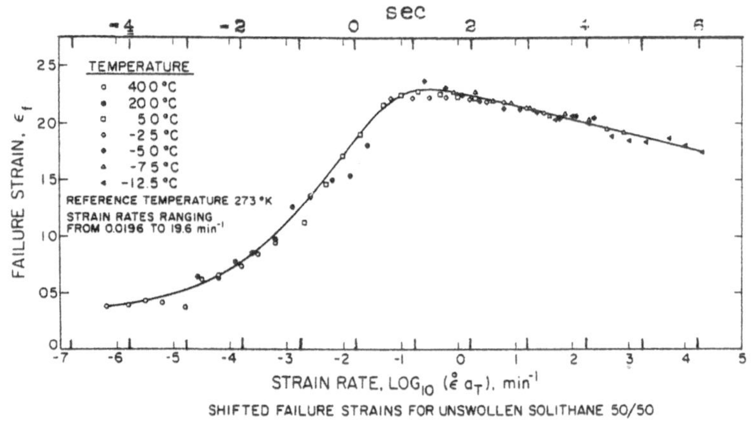

Figure 15. Failure strain plotted vs reduced strain rate (From Mueller and Knauss, Ref. 8)

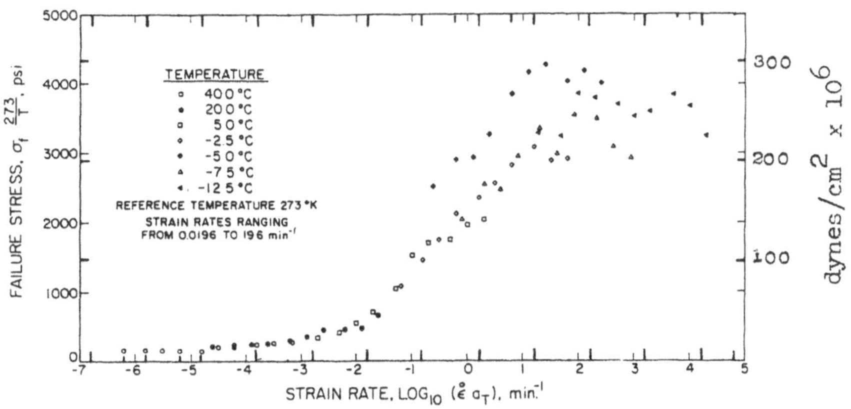

Figure 16. Temperature reduced failure stress vs reduced strain rate (from Mueller and Knauss, Ref. 8).

In spite of this, the failure envelopes are normal. Thus Figure 17 shows the envelopes for several Solithane 113-300 compositions (10). These envelopes can be fitted by the inverse Langevin approximation (11) to the stress-strain curve, and from the curve fit both the number of effective chains per cm^3 Y_e and the number of equivalent random links N can be determined (12). The fit for two compositions is shown in Figure 18 and the results of such an analysis (13) are given in Table II. It can be seen that the chain concentration is almost constant but N increases, i.e. the chains effectively become stiffer as the concentration of prepolymer is increased. This is the only elastomer system we are aware of in which such a change can be effected at constant Y_e.

A change in N is normally associated with a change in the chemical structure of the backbone or with a change in the degree of branching. However, since the chain backbones here are all derived from castor oil, the only change would seem to be the chemical nature and the functionality of the crosslink. In the present case, it appears that the addition of excess isocyanate leads to the formation of allophanates in addition to urethane linkages (A urethane moiety has an active hydrogen which is capable of reacting with isocyanate. The product is an allophanate.) Here, the formation of the allophanate will also lead to branching. Moreover, since an allophanate is more bulky and contains more polar groups than a urethane, it must also confer additional rigidity to the network chain. Both effects evidently appear as the prepolymer content is increased and the decrease in N is simply a reflection of the increased allophanate contents.

TABLE II

Estimates of Y_e and N from Failure Envelopes

Volume ratio of prepolymer/ curing agent	Y_e, moles/cm^3 x 10^{-6}	N, equiv. random links/chain
50/50	410	12
55/45	460	10
60/40	440	7.2
65/35	470	6.6
70/30	490	4.8
75/25	460	4.2
80/20	460	4.2

Swelling measurements were carried out on the same samples listed in Table II. The content was very small in all cases. Values of Y_e calculated from the modified Flory-Ruhner equation were in good agreement with those from the ultimate property data and with values obtained by compressive stress-strain measurements in the swollen state, as measured in several solvents. Table III gives the results and the solubility parameter for the solvents used. These values are apparent values if C_2 in the Mooney-Rivlin plot is not zero or close to it. However, if $C_2 \cong 0$, then Y_e will be independent of v_2 (actually, of $v_2{}^{4/3}$) and an appropriate plot of the data confirms the near zero value, in agreement with Figure 13.

If it is assumed that all network chains are effective, then the average value of $Y_e \approx 440 \times 10^{-6}$ moles/cm^3 can be used to obtain an estimate for the maximum value of the average molecular weight of a network chain, M_c, since for the present castor oil system, chain entanglements are expected to be absent because the molecular weight between chemical crosslinks is so low. Using the relationship $M = \rho/Y_e$ where ρ is the polymer density, ~ 1.00g/cm^3, $M_c \approx$ 2300g/mole. For the real network, M_c is expected to have a somewhat smaller value, since not all of the chains in the gel are effective and hence capable of supporting an equilibrium load. In

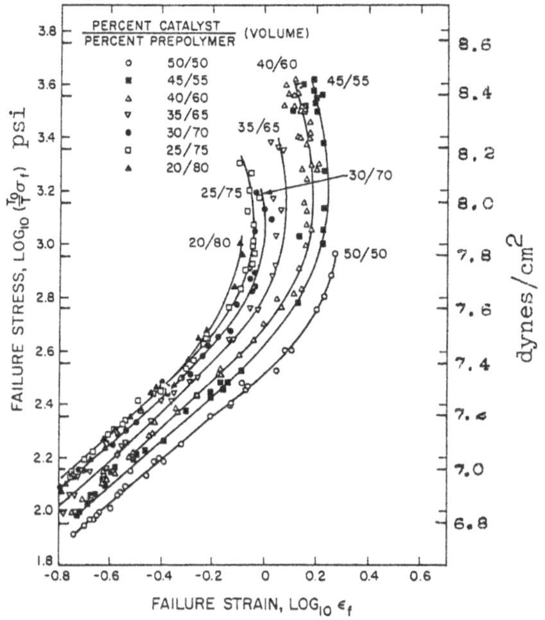

Figure 17. Failure envelopes for several Solithane compositions.

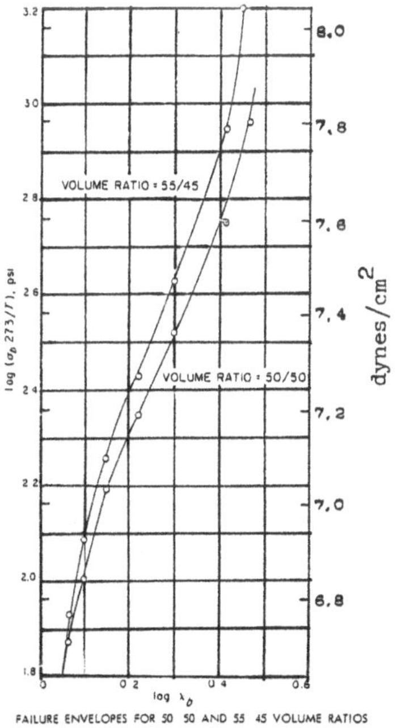

Figure 18. Failure envelopes for two of the Solithane compositions of Figure 17, showing the excellent fit by the inverse Langevin function (line) to the experimental data (points)

TABLE III

Swelling Behavior and Apparent Chain Concentration, in 10^{-6} moles/cm^3, for polyurethane Elastomers

Solvents and their Solubility Parameters in $(cal/cc)^{\frac{1}{2}}$

Volume ratio of prepolymer/ curing agent	Benzene $\delta = 9.22$		Ethylene dichloride $\delta = 9.80$		Cyclohexane $\delta = 8.25$		Heptane $\delta = 7.45$		Isooctane $\delta = 6.70$		Average
	v_2	Y_e	v_2	Y_e	v_2	Y_e	v_2	Y_e	v_2	Y_e	Y_e
50/50	0.353	370	0.380	340	0.660	450	0.771	390	0.794	420	392
55/45	0.398	390	0.418	390	0.667	480	0.798	---	0.813	---	420
60/40	0.438	440	0.457	390	0.692	470	0.813	---	0.836	---	423
65/35	0.462	---	0.481	440	0.706	460	0.816	380	0.863	430	428
70/30	0.492	350	0.510	400	0.721	440	0.827	---	0.848	---	397
75/25	0.510	430	0.527	350	0.733	400	0.833	380	0.868	480	408
80/20	0.527	---	0.542	450	0.745	420	0.839	370	0.943	370	402

principle, the fraction of "active" network capable of permanent elastic deformation can be estimated if the functionality of the reactants, the extent of reaction, and the initial ratio of active groups are known. The value of the fraction of "active" network will be a maximum when this ratio is unity, if all other parameters are held constant. On the other hand, this fraction will be markedly reduced if the ratio differs appreciably from unity. This will occur in practice if the "purity" of the reactants is low, i.e., if either the prepolymer or the curing agent contains appreciable amounts of monofunctional or difunctional molecules.

A minimum value for M_c can be estimated by assuming that all the network chains are effective and, further, that they correspond to the structure obtained from an equimolar reaction of the Prepolymer and Catalyst. Taking the former to be tolylene diisocyanate-capped castor oil and the latter castor oil, i.e., the chain structure of Figure 1, M_c will be 791g/mole. The actual value will be between these two extremes, i.e. between 790 and 2300g/mole.

Based on the manufacturer's specification for equivalent weight and density, the equivalents of NCO + OH per cm^3 can be calculated to be 2.74×10^{-3} for <u>all</u> formulations studied. Since, as just noted, the average value of Y_e is 440×10^{-6} moles/cm^3, this indicates that there are $\frac{1}{2} \times (2.74 \times 10^{-3})/440 \times 10^{-6}$ equivalents incorporated into each chain, or 3.1 reaction sites per chain between crosslinks rather than the single site expected. This value may be compared with the value of 2.9 reaction sites per chain obtained from the ratio of the actual to the theoretical M_c values, 2300/790. Both arguments point to an average functionality which is less than 3. If there are three <u>reaction sites</u> per chain, this implies that there are 3 difunctional molecules present for every trifunctional one and the chains have structure such as shown in Figure 19.

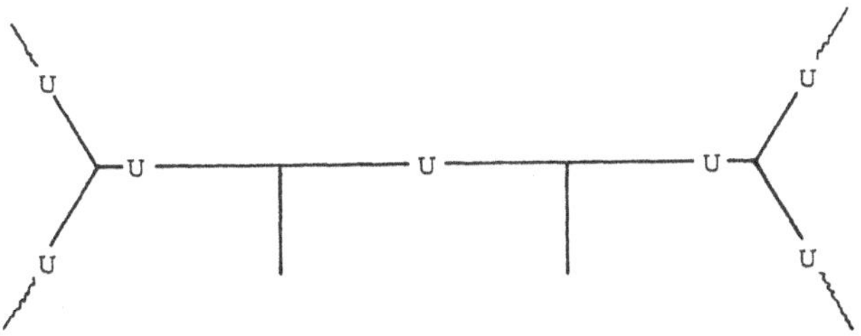

Figure 19. Sketch of the chain structure between crosslinks for Galcit I. Each line segment represents a ricinoleic acid moiety; each U, the di-urethane associated with TDI.

Whatever the actual structure, it is clear that M_c is low and the material is highly crosslinked. Since N=12, the molecular weight of a random link in Galcit I must be about 300 or less. For a gross comparative measure of this value relative to a vinyl-based polymer, the phenyl rings of the structure shown in Figure 16 can be taken to be equivalent to a pair of methylene carbons and the unsaturation can be ignored to give an equivalent average number of carbon bonds of about 100*, or about 8 carbon-carbon bonds/random link. Thus, M_c and the apparent size of a random link are such as to raise doubts as to the applicability of the usual statistically based equations for Galcit I.

The swelling data and Y_e determined from the compressive stress-strain measurements in the swollen state were used to determine the polymer solvent interaction parameters χ_1, and their dependence on v_2. The value of the solubility parameter for the rubber has been deduced from these data to be 11.6 $(cal/cm^3)^{\frac{1}{2}}$.

Table 4. Polymer--Solvent Interaction Parameters

Solvent	v_1	δ_s, $(cal/cm^3)^{\frac{1}{2}}$	χ_1
Benzene	88.9	9.22	$0.69 + 0.63v_2$
Ethylene dichloride	78.8	9.80	$0.70 + 0.60v_2$
Cyclohexane	108	8.25	$0.58 + 0.99v_2$
Heptane	146.5	7.45	$0.16 + 1.8v_2$
Isooctane	165.1	6.90	$0.01 + 2.0v_2$

Multiaxial response. The stress-strain response of Galcit I to multiaxial stress fields has also been investigated (15). These measurements were not made to try to determine the form of the stored energy function, W, but rather to test possible forms, especially in the region where $\lambda \to 1$. The experimental procedure used was combined torsion and tension/compensation on cylindrical rods.

For this deformation, the strain invariants for an incompressible material are:

$$I_1 = \lambda^2 + 2/\lambda + \lambda^2 \psi^2 r^2$$

$$I_2 = 2\lambda + 1/\lambda^2 + \lambda \psi^2 r^2$$

*The most highly crosslinked poly (cis - 1,4 - butadiene) studied by Shen (Chapter 3, this Symposium) has 130 c-c bonds per chain.

where ψ is the twist in radians per deformed length and r is the deformed radius to the material point of interest. Such a biaxial deformation allows a stringent test of any proposed W.

If W is assumed to have the Mooney-Rivlin form, then the observed moment M and the axial or normal stress (per unit undeformed area) can be shown (15) to be given by the following expressions

$$\frac{32M}{\pi d^4 \lambda^2 \psi} = 2C_1 + 2C_2 \lambda^{-1} \quad (2)$$

$$\frac{\sigma}{\lambda - (1 + q/2)\lambda^{-2}} = 2C_1 + 2C_2 \lambda^{-1} \left[\frac{\lambda - (1+q)\lambda^{-2}}{\lambda - (1-q/2)\lambda^{-2}}\right] \quad (3)$$

Where d is the deformed specimen diameter and $q = d^2\psi^2\lambda^3/8$. Note that equation (3) represents the normal Mooney-Rivlin uniaxial tension equation generalized to $\psi > 0$. When the specimen is not twisted, $\psi=0$ and the usual expression results. Thus, a plot of $\sigma/(\lambda - (1+\frac{1}{2}q)\lambda^{-2})$ against $\lambda^{-1}(\lambda - (1+q)\lambda^{-2})/(\lambda - (1+\frac{1}{2}q)\lambda^{-2})$ should be a straight line for all λ and ψ, and a plot of $32\ M/\pi d^4 \lambda^2 \psi$ against λ^{-1} should fall on the same line. The former plot, called the stress plot, has the interesting aspect that the ratio $(\lambda - (1+q)\lambda^{-2})/(\lambda - (1+q/2)\lambda^{-2})$ has a singular point at $\lambda^3=(1+\frac{1}{2}q)$; as λ increases from zero, this ratio goes from $(1+q)/(1+q/2)$ to $+\infty$ and then from $-\infty$ to 1. However, this phenomenon does not impair the usefulness of the plot, as will be seen below. The plot based on equation (2) is called the moment plot.

Specimens of Galcit I of different sizes were tested in tension and compression at various degrees of twist. Specimen sizes ranged from ten centimeters long by three centimeters diameter to two centimeters long by one centimeter diameter. Great care was taken to eliminate possible sources of error in the measurements. Representative moment plots in both tension and compression are illustrated in Figure 20. All of the data can be represented by the straight lines with a small positive slope, i.e., the behavior appears to be Mooney-Rivlin rather than simple Gaussian. The intercepts of the plots at $\lambda=1$ vary slightly, probably because of specimen-to-specimen variations in the crosslink density.

Representative axial stress plots are shown in Figures 21 and 22. The solid lines are drawn through the most accurate data points, using the same slope. The dashed lines are those which are to be expected on the basis of the moment plots. As can be seen, the slopes appear to be in reasonable agreement - the solid lines are drawn parallel to the dashed. It is clear, however, that $C_2 \neq 0$ and that the same values of C_1 and C_2 cannot describe both the moment and the axial force data. Although the discrepancy from the two sets of data is small, we believe that it is real.

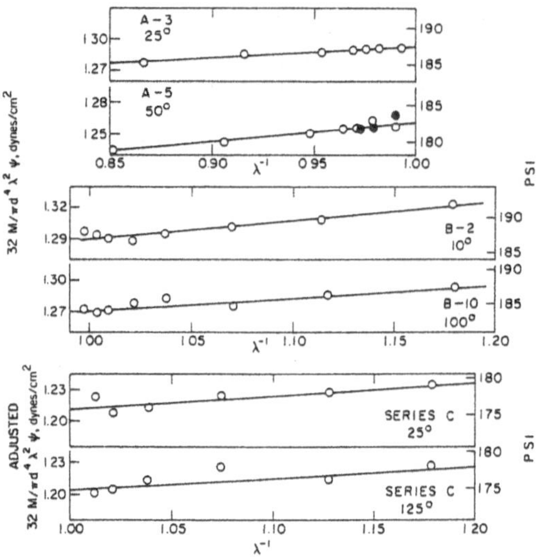

Figure 20.

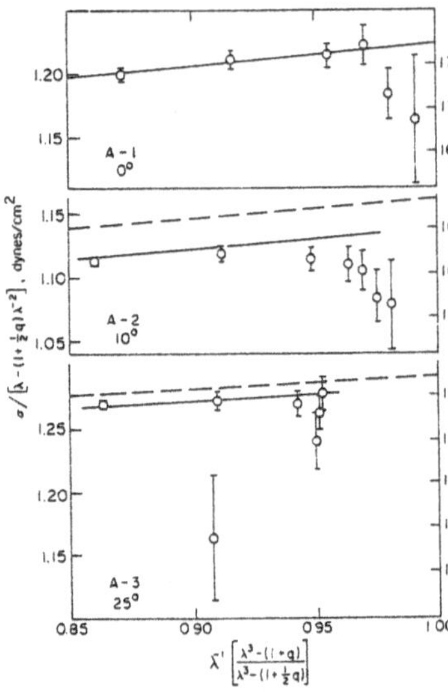

Figure 21. Moment plots for 10 x 3cm specimens in tension at increasing twists. The bar indicates the uncertainty in the point, the largest error occuring at the smallest strain. The solid points for specimen A-5 represents a repeat run at the smaller strains.

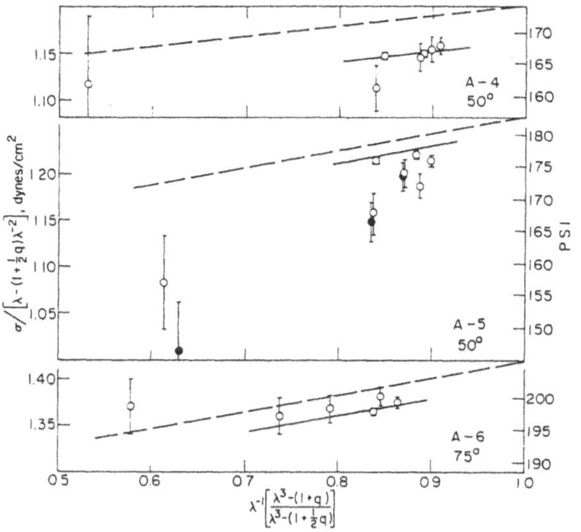

Figure 22. Stress plots for 10 x 3 specimens in compression at increasing twists. Series B, increasing λ at fixed θ; series C, crossplot from data taken at increasing θ but fixed λ.

Thus, the uniaxial data, Figs. 13 and 14, indicate neo-Hookean behavior with $C_2=0$ while the biaxial data obtained at still smaller strains indicate a more complex response. Since the former data are not accurate in this strain region and our experiments could not be carried to large strains without causing the specimens to crack, it is not possible to say whether a discrepancy exists or whether the apparent C_2 value will decrease at still higher strain levels.

In summary, Galcit I has fulfilled the prime purpose of obtaining a reproducible highly birefringent castable elastomer. Several of its properties turn out to be quite unusual:

1. Its small strain modulus reaches a true equilibrium value at room temperature.
2. At the same temperature the stress/strain response, as conventionally measured, appears to follow the Gaussian kinetic theory of rubberlike elasticity, though careful experiments under biaxial conditions show that this is not true for strains less than 10%.
3. Broad changes in the composition of the basic elastomer family do not influence the crosslink density or effective chain concentration but rather change the nature of the crosslink site so that the number of the equivalent random links per chain is changed.
4. M_c is unusually low and the effective number of C-C bonds between crosslinks is even lower.

The origin of the elastomer components is such as to preclude an exact analysis of the results in some instances, the GPC results on the prepolymer being particular disappointing. However, the elastomer is so unusual in its response that experiments with a material based on the pure glyceryl triricinoleate would seem to be highly desirable. Alternatively, low molecular weight synthetic prepolymers based on trimethylolpropane and polypropylene oxide could serve as a useful substitute, though there would always be uncertainty about the length of the branches in the trifunctional starting triol.

Acknowledgement: This paper represents one phase of research performed by the Jet Propulsion Laboratory, California Institute of Technology sponsored by the National Aeronautics and Space Administration, Contract NAS7-100.

REFERENCES

1. San Miguel, A., and Duran, E. Experimental Mechanics, 4 (3), 84 (1964).

2. Duran, E., private communications.

3. Knauss, W. G., "A Cross-Linked Polymer Standard: Report on Polymer Selection," Tech. Rpt. No. AFRPL-TR-65-111, Air Force Rocket Propulsion Lab., Edwards AFB (April 1965).

4. Kuhn, W., Grün, F., and Kallord, F., 101 248 (1942).

5. Treloar, L. R. G., "The Physics of Rubber Elasticity," 2nd Edition, Oxford University Press, London, 1958. Chapter 10.

6. Arenz, R. J., Ferguson, C. W., and Williams, M. L., Experimental Mechanics, 7, 183 (1967).

7. Smith, T. L., J. Polymer Sci. 20, 90 (1956).

8. Mueller, H. K., and Knauss, W. G., Trans. Soc. Rheol., 14, 000 (1970).

9. Landel, R. F., Froelich, D., and Tschoegl, N. W., Further work is in progress and will be reported subsequently.

10. Knauss, W. G., Clauser, J. F., and Landel, R. F., AFRPL-Tr-66-21, Edwards, California, January 1966, MATSCIT PS 66-1, California Institute of Technology, Pasadena, California, January 1966.

11. Reference 5, Chapter 6.

12. Landel, R. F., and Fedors, R. F., Proc. First Int. Conf. on Fracture, Sindai Japan, T. Yokobori, T. Kawasaki and J. L. Swedlow, Edr., I.C.F. Vol. II, 1965, p. 1247. Reprinted in Rubber Chem. and Tech., 40, 1044 (1967).

13. Smith, T. L. and Frederick, J. E., J. Appl. Phys. 10, 2996 (1965).

14. Fedors, R. F., and Landel, R. F., Space Program Summary 37-45, Vol. IV, Jet Propulsion Laboratory, Pasadena, California, June 30, 1967, p. 92.

15. Mancke, R. G., and Landel, R. F., to be submitted. Preliminary results may be found in AFRPL-TR-69-180, Edwards, California, August 1969, CHECIT PL 69-1, California Institute of Technology, Pasadena, California, August 1969.

INHOMOGENEITIES INDUCED BY CROSSLINKING IN THE COURSE OF CROSSLINKING COPOLYMERIZATION

K. Dusek

Institute of Macromolecular Chemistry

Czechoslovak Academy of Sciences, Prague

SUMMARY

Heterogeneous gels are formed as a result of phase separation during copolymerization of 2-hydroxyethyl methacrylate with ethylene dimethacrylate and styrene with divinylbenzene in the presence of various diluents. Phase separation is induced by an increase of the degree of crosslinking and possible change of the polymer-diluent interaction and takes the form of macrosyneresis (deswelling) or microsyneresis (formation of a dispersion). Microsyneresis, if not followed by crosslinking, is a non-equilibrium form of phase separation, occurs in gels of low degrees of crosslinking, and is determined by slow relaxation of such gels. If crosslinking continues, the dispersion becomes fixed and permanent and phase equilibrium is approached again. In the course of crosslinking copolymerization, usually dispersions are formed and phase separation occurs at low polymerization conversions: the theory of equilibrium macrosyneresis can satisfactorily describe the onset of phase separation as well as the phase volume ratio and volume degree of swelling of heterogeneous gels after copolymerization. In homogeneous gels, dilution leads to an increase of the volume degree of swelling, but beyond the separation limit, it decreases with increasing concentration of the diluent.

INTRODUCTION

Crosslinked networks are often regarded as a random collection of network chains with their ends connected in junction points (crosslinks). However, this may not

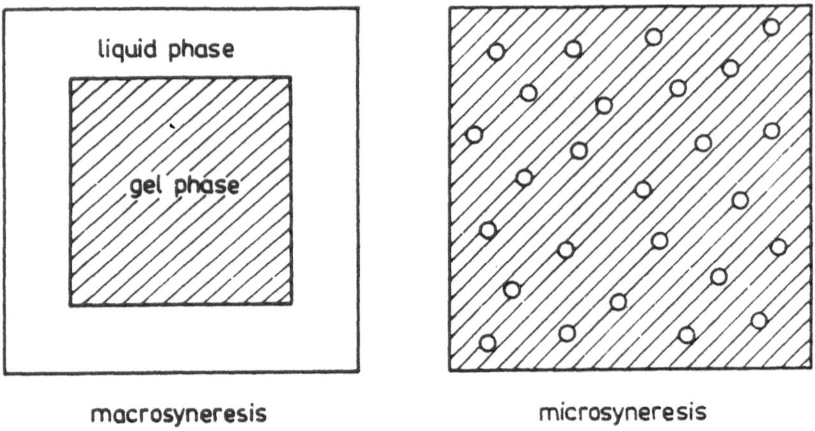

Fig.1
Schematic representation of macro- and microsyneresis

be so for various reasons[1]. It is sometimes desirable from both theoretical and practical points of view to carry out crosslinking in the presence of an inert diluent which is subsequently removed or replaced. The presence of a diluent can become, however, an additional source of inhomogeneities which in certain cases need not be easily detected by naked eye. Formation of networks that are porous in the dry or swollen states is one of the practical aspects of crosslinking in the presence of diluents (see e.g.[2,3]). Phase separation is the main reason for the formation of such inhomogeneities; it can be caused by an increase in the degree of crosslinking (ν-induced syneresis) or by changes in polymer-diluent interactions (χ-induced syneresis). In crosslinking copolymerization (e.g. of vinyl-divinyl type) both effects are usually operative because the monomers act as a special kind of diluent and conversion of monomers into the polymer causes not only an increase in the crosslinking density, but also a change of the composition of the polymerizing system.

Phase separation in networks occurs either in the form of macrosyneresis (deswelling of the gel being crosslinked and formation of continuous liquid and gel phases), or microsyneresis in which the gel and separated liquid phases form a dispersion (e.g. droplets of liquid inside the gel (Fig.1)). In this communication, we will discuss the predictions of the equilibrium theory of macrosyneresis for vinyl-divinyl copolymerization, conditions for microsyneresis and applicability of the equilibrium theory for systems formed by crosslinking copolymerization of styrene and 2-hydroxyethyl methacrylate.

THERMODYNAMICS OF MACROSYNERESIS

If a network is originally swollen to a degree (given by the volume fraction of the network polymer, ϕ_p^o,) which is less than its maximum degree of swelling in the same diluent, ϕ_p, then phase separation starts when the maximum degree of swelling drops to the initial degree of swelling, i.e. when

$$\phi_p = \phi_p^o \tag{1}$$

and when the chemical potentials of each component in both phases are equal

$$\mu_i = \mu_i' \quad . \tag{2}$$

It follows from the statistical theory of rubber elasticity (see e.g.[1]) that the chemical potential of the diluent i in the swollen network is given by

$$(\mu_i - \mu_i^o) = (\Delta\mu_i)_{mix} + RT\nu^* \overline{V}_i [\phi_p^{1/3} (\langle r^2\rangle_d / \langle r^2\rangle_o) - \phi_p/2], \tag{3}$$

where ν^* is the concentration of active network chains in unit dry volume of the gel, $\overline{V}_i$ is the partial molar volume of the diluent i, and $\langle r^2\rangle_d$ and $\langle r^2\rangle_o$ are the mean squares of the end-to-end distances of network chains in the dry and reference states, respectively. The factor $\langle r^2\rangle_d/\langle r^2\rangle_o$ depends on the network structure and especially on the amount of diluent present during crosslinking. At not too high dilutions (excluded volume effects), one may assume the network chains to be essentially unstrained, when they are being crosslinked, and their dimensions to approach the reference dimensions. When the diluent is removed, the chain dimensions shrink correspondingly and it is legitimate to write

$$\langle r^2\rangle_d = \langle r^2\rangle_o (\phi_p)^{2/3}. \tag{4}$$

Inserting relations (1), (2), and (4) into (3) one gets a set of equations that determine the incipience of phase separation. For a binary system, a single equation is obtained

$$(\Delta\mu_1)_{mix}+(\Delta\mu_1)_{net}=\ln(1-\phi_p^o)+\phi_p^o+\chi(\phi_p^o)^2+\nu^*\overline{V}_1\phi_p^o/2=0 \quad (5)$$

which at a given dilution determines the critical values of ν^* or χ.

Continuous crosslinking beyond the critical point leads to a volume change (usually a shrinkage) of the network phase and the dimensions of network chains already formed change correspondingly. When crosslinking has been completed, there exists in the networks a whole spectrum of chain compressions or, possibly, extensions given by the history of the volume change of the network phase superimposed on the chain length distribution given by the Gaussian statistics in a network crosslinked at constant volume. If such a composite network is approximated by a set of interpenetrating networks[4], the maximum degree of swelling of the network phase and the phase volume ratio are described by a set of integral equations having the form (subscript c denotes the incipience of phase separation)

$$\mu_i-\mu_i^o=(\Delta\mu_i)_{mix}+RT\overline{V}_i\left\{\phi_p^{1/3}\left[\nu_c^*(\phi_p^o)^{2/3}+\int_{\nu_c^*}^{\nu^*}\phi_p^{2/3}d\nu^*\right]-\right. \quad (6)$$
$$\left.-\ \nu^*\phi_p/2\right\}$$

The relations given above can readily be applied to crosslinking of existing chains, where the concentration of active network chains, ν^*, is the independent variable. When networks are formed by vinyl-divinyl crosslinking copolymerization, ϕ_p, ϕ_p^o, ν^*, and possibly the interaction parameters χ_{ij} are functions of the polymerization conversion which is now the independent variable of a system characterized by initial concentrations of the mono- and divinyl monomers, the diluent(s), and reactivities of the vinyls. The dependence of ϕ_p^o on the volume conversion of monomer to the crosslinked polymer ξ is simply $\phi_p^o = \xi\phi_p^{oo}$, where ϕ_p^{oo} is the initial volume fraction of monomers. The concentration of network chains usually increases with increasing conversion, depends on the copolymerization parameters r_{ij} of the vinyls involved and on the molecular weight distribution of primary chains and can be obtained from copolymerization and crosslinking statistics using e.g. the extension of the theory of cascade processes by Gordon et al.[5]. When the sol fraction is low, much simpler relations between the overall crosslinking density and conversion[6] can be

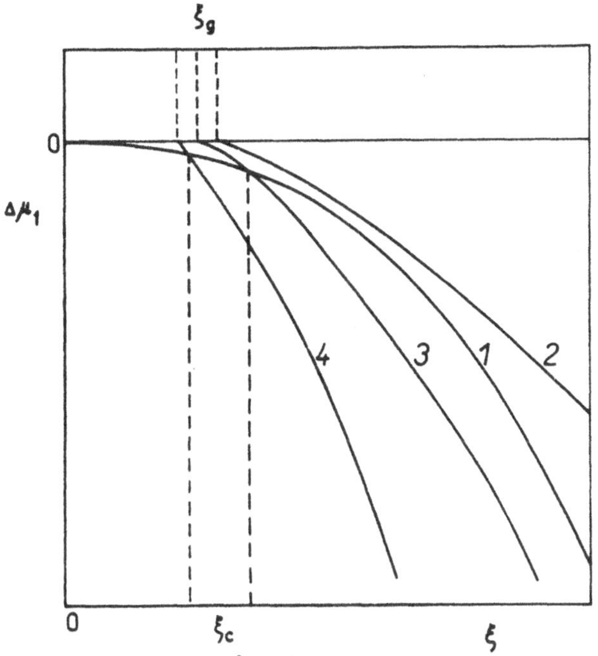

Fig.2
Critical conversion, ξ_c, at incipient phase separation in the copolymerization of a monovinyl and a divinyl monomer (both the monomer and diluent are good solvents - a quasibinary system) given by the intersection of $(\Delta\mu_1)_{mix}$ and $-(\Delta\mu_1)_{net}$. 1 $(\Delta\mu_1)_{mix}$; 2,3,4 $-(\Delta\mu_1)_{net}$ with increasing fraction of the divinyl monomer

used. Yet neither theory does adequately cover the relation between ν^* and ϕ_p^{oo} and ξ and it is, therefore, preferable to work with experimental dependences obtained by measuring the degree of swelling of gels prepared at different initial conditions and isolated at different conversions. This path was followed in this work when real systems were treated.

An example of ν-induced macrosyneresis in a quasibinary system composed of a monomer and a diluent that are both good solvents for the polymer ($\chi = 0.4$) is given on Fig.2. The incipience of phase separation is given by intersection of $(\Delta\mu_1)_{mix}$ and $-(\Delta\mu_1)_{net}$ curves. At a lower concentration of the divinyl monomer, phase separation does not occur at all and on passing to higher concentration it starts already in the vicinity of the gel point. This is because the slope of $(\Delta\mu_1)_{mix}$ falls as a rule steeper than the slope of $-(\Delta\mu_1)_{net}$ unless the vinyls in the divinyl

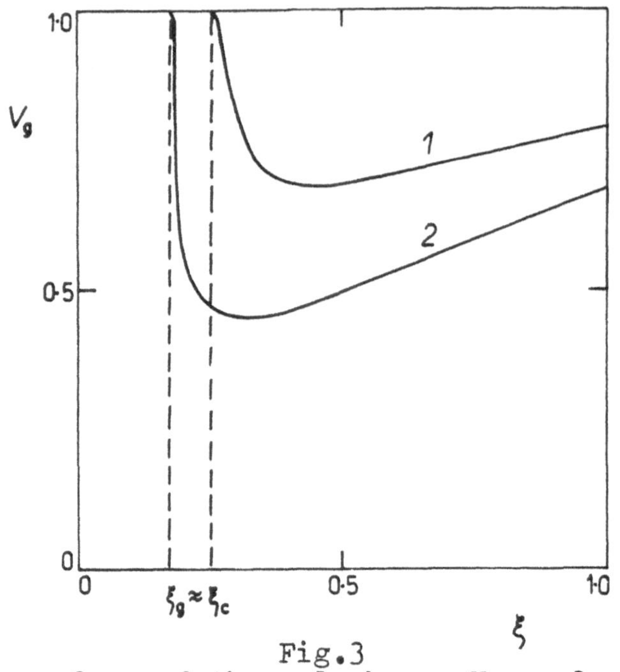

Fig.3

Relative volume of the gel phase, V_g, calculated for the copolymerization of styrene (ST) with divinylbenzene (DVB) in the presence of toluene. 1 : 15% DVB, ϕ_p^{oo}=0.2, 2: 30% DVB, ϕ_p^{oo}=0.33.

monomer have much less reactivity than vinyls in the monovinyl monomer.

Contrary to crosslinking of existing chains, the volume of the network phase need not decrease when the critical point is surpassed. At higher concentrations of the crosslinking agent and higher dilutions, the volume of the network phase first sharply decreases and can slowly increase again as a result of the transfer of monomers into the initially small volume of the network phase. An example calculated according to the modified Eq.(6) is given in Fig.3.

The equilibrium theory predicts that phase separation is promoted by increasing concentration of the divinyl monomer and reactivity of its vinyls, by increasing dilution, molar volume of the diluent, and polymer-solvent interaction parameters. The critical conversion at phase separation is usually located close to the gel point and the volume of the network phase first decrease and may in certain cases increase again

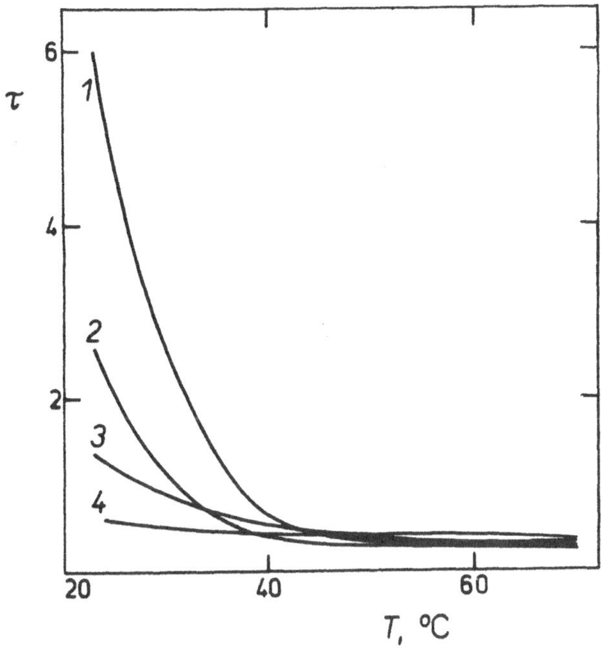

Fig.4
Turbidity induced by a decrease in temperature in poly(2-hydroxyethyl methacrylate) (PHEMA) gels swollen in butanol; concentration of ethylene dimethacrylate in the gel in wt.%: 1: 0.2, 2: 0.5, 3: 1.0, 4: 2.0.

as the polymerization proceeds. These predictions can be subjected to experimental tests.

CONDITIONS FOR MICROSYNERESIS

Phase separation occurring in the course of crosslinking copolymerization usually leads to the formation of dispersions rather than to deswelling. Turbidity develops in gels, and eventually porous structures are formed. To find the reason why microsyneresis has priority over macrosyneresis, we have studied χ-induced microsyneresis in homogeneous gels with constant crosslinking density. Poly(2-hydroxyethyl methacrylate) (PHEMA) gels were swollen in butanol at higher temperature and then rapidly cooled. Butanol is a fairly good solvent for PHEMA at higher temperatures, but becomes a poor one when the temperature is lowered. As a result of cooling, the samples get turbid; the turbidity

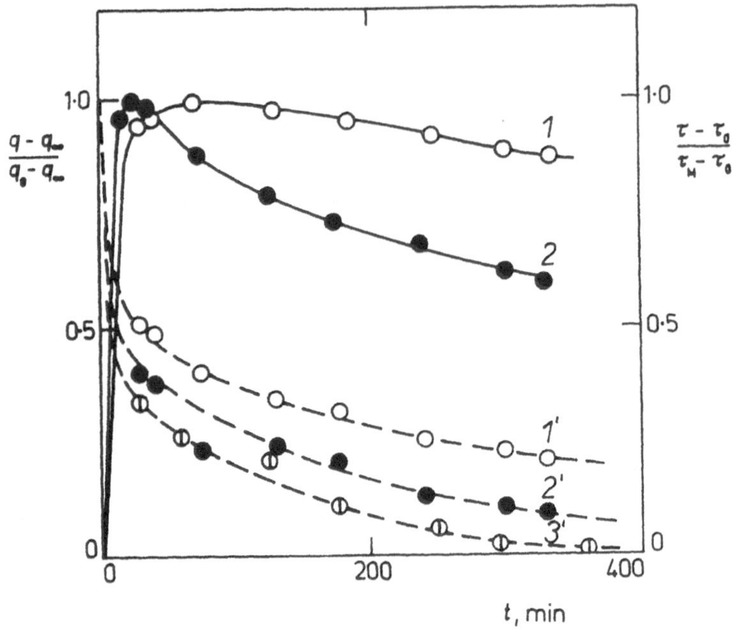

Fig.5
Time decrease of relative turbidity and volume degree of swelling, in PHEMA gels swollen in butanol; q_0 and q_∞ are the initial and equilibrium degrees of swelling, respectively; τ_0 and τ_M are the initial and maximum turbidity, respectively: —— turbidity, ---- degree of swelling; concentration of the crosslinking agent in wt.%: 1: 0.2, 2: 1.0, 3: 5.0.

decreases with increasing crosslinking density (Fig.4). However, the turbidity is not stable and decreases with time. The decay of turbidity is accompanied by deswelling and both changes are faster in gels of higher degree of crosslinking (Fig.5). Highly crosslinked gels (e.g. 4% ethylene dimethacrylate) only deswell without getting turbid. On the contrary, the equilibrium state (disappearance of turbidity) in loosely crosslinked gels can be approached only after several months. The size of the microseparated diluent particles is determined by the interfacial tension and by elastic response of the locally deformed network. Transmission measurements at different wave lengths have shown that the decay of turbidity is accompanied by a decrease in the

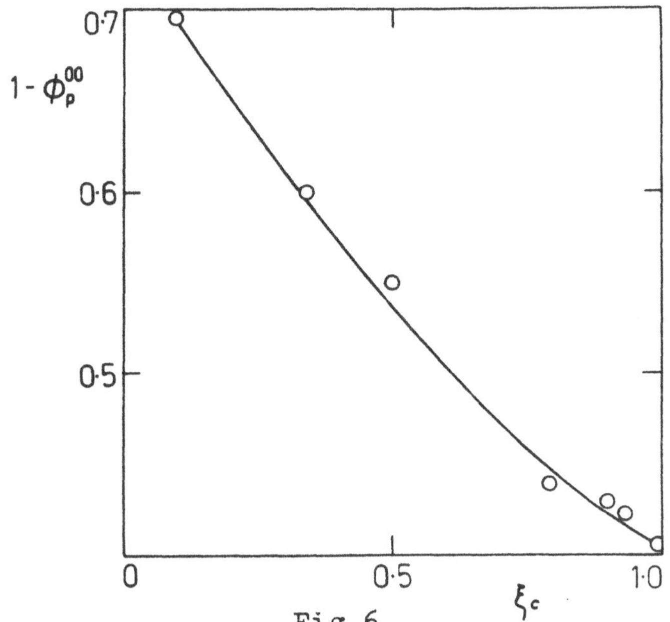

Fig.6
Relation between critical conversion at phase separation and initial volume fraction of water in the copolymerization of HEMA with 0.2% ethylene dimethacrylate in the presence of water

number of particles rather than by their contraction[7]. The formation and instability of dispersions is connected with the very slow relaxation of loosely crosslinked networks[8] and a connection has been established between the stress relaxation[9] and decrease of turbidity in PHEMA gels.

The non-equilibrium nature of microsyneresis can be demonstrated by crosslinking copolymerization of HEMA with small amounts of ethylene dimethacrylate in the presence of water. Water is a poor and HEMA is a better solvent for PHEMA, but water-HEMA systems form a cosolvent mixture with maximum solvent power at 55% HEMA. If the concentration of the crosslinking agent is kept low, phase separation is almost exclusively controlled by polymer-solvent interactions and does not take place at all, when the concentration of water in the system is below 41%. Slightly above this limit, phase separation takes place at high conversions, when the crosslinking has been completed and the turbidity is again not permanent. On increasing the concentration of water, phase separation starts at lower conversions (Fig.6) and the disperse structure becomes fixed by proceeding crosslinking. Therefore, microsyneresis becomes a stable form of phase separation if the polymerization is faster than the decay of turbidity and if the initially non-equilibrium dispersion is fixed by crosslinks.

MICROSYNERESIS AT HIGHER CONCENTRATIONS OF THE DIVINYL MONOMER

When monovinyl monomers are copolymerized with a divinyl monomer (e.g. styrene with divinylbenzene) in the presence of diluents (good solvents, poor solvents, or polymers), microsyneresis usually occurs when the critical concentrations of the divinyl monomer and diluent are surpassed. Preceding sections give an answer to why phase separation usually prefers to take the form of microsyneresis and not that of macrosyneresis. As has been shown by calculation and confirmed experimentally, phase separation takes place in gels of low crosslinking degree (at low conversions) where slow relaxation of the gel gives rise to microsyneresis. However, as copolymerization proceeds, the dispersion becomes fixed by newly formed crosslinks, the relaxation gets faster and equilibrium can be reached between both phases. Application of the equilibrium theory of macrosyneresis to such disperse structures seemed promising and has met with success.

To this end, the copolymerization of styrene with m- and p-divinylbenzene has been studied and the results have been used as input data for calculation of the phase equilibria. Divinylbenzene is a symmetric divinyl monomer and reactivities of its vinyls are interdependent which has been demonstrated by a shift of the infrared absorption band of vinyl at 298-314 mμ with respect to the band of the pendant vinyl or vinyl in styrene[10]. For copolymerization of styrene with p- and m-divinylbenzene, the values of copolymerization parameters found by us[11] were $r_1= 0.15$, $r_2= 1.22$ and $r_1= 0.54$, $r_2= 0.58$, respectively, clearly indicating a higher reactivity and stronger dependence of vinyls in the p-isomer. The increase of the overall concentration of crosslinked units calculated from the copolymerization statistics is shown on Fig.7 for copolymerization of styrene with 15% p-divinylbenzene and compared with the concentration of elastically active chains obtained from swelling data. Whereas the negative deviations at

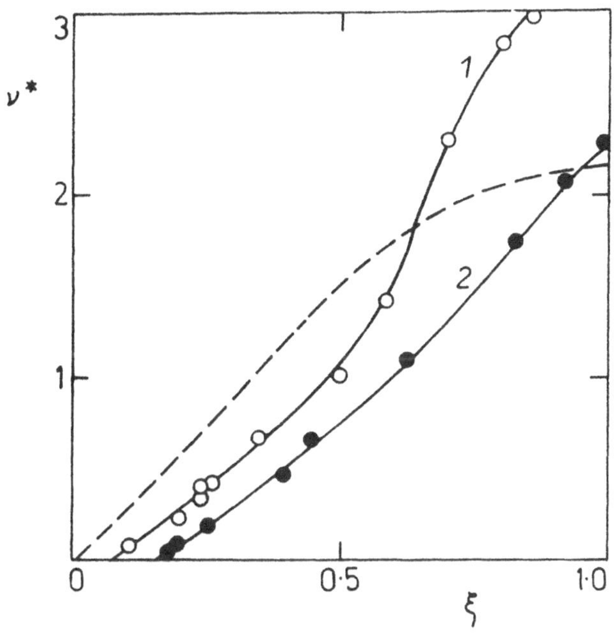

Fig.7
Concentration of elastically active chains in the copolymerization of styrene with 15% p-divinylbenzene; in the absence of a diluent (curve 1) and in the presence of 40% toluene (curve 2); dashed curve corresponds to the theoretical values of the overall concentration of network chains.

low conversions can undoubtedly be ascribed to finite molecular weight of primary chains and cyclization, the high concentration of active chains at high conversions cannot be fully explained by departures from the Gaussian statistics. Instead, one can assume formation of additional entanglements due to a leap-frogging[12] ofnearest crosslinks by network chains. This effect must decrease with dilution, and it does indeed (Fig.7). To calculate the phase separation behavior it was necessary to use experimental dependence of ν^* on conversion and dilution.

The concentration of divinylbenzene and degree of

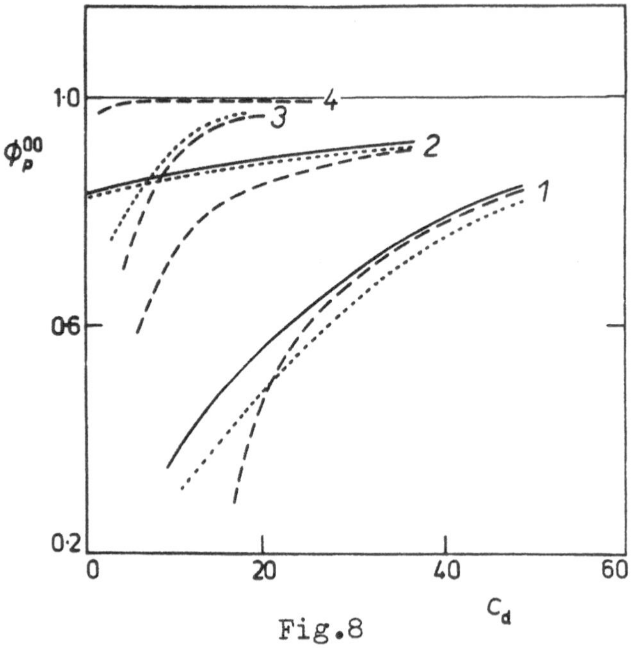

Fig.8

Critical concentrations of the diluent ($1-\phi_p^{00}$) and DVB (c_d,wt.%) for the incipience of phase separation in the copolymerization of ST with DVB. Diluent: 1: toluene (a quasibinary system, χ=0.46); 2: iso-octane, (χ_{13}(styrene-polymer) = 0.46, χ_{23}(iso-octane-polymer)= 1.40, χ_{12} = 0.40); 3: polystyrene mol.wt. 50000 (χ_{23} = 0), 4: polydimethylsiloxane 50cP; —— development of turbidity, ----- formation of networks porous in the dry state, ... theory.

dilution necessary for the appearance of turbidity in the polymerizing system or for obtaining gels porous in the dry state are compared in Fig.8 with the calculated dependence for critical conversion exceeding the conversion at the gel point by 2-10%. The agreement between the cloud points and values calculated from Eq.(5) or two equations obtained from relationship (3) is fairly good. As expected, phase separation is promoted by increasing degree of dilution, increasing concentration of the divinyl monomer, increasing molar volume of the diluent and by passing from a good to a poor solvent for polystyrene.

The volume degree of swelling and phase volume ratio calculated from the integral equations of the form of Eq.(6) agrees fairly well with experimental data when toluene and polystyrene are used as diluents. Thus for 67% toluene ($\chi = 0.46$) and 30% divinylbenzene, the calculated volume fraction of the network phase at full conversion was 0.55-0.60, which is in agreement with the experimentally observed value 0.53. The observed and calculated toluene regain as 2.25 and 2.70 ml/g, respectively. When the monomers are diluted with more than about 20% iso-octane (a poor solvent), the agreement between calculated and experimental values of phase volume ratio becomes poorer. In that case, the monomer-diluent mixture is so poor that phase separation takes place already before the gel point has been reached.

The dependence of the phase separation on the molar volume of the diluent deserves further comment. In a series of polystyrenes, the appearance of macroporosity after extraction of the gel(development of turbidity cannot be observed because of small differences in refractive indexes of both phases) depends on the molecular weight of polystyrene (Fig.9) and on the concentration of divinylbenzene (Fig.10). Since the segments of the diluent and of network chains are identical, the interaction parameter of network chain and diluent segments is zero and the separation effect is due to the large molar volume of the diluent which causes polystyrene to be almost completely excluded from the gel phase. Combining the high molar volume and chemical dissimilarity of network chain and diluent segments, one arrives at diluents that are the most powerful with respect to phase separation. Polydimethylsiloxanes serve as a good example: phase separation is caused by 0.5-1% of these polymers. In that case however, phase separation is governed mainly by polymer-polymer incompatibility and is not much sensitive to the concentration of the divinyl monomer.

The equilibrium theory predicts for some cases an increase following a sharp decrease in the volume of the network phase (Fig.3). The volume increase forces the network chains already formed to become strained and the ratio $\langle r^2\rangle_d/\langle r^2\rangle_o$ may become greater than unity. The volume fraction of the network phase is small after the onset of phase separation and monomers contained in the liquid phase are continuously transferred into

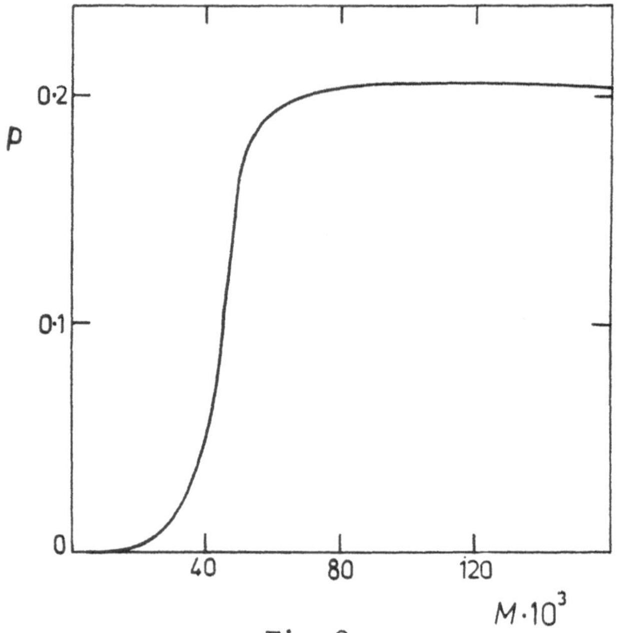

Fig.9
Volume fraction of pores, p, in dry copolymers of ST with 10% DVB prepared in the presence of 20% polystyrene of different molecular weight, M. Ref. (2).

the network phase as the copolymerization proceeds; the volume of the network phase increases while its degree of swelling decreases. The diluent has thus a dual effect on the volume degree of swelling. If the diluent concentration is less than critical, the volume degree of swelling increases with increasing dilution, but beyond the critical point the effect is opposite. Such a dual effect of the diluent is predicted by the theory, and an experimental evidence is given in Fig.11. It is worthwhile to mention that at high dilutions the volume degree of swelling can be lower than that of a network with the same concentration of crosslinks, but prepared in the absence of any diluent.

CONCLUSIONS

It has been demonstrated that crosslinking copolymerization in the presence of diluents can be accom-

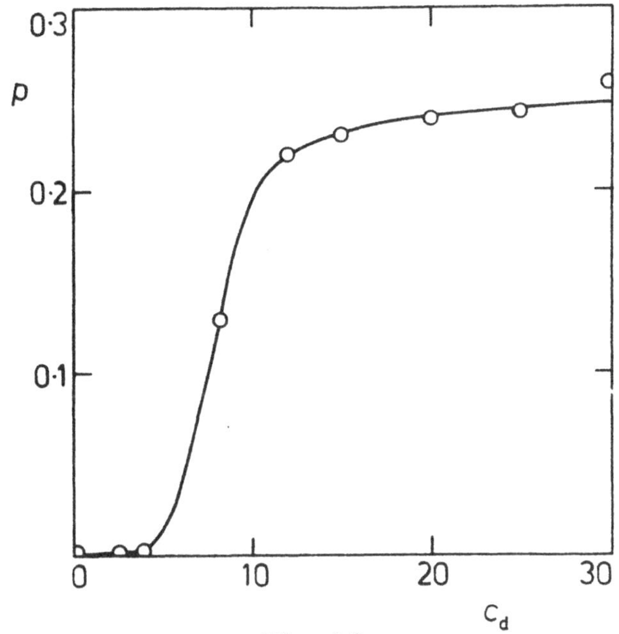

Fig.10
Volume fraction of pores, p, in dry ST-DVB copolymers prepared in the presence of 20% polystyrene, (M = 87000) in relation to the concentration of divinylbenzene, c_d. Ref. (2).

panied by phase separation leading to inhomogeneous networks. Phase separation occurs mainly in the form of microsyneresis; although it is of non-equilibrium nature, the incipience of phase separation and equilibrium properties of the heterogeneous gels can be described by the equilibrium theory of phase separation, because the originally non-equilibrium structure is readily fixed by proceeding crosslinking. Heterogeneous gels are not transparent in the swollen state. It should be noted, however, that the turbidity or porosity sometimes disappears as a result of capillary contraction when the gels are dried out (cf.[3]). The disappearance of turbidity is reversible and is characteristic of gels with lower crosslinking density whereas in highly crosslinked networks, the structure is rigid enough to prevent capillary contraction. Transparence of networks is thus not a sufficient quarantee for their homogeneity.

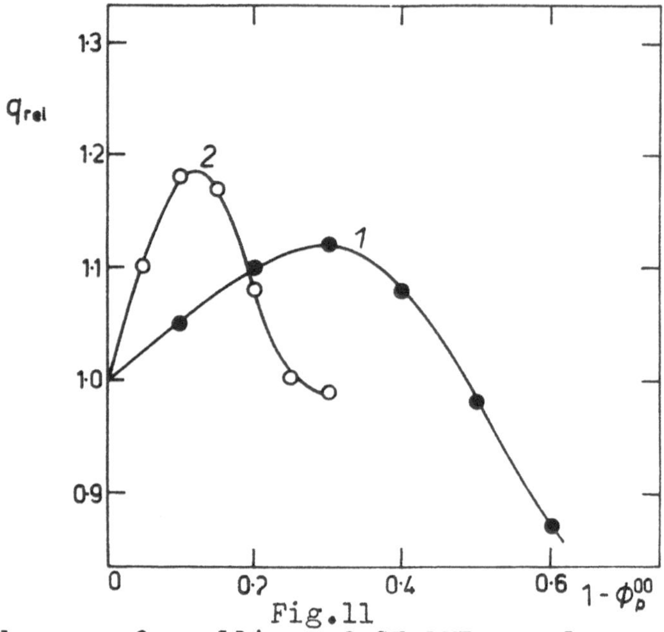

Fig.11
Volume degree of swelling of ST-DVB copolymers related to the swelling of the copolymer prepared in the absence of diluent, q_{rel}; 1: 20% DVB, diluent: iso-octane, 2: 10% DVB, diluent: polystyrene (M = 80000).

REFERENCES

1 K.Dušek and W.Prins, Adv.Polymer Sci.,6,1(1969).
2 J.Seidl, J.Malinský, K.Dušek, and W.Heitz, Adv.Polymer Sci., 5,113(1967).
3 K.A.Kun and R.Kunin, J.Polymer Sci., Pt.A-1,6,2689 (1968).
4 K.Dušek, J.Polymer Sci., Pt.C, 16,1289(1967).
5 M.Gordon and G.N.Malcolm, Proc.Roy.Soc.,A295,29(1966).
6 K.Dušek, Collect.Czech.Chem.Commun.,34,1891(1969).
7 K.Dušek and B.Sedláček, Collect.Czech.Chem.Commun., 34,136(1969).
8 R.Chasset and P.Thirion, Proc.Internat.Conf.Non-Cryst.Solids, North-Holland Publ.Co, Amsterdam 1965, p.345.
9 M.Ilavský and W.Prins, Macromolecules, in press.
10 D.J.Worsfold, J.G.Zilliox, and P.Rempp, Can.J.Chem. 47, 3379(1969).
11 J.Malinský, J.Klaban, and K.Dušek, Collect.Czech. Chem.Commun., 34, 711 (1969).
12 T.Alfrey and W.G.Lloyd, J.Polymer Sci., 62,159 (1962).

THE SWELLING OF NONUNIFORMLY CROSSLINKED POLYMERS IN SOLVENTS

Z. Rigbi

Technion - Israel Institute of Technology

Department of Mechanics, Haifa, Israel

SUMMARY

The mechanics of swelling of a heterogeneously crosslinked polymer is considered by means of a model consisting of a spherical core of a highly crosslinked material within a shell of a lower degree of crosslinking. Use is made of the equations relating the stresses developed in each of the parts with the relevant extension ratios and the strain energy function, in order to determine the stress distributions. The dependence of the strain energy function on the first and third invariants of the extension ratios is discussed and derived for a swollen, crosslinked elastomer.

Is is shown that the average degree of crosslinking obtained from swelling is neither the arithmetic mean, nor the weighted volume average of the crosslinking densities of the two phases.

When a crosslinked polymer is placed in a liquid which is a solvent for the uncrosslinked polymer, it swells to an extent which is dependent upon the interaction between the polymer and the solvent, defined by a swelling potential, and the degree of crosslinking, which is a measure of the elastic potential. Equilibrium is reached when these two potentials, which act in opposite directions, are equal.

In his fundamental work, based on that of Meyer[1], Flory[2] extended the thermodynamics of polymer solutions to cover swollen, crosslinked polymers. The expressions developed include the number of network chains per unit volume or, in terms of the density

ρ of the polymer and $\tilde{M}$, the mean molecular weight between crosslinks, $\rho/\tilde{M}$. The mean used here is the number average, and it is not known in many systems whether this represents a single value, or a narrow or wide distribution. In other systems, such as divinylbenzene-styrene copolymers, the distribution of crosslinks has been studied and largely determined by Flory[3] and Stockmayer[4]. However, it is known that when the chemical species being copolymerized are of very different character, such as in polyester-styrene resins and alkyd resins, the calculated distributions may not be similar to those actually established.

The concept of the "gel point", when a very rapid build-up of the viscosity of crosslinking systems takes place and is followed by relative stabilization of polymer segments, implies that in crosslinked systems, molecular weights between crosslinks may vary widely, possibly even in a bimodal distribution with means widely separated. The results of experimental work by Cuthrell[5] may be interpreted as an indication that such nonhomogeneous crosslinking with a mode of higher crosslink density located within small, closed domains dispersed throughout a "continuous phase" of material of low crosslink density, or vice-versa, may occur preferentially in systems in which crosslinking takes place in the presence of large proportions of a diluent[6]. Similar distribution of molecular weight were postulated by Pinner[7] in discussing the crosslinking of pvc with allyl esters by ionising radiation.

Stein[8] described a method of studying the nonhomogeneity of crosslinking of rubber using light scattering, but other than this paper, there appear to have been no developments of even moderate success in dealing with this problem. As stated by Dusek and Prins[6], ".....a network is insufficiently described if only the numer of network chains and their length is specified. In addition, knowledge of the *topography of the chain length distribution* is required." (Authors' italics).

Two studies by the present author[9,10], dealing with the influence of rigid inclusions in crosslinked systems indicate how the effect of a non-uniform distribution of crosslink density of swelling may be studied. The methods of these papers are extended in the following paragraphs in an attempt to consider this problem.

Strain energy function. In order to solve the hyperelastic problem presented in a later section, it was first necessary to develop a strain energy function for elastomers swollen in solvents.

The previous formulation of the strain energy function proposed by the present author[9] has been criticised by Treloar[11] and Flory[12]. Treloar asserts that the function given by Gumbrell, Mullins and Rivlin[13] for the strain energy function, namely

$$W_1 = C_{dry}\ \phi_2^{\ 1/3}\ (I_1-3) \qquad (1)$$

where I_1 is the first invariant of the extension ratios $(\lambda_1^2+\lambda_2^2+\lambda_3^2)$, $C_{dry} = \rho RT/2\tilde{M}$, and ϕ_2 is the ratio of the volume of dry rubber to the swollen gel, is correct for a swollen rubber containing a fixed amount of solvent, and that only under these conditions is $\partial W/\partial I_2=0$. The situation described in my previous papers and in the present is such that the solvent/polymer system is in equilibrium with solvent surrounding it, and in this case, Treloar suggests that it is possible $\partial W/\partial I_2 \neq 0$. However, experimental verification is still not available and it is believed that because $\partial W/\partial I_2$ is very small for swollen elastomers[14], its effect on the strain energy function derived cannot be substantial.

Flory[12] draws attention to the fact that the interaction constant χ is a function of ϕ_2 and that the strain energy function derived in Ref.3 must be modified accordingly, although Gee's results[15] indicate that for one system χ is independent of ϕ_2. Experimentation of both the experimental data by Eichinger and Flory[16] and their theoretical reasoning shows that the constant χ may be written as $\chi_1 + \chi_2\phi_2$, a proposal made earlier on empirical grounds by Tompa[17]. Using this form, it is possible to derive the following expression for the strain energy dependence upon I_3, using the same arguments as in reference[9], assuming a crosslink functionality of 4:

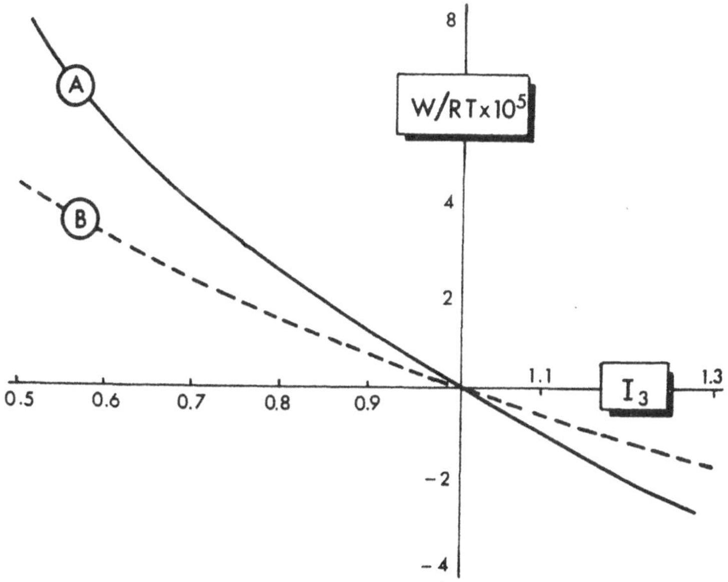

Fig.1: Strain-energy dependence on the third invariant of the extension ratios of a swollen natural rubber/benzene gel. A- as first plotted by Rigbi (Ref.9). B- based on constants obtained from Eichinger and Flory (Ref.16).

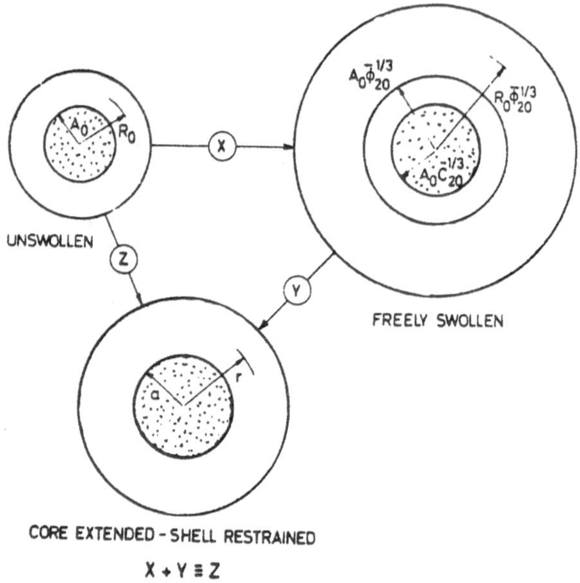

Fig.2: Scheme for calculating the mechanics of swelling of attached elastomer shell and core.

$$-\left(\frac{V_1^o}{RT}\right)W_3 = (I_3^{1/2} - \phi_{20})\ln(1-\phi_{20}I_3^{-1/2}) - (1-\phi_{20})\ln(1-\phi_{20})$$
$$- \chi_1\phi_{20}^2(I_3^{-1/2}-1) - \frac{1}{2}\chi_2\phi_{20}^3(I_3^{-1}-1) \quad (2)$$
$$- \frac{\rho V_1^o}{2M}\left[\phi_{20}^{1/3}(3I_3^{1/3} + 6I_3^{1/6}-9) - \frac{1}{2}\phi_{20}\ln I_3\right]$$

A plot of this function, compared with the function previously derived, appears in Fig.1. The partial derivative of this strain energy function with respect to I_3 at $I_3 = 1$ can be simply reduced to

$$\partial W/\partial I_3 = -\phi_{20}^{1/3}\cdot\rho RT/2\tilde{M} \quad (3)$$

and for the purposes of calculation, this will be assumed to be the value over the whole range. Examination of Fig.1 will show that this assumption is not likely to introduce large errors.

<u>Swelling of a highly crosslinked core surrounded by a loosely crosslinked shell.</u> As our next step in the establishment of a model for the swelling of the system, we now consider a spherical core of highly crosslinked polymer surrounded by spherical shell of lightly crosslinked material. From the point of view of the description of the system, the reverse situation (with the highly crosslinked material surrounding a lightly crosslinked core) is equally likely,

but rather less tractable analytically. No loss of generality in these considerations is envisaged by the restriction to the particular model selected.

The composite sphere, with core and shell inseparable at their common surface, is now allowed to swell to equilibrium in a solvent The final situation, as shown in Fig.2, may however be reached by a different route: this is, by allowing the core and the shell to swell freely and separately, each achieving its equilibrium radius. The subsequent application of a centrally directed tension f between the surface of the core and the inner surface of the shell to draw them together results in the identical geometry. It is now possible, using the methods described by Eringen[18], to determine the states of stress in both core and shell. These will result in a (hydrostatic) tension in the core, and a hydrostatic compression, together with a complicated deviatoric stress picture in the shell.

Hart-Smith and Crisp[19] give a general expression for the stresses developed in a system such as the one under discussion. This is

$$\sigma_i = \lambda_i^2\zeta + \lambda_i^2(\lambda_j^2 + \lambda_k^2)\xi + p$$

$$\zeta = \frac{2}{\sqrt{I_3}}\,\frac{\partial W}{\partial I_1} \qquad \xi = \frac{2}{\sqrt{I_3}}\,\frac{\partial W}{\partial I_2} \tag{4}$$

$$p = 2\sqrt{I_3}\,\frac{\partial W}{\partial I_3}$$

$$(i,j,k, = 1,2,3)$$

where λ is the extension ratio along the i-axis. Introducing the proper extension ratios, in the radial (r) and tangential (θ) directions, we obtain

$$I_1 = \lambda_r^2 + 2\lambda^2{}_\theta$$

$$I_2 = \lambda_r^4 + 2\lambda_r^2\lambda^2{}_\theta \tag{5}$$

$$I_3 = \lambda_r^2\lambda^4{}_\theta$$

We note that I_3 is the square of the ratio of the final to the initial elementary volume at a given point. In the prevailing situation, this can be written as $\lambda_r^2\lambda^4{}_\theta = (\phi_{20}/\phi_2)^2$ so that

$$I_1 = \lambda_r^2 + \frac{2}{\lambda_r}\,\frac{\phi_{20}}{\phi_2}$$

$$I_2 = \lambda_r^4 + 2\lambda_r\,\frac{\phi_{20}}{\phi_2} \tag{5a}$$

$$I_3 = \left\{\frac{\phi_{20}}{\phi_2}\right\}^2$$

Here ϕ_{20} is the value of ϕ_2 at zero pressure.

Upon carrying out the necessary operations, we obtain for the stresses σ in the radial and tangential directions,

For the shell:

$$\left(\frac{\tilde{M}}{\rho}\right)\frac{\sigma_r}{RT} = \frac{\phi_{20}}{\phi_2^{2/3}}\lambda_\theta^{-4} - \frac{\phi_{20}^{4/3}}{\phi_2} \tag{6}$$

$$\left(\frac{\tilde{M}}{\rho}\right)\frac{\sigma_\theta}{RT} = \frac{\phi_2^{4/3}}{\phi_{20}}\lambda_\theta^{2} - \frac{\phi_{20}^{4/3}}{\phi_2} \tag{7}$$

$$\left(\frac{\tilde{M}}{\rho}\right)\frac{\tilde{p}}{RT} = \frac{1}{3}\frac{\phi_{20}}{\phi_2^{2/3}}\left(\lambda_\theta^{-4} + 2\frac{\phi_2^2}{\phi_{20}^2}\lambda_\theta^2\right) - \frac{\phi_{20}^{4/3}}{\phi_2} \tag{8}$$

while Flory's equation also holds at the same time

$$\frac{\tilde{p}V_1^\circ}{RT} = \ell n(1-\phi_2) + \phi_2 + \chi_1\phi_2^2 + \chi_2\phi_2^3 + \frac{\rho V_1^\circ}{\tilde{M}}\left(\phi_2^{1/3} - \frac{\phi_2}{2}\right) \tag{9}$$

The boundary conditions for this situation are

$$\text{(1)}\quad \text{at } r = a \qquad \sigma_r = -f \qquad \lambda_\theta = a\phi_{20}^{1/3}/A_o \tag{10}$$

$$\text{(2)}\quad \text{at } r = b \qquad \sigma_r = 0 \qquad \lambda_\theta = b\phi_{20}^{1/3}/B_o$$

It will be noted that when $\phi_2 = \phi_{20}$, that is a situation of uniform swell exists, we have $\lambda_\theta = \lambda_r = 1$ and $\tilde{p} = \sigma_r = \sigma_\theta = 0$ as would be expected. If we use a series approximation for the logarithm, which is very accurate at high degrees of swelling, we may combine (8) and (9) to give

$$\lambda_\theta^{-4} + 2\frac{\phi_2^2}{\phi_{20}^2}\lambda_\theta^2 = 3\left(\frac{\phi_{20}}{\phi_2}\right)^{1/3} + 3\frac{\phi_2}{\phi_{20}} - \frac{3}{2}\frac{\phi_2^{5/3}}{\phi_{20}}$$
$$- \frac{3\tilde{M}}{\rho V_1^\circ}\cdot\frac{\phi_2^{2/3}}{\phi_{20}}\left[(1/2 - \chi_1)\phi_2^2 + (1/3 - \chi_2)\phi_2^3\right] \tag{11}$$

It is interesting to note the relationship between λ_θ and ϕ_2 is independent of the temperature.

The shell must be in a state of equilibrium established according to

$$\frac{d\sigma}{dr} = \frac{2}{r}(\sigma_\theta - \sigma_r) \tag{12}$$

The mutual interaction between the core and the shell subjects the former to a hydrostatic tension equal to σ_r, the radial stress developed in the latter. Under this tension, the swelling ratio in the core, c_2, will be

$$\frac{fV_1^o}{RT} = \ell n(1-c_2) + c_2 + \chi_1 c_2^2 + \chi_2 c_2^3 + \frac{\rho V_1^o}{M_c}(c_2^{1/3} - \frac{c_2}{2}) \tag{13}$$

where

$$f = 0 \quad \text{for } c_2 = c_{20}; \qquad f = \sigma_r \text{ when } \lambda_\theta = (c_{20}/c_2)^{1/3} \tag{14}$$

and

$$c_2 = (A_o/a) \tag{15}$$

Equations (8), (9), (12) and (13), taken with the boundary conditions (10) and (13) and the geometry of the core/shell system before swelling, are sufficient to determine the geometry of the core/shell system after swelling.

A typical example. In order to demonstrate the above in figures, the following case has been assumed:

System: Natural Rubber - Benzene

$\chi_1 = 0.4$ $\chi_2 = 0.11$ $V_1^o = 87$ $\rho = 0.92$

$\tilde{M} = 7833$ $M_c = 527$

$\phi_{20} = 0.18$ $c_{20} = 0.506$

Utilising a computer program which was written for this purpose, $\tilde{p}$ was calculated from equation (9) for a range of values of ϕ_2, and using these values, λ_θ was determined from equation (8), and corresponding values of σ_r and σ_θ were calculated from (6) and (7) respectively. These are shown plotted as functions of λ_θ on fig.3. A curve relating $f(=\sigma_r)$ to $(c_{20}/c_2)^{1/3}$ was plotted and the intersection of these two curves determines the value of λ_θ and σ_r at the interface. In this particular case, the values obtained are

$$c_2 = 0.495 \qquad \lambda_\theta = 0.72$$

The initial situation (refer to fig.1) is that of a core of radius A_o in a shell of the same internal radius. Free swelling of the core would give a radius of 1.26 A_o which would then be increased by the pull of the shell to give a radius of 1.27 A_o. The figures would be more impressive if it were stated that under the influence of the shell, the core would have a volume of 1.498 times its original volume, while its freely swollen volume is only 1.465 times the original. The internal radius of the shell, if allowed to swell freely, would have increased to a value of 1.765 A_o, whereas under restraint from the core, it increased to only 1.27 A_o. The stress σ_r at the interface 2.38 × 10^{-4} RT dynes/sq.cm.

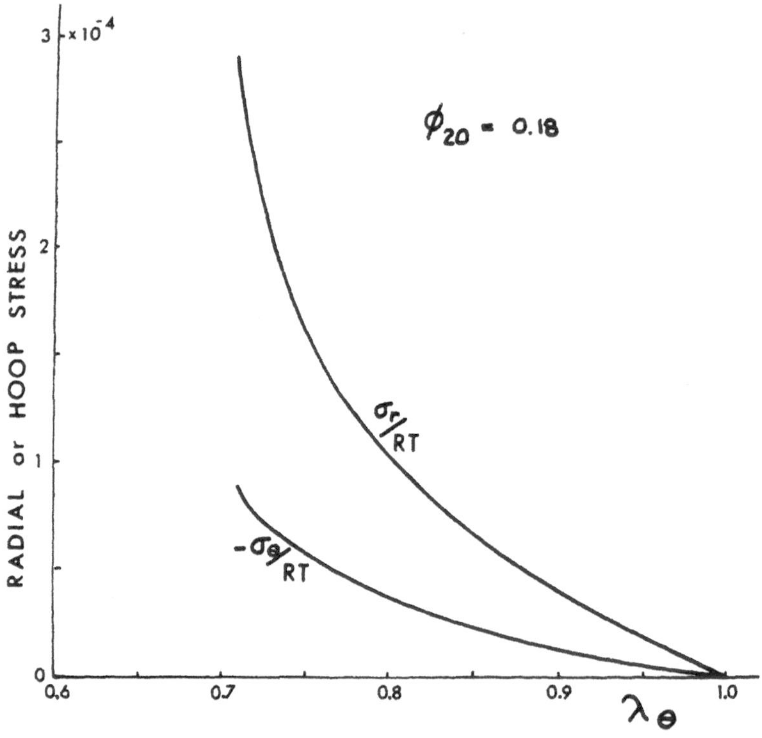

Fig.3: Radial and hoop stresses as functions of the hoop extension ratio. Dimensions of the stress are in cc/mole.

The stress distribution within the sphere must be determined if the influence of the core on the total swelling must be calculated. No doubt, a suitable computer program may be established for this purpose, but it has been found relatively simple to carry out the operations graphically. Thus we rewrite (12) as follows

$$r\,\frac{d\theta}{dr} = \frac{2(\sigma_\theta - \sigma_r)}{d\sigma_r/d\lambda_\theta} \tag{12a}$$

The numerator of the r.h.s. of this equation is easily determined from fig.3 as a function of λ_θ. The denominator can also be determined using any method, such as that of differences. A curve showing $r d\lambda_\theta/dr$ as a function of λ_θ can therefore be plotted, as in fig. 4. Now, except for the region where $\lambda_\theta > 0.85$, the line is

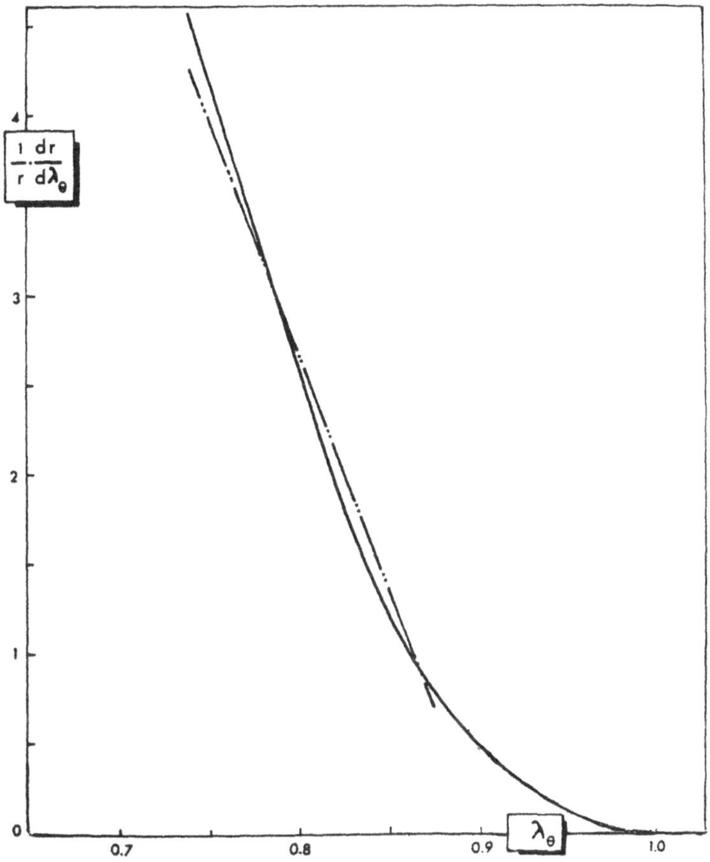

Fig.4: (See text)

virtually straight, and can be approximated by

$$\frac{1}{r}\frac{dr}{d\lambda_\theta} = \frac{8}{3}\,(9-10\,\lambda_\theta) \tag{16}$$

Integration, subject to the proper boundary condition results in

$$\ln(r/A_o) = \frac{8}{3}(9-5\lambda_\theta)\lambda_\theta - 9.769 \tag{17}$$

over the range of $\lambda_\theta = 0.72$ to 0.85. For values of $\lambda_\theta > 0.85$, it was found simpler to work by differences. The results are shown in fig.5 where both the freely swollen radius (R) and the corresponding restrained radius (r) are shown as functions of the initial

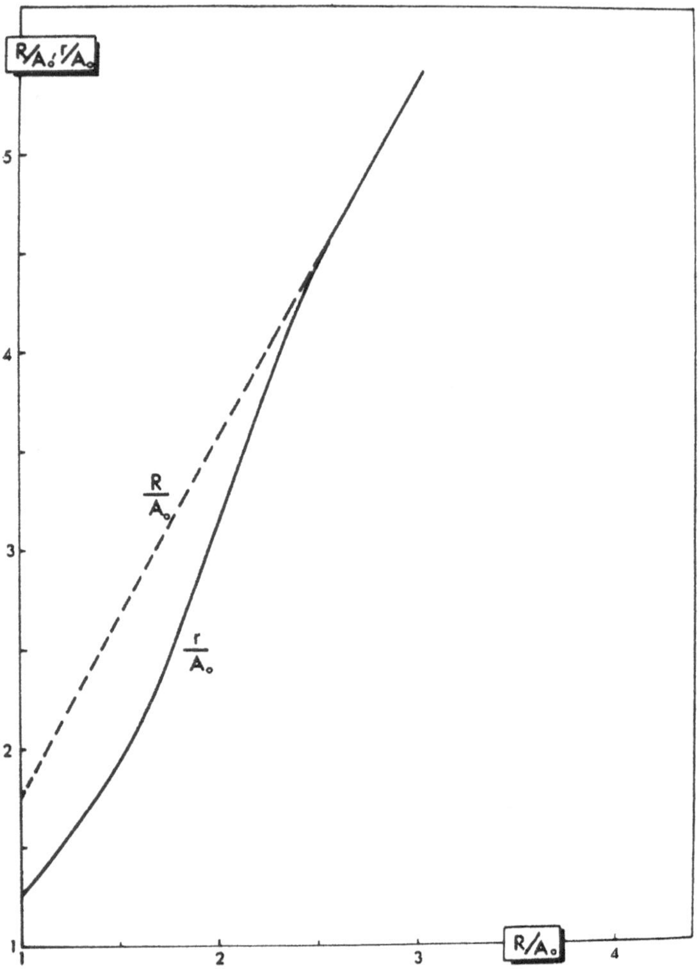

Fig.5: Free and restrained swollen radii as functions of initial position radius.

radius ($R_o = R\phi_{20}^{1/3}$). It will be seen that there is a maximum restraint at the surface of the core, but that the restraint falls rapidly as the distance from the core is increased. At a distance of $2\frac{1}{2}$ radii, the difference of swelling due to the restraint imposed by the core virtually vanishes.

We can now proceed to compare the swollen volume obtained by assuming

(a) Two different mean molecular weights

(b) A single volume-average molecular weight for the core/shell structure.

As above, let $\tilde{M} = 7833$ and $M_c = 527$, and let the outer radius of the shell $B_0 = 2A_0$. The swollen volume of the core in benzene will be 1.498 times the initial volume, while the shell will swell, in accordance with fig.5, to give a radius $b = 3.22\ A_0$. The total volume of the sphere (core and shell) therefore equals $4.19 \times (3.22A_0)^3 = 135.59\ A_0{}^3$. The average dry rubber content in this swollen sphere will be $4.19\ (2A_0)^3 = 33.52\ A_0{}^3$ giving a ratio of $\phi_{20} = 0.247$.

This would correspond to a "swelling-average" molecular weight between crosslinks of 4687. The "volume-average" molecular weight between crosslinks is $(527 + 7 \times 7833)/8 = 6920$. On the other hand, if our compound sphere were able to accomodate free swelling of each of its parts (which is a geometrical impossibility) the dry rubber content which would result would be 0.172, to be compared with the "average" of 0.247.

Conclusion: I have shown that the total swelling of a system comprising a core of highly crosslinked elastomer and a tightly bound shell of more loosely crosslinked elastomer of the same chemical structure is lower than can be expected from the crosslink density of the separate components. Where a real crosslinked polymer can be assumed to consist of a "dispersed phase" of highly crosslinked volumes within a matrix of lower crosslink density, the "average" as determined by swelling does not correspond to a numer average over the whole volume, but must be smaller than it.

BIBLIOGRAPHY

1. H.K. Meyer, *Z. Phys. Chem.*, B44, 383 (1939).

2. P.J. Flory, Principles of Polymer Chemistry, Cornell University Press, 1953.

3. P.J. Flory, *J. Amer. Chem. Soc.*, 63, 3097 (1941).

4. W.H. Stockmayer, *J. Chem. Phys.*, 12, 125 (1944).

5. R.E. Cuthrell, *J. Appl. Polym. Sci.*, 12. 1263 (1968).

6. K. Dusek and W. Prins, *Adv. Polym. Sci.*, 6, 1 (1969).

7. S.H. Pinner, *Plastics*, 25, 35 (1960).

8. R.S. Stein, *Polym. Letters*, 7, 657 (1969).

9. Z. Rigbi, *Int. J. Engng. Sci.*, 7, 1163 (1969).

10. Z. Rigbi, paper read at Int. Rubber Conf., Paris, June 1970.

11. L.R.G. Treloar, private communication dated September 30, 1968.

12. P.J. Flory, private communication dated August 27, 1969.

13. S. Gumbrell, L. Mullins and R.S. Rivlin, *Trans. Faraday Soc.*, 48, 200 (1952).

14. A. Ciferri and P.J. Flory, *J. Appl. Phys.*, 30, 1498 (1969).

15. G. Gee, *J. Chem. Soc.*, p.280, (1947).

16. B.E. Eichinger and P.J. Flory, *Trans. Faraday Soc.*, 64. 2035 (1968),

17. H. Tompa, *C.R. II Reunion Chem. Phys. (Paris)* 1962, p. 163.

18. A.C. Eringen, Non-Linear Theory of Continuous Media, McGraw-Hill, 1962.

19. L.J. Hart-Smith and J.D.C. Crisp, *Int. J. Engng. Sci.*, 5, 1 (1967).

BIREFRINGENCE ANALYSIS OF INHOMOGENEOUS SWELLING IN FILLED ELASTOMERS

T. Kotani* and S. S. Sternstein

Materials Division, Rensselaer Polytechnic Institute

Troy, New York 12181

SUMMARY

The inhomogeneous swelling of an elastomeric matrix containing isolated, spherical filler particles has been described in detail elsewhere [7]. In this paper, the birefringence resulting from the spherically symmetric swelling field is considered in detail. Equations are derived that predict the retardation of polarized light that would be observed in a plane. Experimental results on crosslinked natural rubber containing bonded glass spheres and swollen in benzene and o-dichlorobenzene are given.

Observations on the interfacial retardation, decay of retardation with distance from the filler, effect of filler size, and the dependence on Flory-Huggins χ parameter are in quantitative agreement with the theory. Additional data on the effect of crosslink density indicate that cohesive failure of elastomer at low crosslinking and non-Gaussian network statistics and/or failure at high crosslinking must be taken into consideration.

INTRODUCTION

The compounding of fillers into rubber is a widely used means of reinforcement that improves modulus, tensile strength, and wear resistance. While the mechanism by which reinforcement occurs is still obscure, the heterogeneous stress field in the elastomeric matrix must be a significant factor that requires additional

* Present address: Japan Synthetic Rubber Co.

characterization [1,2].

In non-filled vulcanizates of homopolymers and random copolymers, physical properties such as elongation at break and strength are reasonably understood as functions of crosslink density and the chemical composition of the elastomer. However, the measurement of crosslink density in filled elastomers is a difficult problem that has received much attention [3,4].

A common technique for measuring crosslink density is by equilibrium swelling measurements. In filled elastomers, the swelling is anomalous in that simple correction for the volume fraction of elastomer is insufficient to account for the observed swelling diminution. Scanlan [5] has suggested that the interfacial swelling restriction of elastomer which adheres to the filler surface may be an important factor. Additional calculations by Rigbi [6] have followed a similar line of reasoning, but invoke unnecessary approximations in the formulation and solution for the stress and deformation fields.

Recently, Sternstein [7] has formulated the problem of inhomogeneous swelling about isolated spherical filler particles with generality, the theory being limited only by the extent to which the swelling thermodynamics are known. In this paper, an experimental study of the inhomogeneous swelling field by birefringence techniques is shown to be in quantitative agreement with the theory.

INHOMOGENEOUS SWELLING

A formulation for the inhomogeneous swelling of an elastomer matrix containing a spherical inclusion has been obtained for arbitrary free energy of mixing and network strain energy functions, rigid or soft inclusion, and finite or infinite matrix size [7]. Calculations have been performed using the Flory-Huggins theory and Gaussian network theory to describe the free energy of mixing and network strain energy function, respectively, for a rigid inclusion in an infinite matrix, a situation which corresponds to the

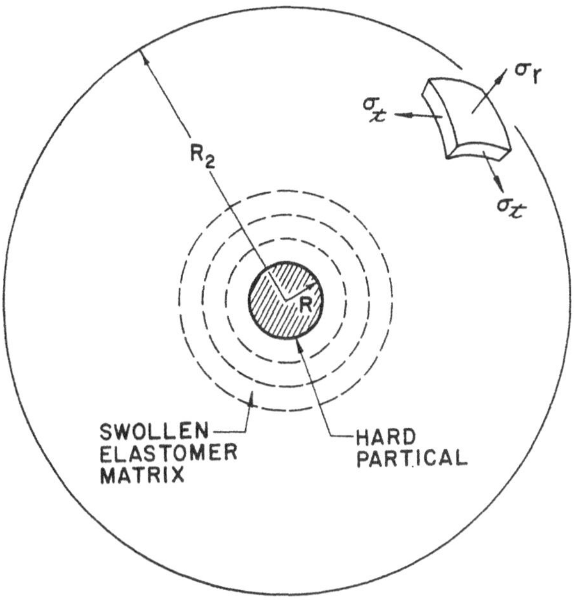

Figure 1. The spherically symmetric swelling field about a rigid inclusion.

experiments reported here.

The spherical symmetry of the stress field produced by swelling is illustrated in Fig. 1, and is characterized by a radial stress (σ_r) and two equal tangential stresses (σ_t), which represent the three physical, principal components of the stress tensor. Similar symmetry applies to the deformation field which is characterized by a radial extension ratio (λ_r) and two equal tangential extension ratios (λ_t).

Typical behavior of the deformation field as a function of distance from the inclusion is shown in Fig. 2. The extension ratios are defined on the basis of swollen, stressed state relative to unswollen, unstressed state. The associated stress field is shown in Fig. 3, and the local volume fraction of elastomer, $v_2 = 1/\lambda_r\lambda_t^2$, is given in Fig. 4. In Figs. 2-4, radius r represents the reduced distance from the center of the rigid inclusion, whose radius is taken to be unity.

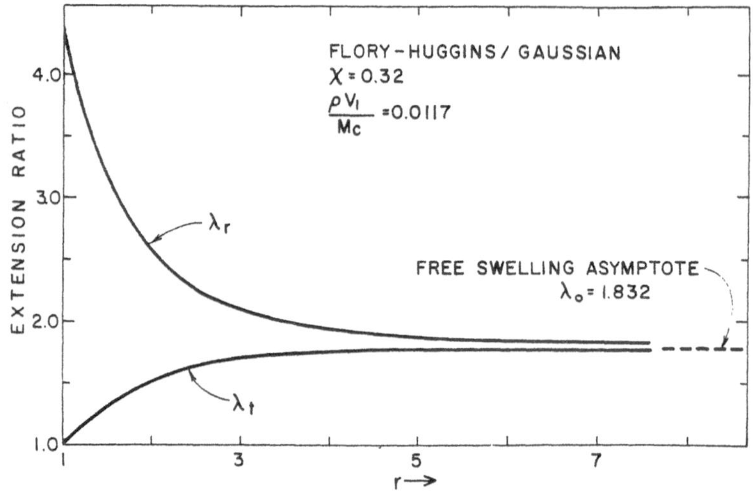

Figure 2. The deformation field produced by inhomogeneous swelling for the parameters shown. Radius r represents the reduced distance from the inclusion surface.

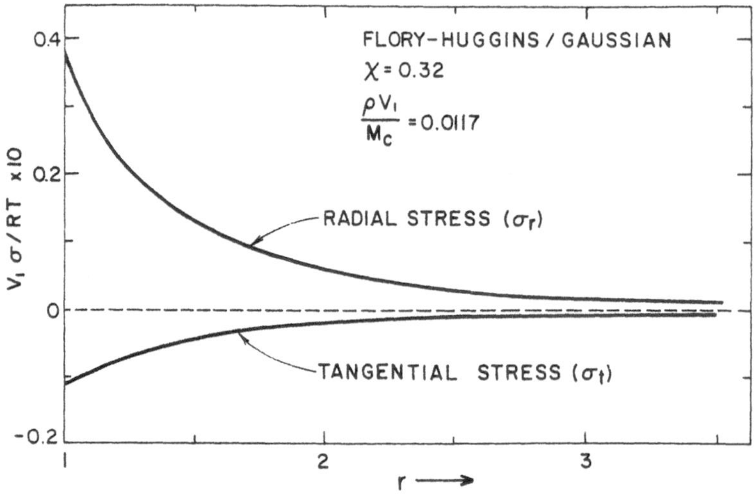

Figure 3. The stress field associated with the deformations shown in Fig. 2.

These results are for a Flory-Huggins $\chi = 0.32$ and a crosslink constant $\rho V_1/M_C = 0.0117$, where ρ is the density of dry rubber, V_1 is the molar volume of the solvent, and M_C is the molecular weight between crosslinks. The calculations are based on the assumption

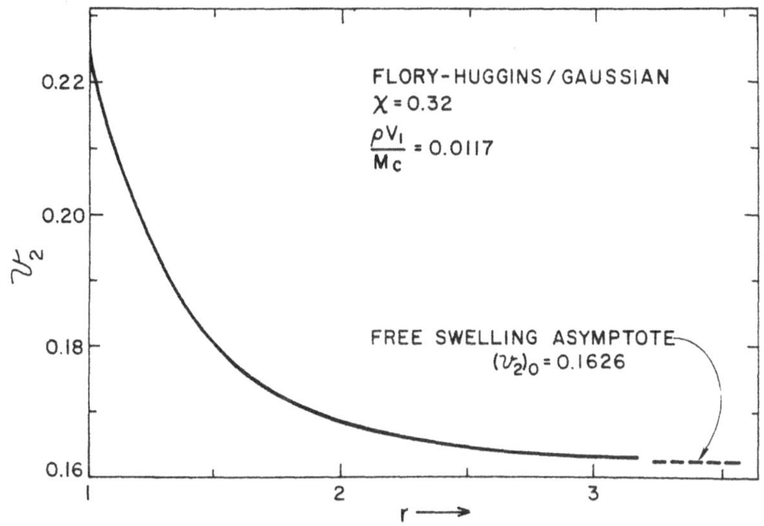

Figure 4. The local volume fraction of elastomer associated with the deformations in Fig. 2.

that perfect adhesion exists between the matrix and inclusion. Imperfect adhesion or partial failure of the interface or surrounding matrix as a result of the swelling would diminish the values shown in Figs. 2-4. Also, a non-rigid inclusion would result in lower values for σ_r, λ_r, and v_2 in the interfacial region. Thus, for a given χ and $\rho V_1/M_c$, Figs. 2 and 3 represent the maximum values of stress and deformation that can result from swelling restriction by an isolated, spherical filler particle, provided that Flory-Huggins/Gaussian swelling behavior is applicable. For this reason, an experimental verification of the theory is desirable.

THE RETARDATION INTEGRAL

The anisotropic deformation state shown in Fig. 2 results in a birefringence field which can be calculated and compared with experimental results. The optical indicatrix possesses the same symmetry as the deformation field, and is characterized by principal refractive indices $n_1 = n_r$ in the radial direction and $n_2 = n_3 = n_t$ in any direction tangent to a surface of constant radius, both of

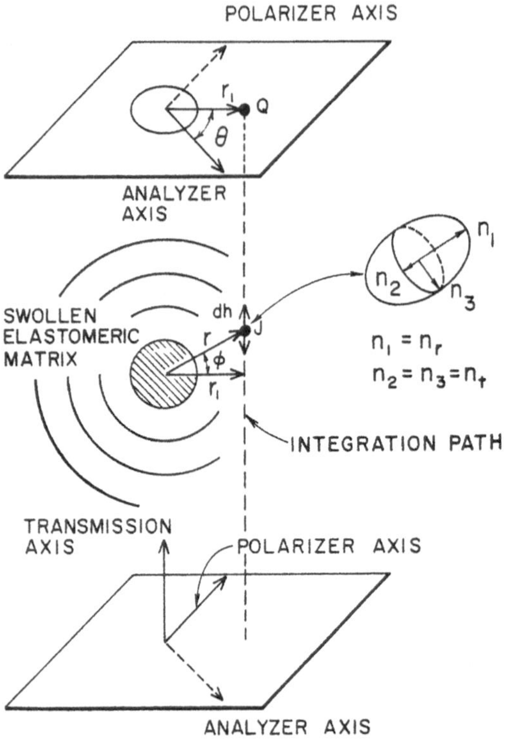

Figure 5. Schematic of the geometry used to derive the retardation integral.

which are functions of the radius (r) of the point under consideration.

With reference to Fig. 5, it is required to calculate the retardation of light at a point Q which is at a distance r_1 from the center of projection of the inclusion onto the polarizer-analyzer plane, and at an angle θ with respect to the analyzer axis. The retardation measured at Q represents the additive retardation of all matrix points which lie on a line parallel to the transmission axis and passing through Q. Holding r_1 and θ constant, an integration on ϕ, the angle made by the radius vector of any matrix point with respect to a plane parallel to the polarizer-analyzer plane, is required to obtain the observable retardation. By varying the parameters r_1 and θ, the entire birefringence pattern can, in principle, be determined.

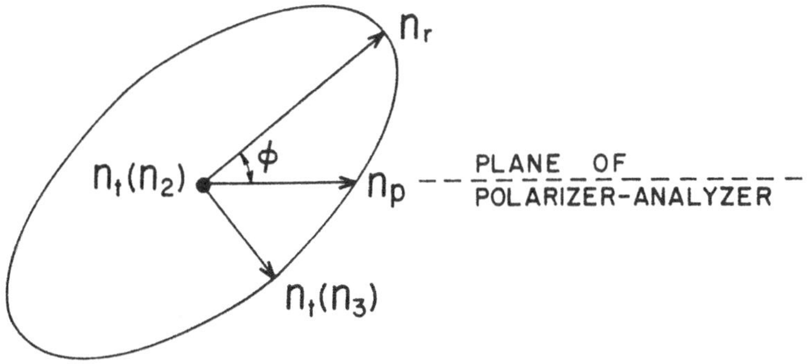

Figure 6. A cross section of the refractive index ellipsoid.

It can be shown that the intensity of light which is transmitted through any matrix point is proportional to $\sin^2 2\theta$, giving zero intensity along the polarizer and analyzer axes (θ = 0 and $\pi/2$), and maximum intensity at $\theta = \pi/4$. Also, the four quadrants defined by the polarizer and analyzer axes are identical, and the birefringence pattern is invariant upon rotation of the polarizer-analyzer pair because of the projected symmetry of the spherically symmetric matrix. For these reasons, it will suffice to calculate the retardations along the brightest line, namely, the locus of points Q which makes an angle $\theta = \pi/4$ with respect to the analyzer axis.

Since the optical indicatrix is circularly symmetric about the major axis n_r, one of the minor axes, n_t (n_2) can be chosen such that it lies in a plane parallel to the polarizer-analyzer plane. Taking a cross-section through the optical indicatrix gives an ellipse whose major axis is n_r and minor axis is the remaining n_t (n_3). As shown in Fig. 6, the refractive index in the polarizer-analyzer plane n_p is given by the elliptical formula

$$\frac{(n_p \cos \phi)^2}{n_r^2} + \frac{(n_p \sin \phi)^2}{n_t^2} = 1$$

where, from Fig. 5,

$$\cos \phi = r_1/r \qquad \text{and} \qquad \sin \phi = (r^2 - r_1^2)^{1/2}/r$$

Solving for n_p,

$$n_p = n_r n_t [r/r_1] \left[\frac{1}{(n_t^2 - n_r^2) + (r/r_1)^2 n_r^2} \right]^{1/2} \quad (1)$$

Since n_p and n_t (n_2) lie in the polarizer-analyzer plane and are perpendicular to each other, the local birefringence at the point J (see Fig. 5) is given by $(n_p - n_t)$:

$$n_p - n_t = n_t \left[n_r(r/r_1) \left(\frac{1}{(n_t^2 - n_r^2) + (r/r_1)^2 n_r^2} \right)^{1/2} - 1 \right] \quad (2)$$

where the refractive indices n_r and n_t are related to the extension ratios λ_r and λ_t in the next section. The differential thickness dh, measured in the transmission direction at the point J, is obtained by noting that r_1 is constant along the path and $\tan \phi = (r^2 - r_1^2)^{1/2}/r_1$. Thus

$$dh = d[(r^2 - r_1^2)^{1/2}] = d(r_1 \tan \phi) = r_1 \sec^2 \phi \, d\phi \quad (3)$$

or equivalently,

$$dh = \frac{(r/r_1)\, dr}{[(r/r_1)^2 - 1]^{1/2}} \quad (4)$$

Because the radius, r, is expressed as a reduced distance (relative to unit radius inclusion) in Figs. 2-4, the physical path length is given by R dh, where R is a scale factor representing the actual radius of the inclusion. Thus, the differential retardation, dP, caused by an element of swollen rubber at point J is given by

$$dP = v_2(n_p - n_t) R \, dh$$

where v_2, the local volume fraction of elastomer, is included to account for composition variations along the path of integration.

The observable retardation at point Q is obtained by summing the contributions dP, giving

$$P/R = \int v_2(n_p - n_t)\, dh \quad (5)$$

where $(n_p - n_t)$ is given by eq. 2 and dh by eq. 3 or eq. 4, depending on the choice of integration variable. For an infinite matrix, the limits of integration on ϕ are $-\pi/2$ to $+\pi/2$, with the symmetry allowing the interval to be reduced to zero to $\pi/2$ and the result doubled. For the case where r is chosen as the integration variable,

$$P/R = 2\int_{r_1}^{\infty} v_2(n_p - n_t)\ \frac{(r/r_1)}{[(r/r_1)^2 - 1]^{1/2}}\,dr \tag{6}$$

where v_2 and $(n_p - n_t)$ are functions of r, but not of r_1.

The integration of eq. 6 by any finite difference technique gives some initial difficulty due to the singular behavior of dh in eq. 4 at the point $r = r_1$. This problem is easily circumvented by initiating the integration using incremental values of ϕ, since eq. 3 is well behaved for $\phi = 0$, and subsequently switching to integration with respect to r, say when ϕ achieves a value such that $r/r_1 > 1.02$. It is expedient to use r as the integration variable when possible, since the swelling deformations, e.g. Fig. 2, are generated numerically at equal increments of r. The infinite upper limit of eq. 6 poses no problem since the integrand approaches zero (i.e. $n_r \to n_t$) at large r, which is to be expected for isotropic, homogeneous swelling.

BIREFRINGENCE THEORY

The homogeneous stretching of a swollen rubber results in a birefringence $(n_1 - n_2)$ that, according to Kuhn and Grün [8], is given by

$$(n_1 - n_2) = K\nu v_2(\lambda_1^2 - \lambda_2^2) \tag{7}$$

where

$$K = \frac{2\pi}{45}\,\frac{(\bar{n} + 2)^2}{\bar{n}}\,(\alpha_{\parallel} - \alpha_{\perp})_{seg} \tag{7a}$$

and $\bar{n}$ is the mean refractive index $[(n_1 + n_2 + n_3)/3]$, $(\alpha_{\parallel} - \alpha_{\perp})_{seg}$ is the anisotropic polarizability per mole of segments, $\nu(=\rho/M_c)$

is the crosslink density in the dry state, and λ_1 and λ_2 are the extension ratios (referred here to unswollen, unstressed state) in the plane whose birefringence is $(n_1 - n_2)$. The factor v_2 is introduced for the correction of the crosslink density to unit volume of swollen elastomer.

The optical constant K has been studied for various polymer-solvent systems by Kuhn and Grun [8], Treloar [10], Hermans [9], Stein and Tobolsky [11], Furukawa et. al. [12], and others. Since the problem of concern here involves inhomogeneous swelling, the dependence of K on concentration (v_2) requires consideration. For a given polymer-solvent system, the effect of v_2 on $(\alpha_{\parallel} - \alpha_{\perp})_{seg}$ is probably small and can be neglected for the purpose of this work [13]. The effect of concentration on mean refractive index can be represented adequately by the formula

$$\bar{n} = n_s + v_2(n_e - n_s) \tag{8}$$

where n_s and n_e are the refractive indices of the pure solvent and dry elastomer, respectively, This equation has been verified by refractive index measurements on the systems natural rubber (NR) - benzene and NR-o-dichlorobenzene which were swollen to various levels of v_2.

Making the usual assumption that the birefringence arising from the deformation of swollen elastomer is a result of the orientation of optically anisotropic segments, then the mean macroscopic polarizability remains constant, $(\alpha_1 + \alpha_2 + \alpha_3) = 3\bar{\alpha}$, where α_i is the polarizability along the i principal axis. Using the linear approximation of the Lorenz-Lorentz equation, one obtains

$$\Delta n_1 + \Delta n_2 + \Delta n_3 = 0 \tag{9}$$

where $\Delta n_i = n_i - \bar{n}$. Applying eqs. 7 and 9 to the spherically symmetric swelling field, and solving for the refractive indices n_r and n_t:

$$n_r = 2K\nu v_2(\lambda_r^2 - \lambda_t^2)/3 + \bar{n} \tag{10}$$

$$n_t = -K\nu v_2(\lambda_r^2 - \lambda_t^2)/3 + \bar{n} \tag{11}$$

Table 1. Values of the Retardation Integral, P/R for Natural Rubber: $\nu = 1.04 \times 10^{-4}$ mole/cc, K(dry) = 4.98 cc/mole, $n_e = 1.518$

Solvent	Benzene	o-Dichlorobenzene
n_s	1.498	1.551
χ	0.44	0.32
r_1	P/R ($\times 10^3$)	P/R ($\times 10^3$)
1.00	2.65	3.16
1.20	1.98	2.38
1.40	1.54	1.87
1.60	1.23	1.51
1.80	1.01	1.24
2.00	0.84	1.04
2.20	0.71	0.89
2.50	0.56	0.71

where the value of K is a function of v_2 through eqs. 7a and 8, and v_2, λ_r, and λ_t depend on the distance (r) from the filler surface.

THEORETICAL RETARDATION VALUES

The numerical evaluation of the retardation integral, eq. 6, in conjunction with eqs. 2, 7a, 8, 10, and 11 has been coupled to the procedure used to obtain the inhomogeneous swelling field, for example Figs. 2-4. The retardations are given in Table 1 as functions of distance from the filler surface ($r_1 = 1$) for two swelling agents, together with the appropriate physical constants at 30°C. The maximum retardation of polarized light occurs at the particle-matrix interface and this value is given in Table 2 as a function of crosslink density for the natural rubber-benzene system. The effects of crosslink density on the interfacial swelling, radial stress, and radial extension ratio are given in detail elsewhere [7].

EXPERIMENTAL PROCEDURE

Spherical glass beads of various diameters were washed with carbon tetrachloride and then acetone to remove surface oils and dust. The dried beads were treated with 10% sodium hydroxide

Table 2. Effect of Crosslink Density on the Maximum Retardation (at r_1 = 1.0): Natural Rubber - Benzene (Physical constants as given in Table 1).

ν(mole/cc)	P/R (x10³) at r_1 = 1.00
1.11×10^{-5}	0.59
1.04×10^{-4}	2.65
2.22×10^{-4}	4.32
1.11×10^{-3}	10.84

solution for six hours, washed thoroughly, and redried. To insure surface adhesion to the rubber, the beads were coated with an epoxy prepolymer-polyamine hardener mixture and partially cured at room temperature. A suspension of the bead-precoat mixture was prepared in carbon tetrachloride and additional polyamine added. Curing was then completed and the solvent removed.

About 0.02 phr of beads and 0.5 phr of antioxidant 425 were mixed with pale crepe and various quantities of dicumyl peroxide on three inch open rollers. The mixtures were press-cured for 30 minutes at 160°C in sheets of one to two mm thickness.

The crosslink densities were calculated from swelling measurements in benzene and o-dichlorobenzene through the use of the Flory-Rehner equation [14,15]. For the natural rubber-benzene system, a value of χ = 0.44 was used, and for NR-o-dichlorobenzene, χ = 0.32. Consistent crosslink densities were obtained from both solvents. The swelling ratios were determined from weight gain measurements, and the effects of 0.02 phr glass beads on the swelling were neglected.

For the birefringence measurements, a trough was constructed for the microscope stage which allowed swelling of the sample and subsequent measurement of the retardations with the sample immersed. Polarized blue light (486 mμ) was passed through the swollen sample and the retardations around various isolated glass beads were measured along a line at 45 degrees to the crossed polarizer-analyzer axes by means of a Berek compensator.

For a perfectly symmetric swelling field, rotation of the sample stage would leave the observed birefringence field unaltered in direction and intensity, and the retardation at any point unchanged. For each radial distance (r_1 in Fig. 5) from the filler particle six retardations were measured, corresponding to 60 degrees rotational increments of the sample stage. The average of the six readings was adopted as a datum. The various retardations differed at most by ±0.5 degrees on the Berek compensator setting (with a compensator constant of 3.936). This represents the combined errors of slight deviations from symmetric swelling and the usual experimental reproducibility of retardation measurements by compensator techniques.

RESULTS AND DISCUSSION

The light intensity pattern produced by swelling in the vicinity of a rigid spherical inclusion, when viewed with polarized light and crossed polarizer-analyzer axes, is shown in Fig. 7. The complete extinction along the polarizer axis (vertical) and analyzer axis (horizontal), approximate symmetry of the four quadrants, and invariance of the pattern when the sample stage is rotated are as expected from theoretical considerations. This pattern is typical of those samples which retained their interfacial integrity upon swelling.

In those cases where the crosslink density was insufficient to produce a network structure capable of withstanding the high radial extension ratio at the interface, a cohesive failure of the elastomer in the interfacial region developed. This is shown in Fig. 8 by the dark ring which separates two regions of higher intensity along a line at 45 degrees to the horizontal (this is most clearly seen in the first quadrant). The minimum crosslink density required to prevent failure upon swelling depends on the swelling agent. Thus, a value of $\nu = 1 \times 10^{-4}$ mole/cc is adequate to prevent cohesive failure with benzene as the solvent, but not with hexane (as shown in Fig. 8).

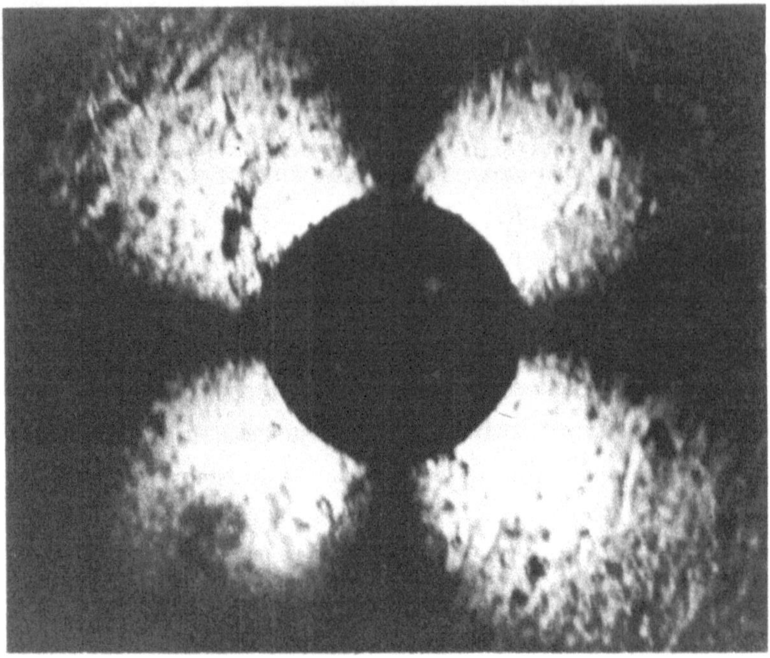

Figure 7. Light intensity pattern produced by swelling around a spherical inclusion (R ∿ 100μ). Natural Rubber - Benzene. (Crossed polarizer-analyzer).

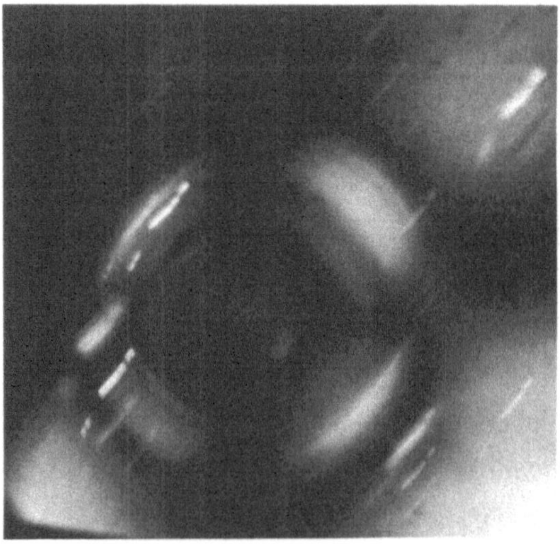

Figure 8. Birefringence pattern produced by swelling when cohesive failure of the matrix occurs. Note the two light regions separated by a dark ring in the first quadrant. Natural Rubber - Hexane.

Table 3. Effect of Particle Size on Maximum Retardation, Natural Rubber - Benzene, $\nu = 1.04 \times 10^{-4}$ mole/cc

Particle Radius R (μ)	Measured Retardation at $r_1 = 1.0$ (mμ)	P/R ($\times 10^3$)
142	377	2.67
118	316	2.68
100	262	2.62
95	262	2.76
71	190	2.68
47	128	2.72
	Theoretical (from Table 1)	2.65

Measurements of the maximum retardation (P at $r_1 = 1.0$) using different radii filler particles in the natural rubber-benzene system are given in Table 3. The constancy of P/R and agreement with theory further indicate no appreciable adhesive or cohesive failure in the interfacial region, since such failure would significantly lower the retardation value by creating local zones of nearly zero stress and birefringence.

The results given in Table 3 also suggest that there is no significant geometrical perturbation of the spherically symmetric field. Such perturbations would arise from particle-particle interactions or the effect of the finite thickness samples. For a 2 mm thick sample, a 200μ diameter particle gives an aspect ratio of only 10/1. However, any given particle could be close enough to the surface of the sample to significantly alter the swelling field from that given by the infinite matrix solution.

Inspection of Fig. 2 and eqs. 2, 6, 10, and 11 shows that the smaller values of r_1 (say $r_1 < 3$) result in values of retardation (P/R) which are determined primarily by the highly birefringent region within several radii of the particle surface, with little contribution to the integral from r = 3 to infinity. Conversely, for larger r_1 (say $r_1 > 3$) the highly birefringent region is removed from the integration (note the limits of integration in eq. 6) and the retardation is numerically sensitive to the asymptotic region

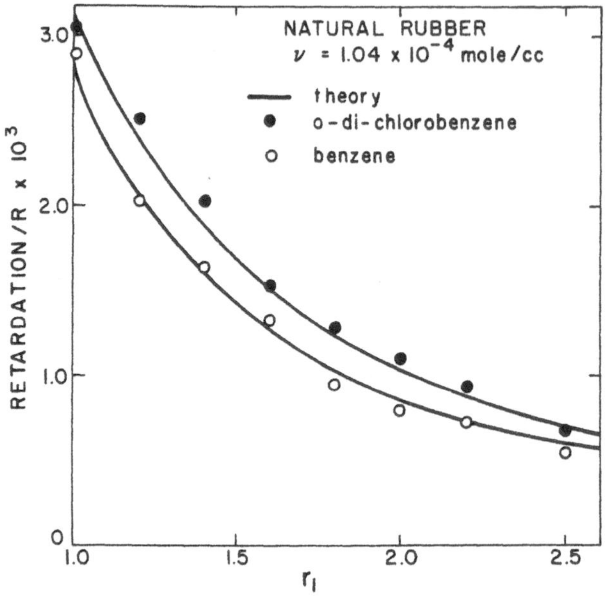

Figure 9. Experimental and theoretical retardations as a function of distance from the inclusion.

of swelling. It can be concluded that values of retardation for $r_1 < 3$ are not overly dependent on the sample thickness and associated perturbations of the swelling field. The results in Table 3 substantiate this.

The dependence of measured retardation with distance (r_1) is shown in Fig. 9 for the systems NR-benzene and NR-o-dichlorobenzene. These data were obtained using particles of ca. 100μ radius. The theoretical curves represent the results given in Table 1. Since the physical constants used in the theory were obtained independently of the inhomogeneous swelling experiments, the agreement in Fig. 9 is quantitative and contains no adjustable parameters. It is concluded that the inhomogeneous swelling theory [7] describes correctly the deformation and stress fields about an isolated, spherical filler particle.

Experimental retardations have also been obtained on samples

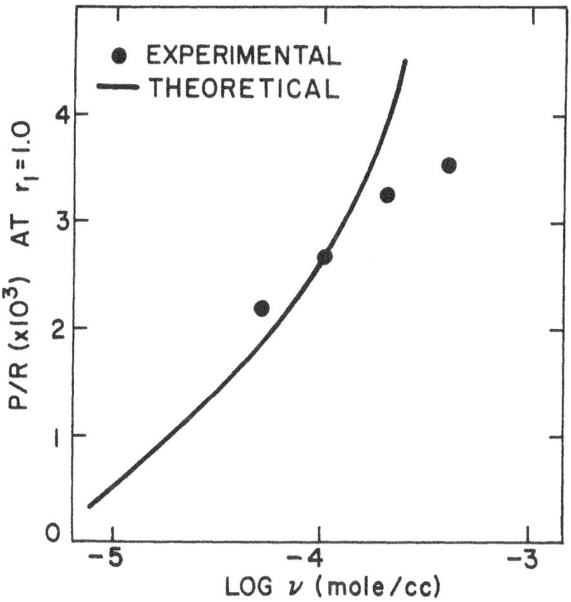

Figure 10. The effect of crosslink density on the maximum retardation for the system natural rubber - benzene.

with higher crosslink density than shown in Fig. 9. The maximum retardations are shown as a function of crosslink density and compared with the theoretical values (Table 2) in Fig. 10. The deviation from the theoretical curve may be attributed to several factors. First, the calculations are based on Flory-Huggins/ Gaussian swelling behavior which may not be an adequate description of the thermodynamics in the interfacial region. In particular, deviations from Gaussian network behavior should increase with increasing crosslink density, especially in the highly deformed and stressed interfacial region. Additional calculations using non-Gaussian network statistics will be made to assess the contribution of this factor to the retardation integral.

Second, some local cohesive failure of the elastomer that is not resolvable optically may be occuring in the interfacial region. Previous calculations [7] have shown that the interfacial radial stress increases but that the radial extension ratio decreases with increasing crosslink density. The failure characteristics of swollen elastomer in the particular triaxial stress field considered here are not known at present. The negative deviations in Fig. 10 suggest that microfailure may be a significant factor. If this is true, then the stress induced failure at high crosslink density and lack of network integrity at low crosslink density may limit the range of applicability of the theory insofar as this factor is concerned.

Light scattering studies provide another experimental technique for analysis of swelling in filled elastomers. In a recent study, Stein and coworkers [16] have demonstrated that the inhomogeneous swelling theory is also in agreement with their measurements. Stein [13] has noted that light scattering may provide a very sensitive technique for detecting the onset of localized failures because of the major changes in the scattering pattern which occur when cavities initiate.

The authors hope that this work may provide additional experimental and theoretical bases by which the characterization of filled elastomers can be improved. Ultimately, a complete analysis of interfacial pheomena should provide a basis for understanding of the filler reinforcement mechanism.

ACKNOWLEDGEMENTS

The authors are grateful for a General Electric Company grant which supported one of them (T.K.) as a post-doctoral research associate. One of us (S.S.S.) wishes to express his gratitude to Professor Richard S. Stein for discussions which initiated our interest in inhomogeneous swelling phenomena.

This work was performed in the Materials Research Center which is supported by a grant from the National Aeronautics and Space Administration.

The natural rubber crepe samples were obtained through the kind assistance of Dr. S. L. Aggarwal of the General Tire and Rubber Co.

REFERENCES

1. K. Fujimoto, Kogyozairyo 16, No. 8, 35 (1969).

2. G. Kraus, Rubber Chem. Technol. 38, 1070 (1965).

3. G. Kraus, Rubber Chem. Technol 37, 6 (1964); J. Appl. Polymer Sci. 7, 861 (1963).

4. W. R. Krigbaum, R. J. Roe, Rubber Chem. Technol. 38, 1039 (1965).

5. J. Scanlan, Rev. Gen. Caoutch. 41, 514 (1964).

6. Z. Rigbi, Technion (Israel Institute of Technology) Reports TDM 69-11 (Nov. 1969); TDM 70-3 (March 1970); see also paper by Rigbi, this symposium.

7. S. S. Sternstein, J. Macromolecular Sci., Pt. B, Physics, In Press.

8. W. Kuhn, F. Grün, Kolloid Z, 101, 248 (1942).

9. J. J. Hermans, Kolloid Z, 103, 210 (1943).

10. L. R. G. Treloar, The Physics of Rubber Elasticity, 2nd Ed., Oxford. 1958.

11. R. S. Stein, A. V. Tobolsky, Textile Research J. 18, 201, 302 (1948).

12. J. Furukawa, S. Yamashita, T. Kotani, T. Kawashima, J. Appl. Polymer Sci. 13, 2527 (1969).

13. R. S. Stein, University of Massachusetts, Private communication.

14. P. J. Flory, J. Chem. Phys. 18, 108 (1950).

15. P. J. Flory, J. Rehner, J. Chem. Phys. 4, 521 (1943).

16. C. Picot, M. Fukuda, C. Chou, and R. S. Stein, J. Macromolecular Sci. Pt. B., Physics, In Press; see also article in this symposium

DEPOLARIZED LIGHT SCATTERING FROM SWOLLEN-FILLED RUBBER[‡]

C. Picot[*], M. Fukuda[+], C. Chou, and R. S. Stein[ø]

Polymer Research Institute and Department of Chemistry

University of Massachusetts, Amherst, Massachusetts 01002

SUMMARY

When rubber containing glass sphere filler particles swells, the swelling is inhomogeneous in that tangential strain is not permitted at the surface of the filler particle providing the rubber remains attached to this particle. Consequently, the degree of swelling in the immediate vicinity of the filler particle will be less than in the bulk of the rubber and there will be a gradient of swelling with distance away from the particle. This leads to a stress gradient in which the radial and tangential component of stress are different. The theory of this effect has been developed by Sternstein. The rubber will become birefringent as a result of this stress gradient leading to a difference between radial and tangential refractive indices. Wilkes and Stein have developed a

[*]Present Address: C.R.M., 6 rue Boussingault, 67 Strasbourg, France.

[+]Present Address: Central Research Laboratory, Denki Kagaku Kogyo Company, Tokyo, Japan.

[‡]Supported in part by a contract with the Office of Naval Research and in part by grants from the General Tire and Rubber Company, the Petroleum Research Fund of the American Chemical Society and the National Institute of Health. This material has been submitted for publication to J. Macromolecular Science - Physics.

[ø]To whom correspondence should be sent.

theory for scattering from material containing such a spherically symmetrical refractive index difference gradient. The equations for this gradient from Sternstein's theory were inserted in the Wilkes-Stein theory and the H_v light scattering patterns obtained by observing the sample between cross polaroids has been calculated. Experimental patterns have been obtained using synthetic cis-1,4-polyisoprene rubber filled with small glass spheres. The patterns are in agreement with the theoretically calculated ones and verify the theory of Sternstein. With a dispersion of glass beads of different sizes, the higher order scattering maxima must smooth out, but if the scattering is obtained from single large glass beads these maxima may be resolved.

INTRODUCTION

In recent publications,[1,2] Stein and Wilkes have pointed out that light scattering can originate from deformed regions surrounding voids and inclusions in solid high polymers. As it arises from birefringence gradients in such strained regions, the light scattering will be depolarized.

Rubbers containing fillers such as glass spheres are inhomogeneous and scatter light. However, if the glass and rubber are isotropic, the scattering will be principally of the V_v type (vertically polarized scattered light from vertically polarized incident light) and the depolarized H_v component (horizontally polarized scattered light) will be weak.

When such a filled rubber is swollen, however, stresses result because of inhomogeneity of swelling. Assuming that the rubber remains firmly bound to the sphere during swelling, a tangential strain is not possible at the rubber-glass interface but a radial strain can occur as a result of swelling. Thus, the rubber will be subjected to a biaxial strain in the vicinity of the glass sphere with a resulting biaxial stress. This will result in a uniaxial birefringence having its principal axis along the radial direction and which will asymptotically approach zero with increasing distance from the center of the sphere. This birefringence gradient fulfills the requirements of the Stein-Wilkes theory and will lead to an enhanced H_v scattering component.

STERNSTEIN THEORY

In recent calculations Sternstein[3] has worked out the problem of swelling of rubber containing spherical filler particles subjected to the boundary conditions that the tangential strain is zero at the surface of the particle. He has obtained expressions

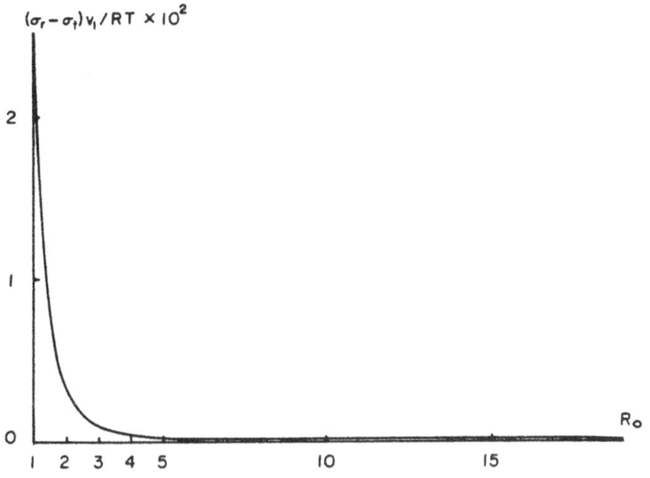

Figure 1. Variation of $(\sigma_r - \sigma_t)\ V_1/RT$ vs. R_o where V_1 is the molar volume of solvent in the swollen state.

for the radial and tangential stress and strain as a function of distance r away from the center of the particle. Figure 1 gives an example of his results of the variation of the anisotropic stress $(\sigma_r - \sigma_t) = \sigma(R_o)$, as a function of the reduced variable $R_o = r/r_o$ where r_o is the size of the particle. This variation corresponds to a volume fraction of the rubber in the swollen state $v_2 = 0.25$. It is then possible to relate the stress to the resulting optical birefringence.

STRAIN-BIREFRINGENCE RELATIONSHIP

According to Kuhn and Grün theory, indeed, the birefringence of an unswollen rubber network submitted to a homogeneous strain can be expressed by the relationship:

$$\Delta n = \frac{2}{45}\pi \frac{(\bar{n}^2 + 2)^2}{\bar{n}} Y_e\ (b_1 - b_2)\ (\lambda_1^2 - \lambda_2^2) \qquad (1)$$

where Δn is the difference of the refractive indices along and perpendicular to the elongation directions, $\bar{n}$ is the average refractive index, Y_e is the number of effective network chains per cm^3, b_1 and b_2 are the polarizabilities parallel and perpendicular to the direction of a link and λ_1 and λ_2 are the principal extension ratios.

In the case of a swollen network, this relationship becomes:

$$\Delta n' = \frac{2}{45}\pi \frac{(\bar{n}^2 + 2)^2}{\bar{n}} Y_e (b_1 - b_2) v_2^{1/3} (\lambda_1^2 - \lambda_2^2) \quad (2)$$

where v_2 is the volume fraction of rubber in the swollen state.

Now, the anisotropic stress $(\sigma_1 - \sigma_2)$ is related to the extension ratios by

$$\sigma_1 - \sigma_2 = kT\, Y_e\, v_2^{1/3} (\lambda_1^2 - \lambda_2^2) \quad (3)$$

where k is Boltzmann's constant and T is the absolute temperature.

In our particular case, as we are dealing with a spherically symmetrical distribution of the stresses, we can identify $\sigma_1 - \sigma_2$ with $\sigma_r - \sigma_t$ and in the same way $\Delta n'$ with $n_r - n_t$, n_r and n_t being the radial and tangential refractive indices. Thus from (2) and (3)

$$\frac{n_r - n_t}{\sigma_r - \sigma_t} = \frac{(\bar{n}^2 + 2)^2}{\bar{n}} \frac{2\pi}{45kT} (b_1 - b_2) \quad (4)$$

which is nothing else than the stress optical coefficient, C. In the case of natural rubber, $C = 2.38 \cdot 10^{-10}$ cm^2/dyne. It is possible that this value may be affected by solvent as indicated by studies of Gent[5] and Nagai[6] on polyisoprene and polybutadiene. Since such an effect depends upon v_2 which varies with radius, it may be that the stress-optical coefficient varies to some extent with radius. This possibility was neglected in the studies reported here.

A direct test of this birefringence variation has been carried out by Kotani and Sternstein[7] by observation of the polarization pattern between crossed polaroids using a microscope. They have found a distribution of birefringence which quantitatively agrees with the prediction of their theory.

LIGHT SCATTERING CALCULATIONS

The amplitude of light scattered at small angles can be calculated by using the expression derived by Stein and Wilkes[1] for symmetrical regions where the optic axis is directed along the radius. For crossed polarizer and analyzer (H_V), it can be written:

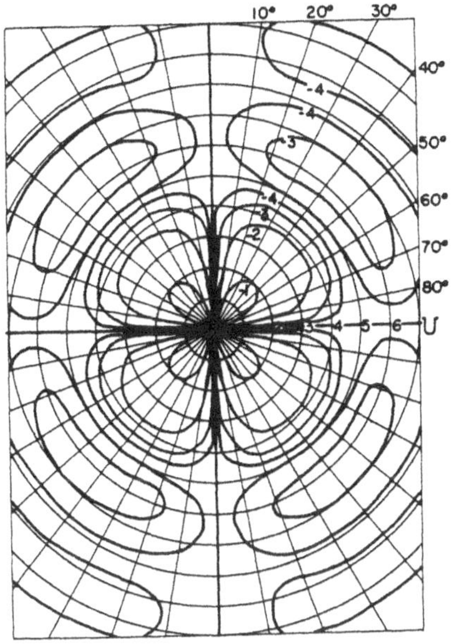

Figure 2. (a) Theoretical H_V small angle light scattering pattern resulting from the anisotropic surrounding of particles embedded in a swollen crosslinked polymer network (v_2 = 0.25).

$$(E_S)_{H_V} = C' \cos\rho_2 \sin\mu \cos\mu \frac{r_o^3}{U^3} \int_{R_o=0}^{\infty} [n_r (R_o) - n_t (R_o)] \left[3 \cos(UR_o) + UR_o \sin (UR_o) - 3 \frac{\sin (UR_o)}{UR_o}\right] dR_o \quad (5)$$

where C' is a numerical coefficient, $\cos\rho_2 = \cos\theta/[\cos^2\theta + \sin^2\theta \sin^2\mu]^{1/2}$, θ and μ are the scattering and azimuthal angles defining the scattering direction, $U = (4\pi r_o/\lambda) \sin (\theta/2)$ and λ is the light wavelength in the medium.

By substituting into (5) the value of $(n_r - n_t)$ as a function of $(\sigma_r - \sigma_t)$ for $r > 1$ and taking $(n_r - n_t) = 0$ for r less than one and using Sternstein's values, we calculated the scattered intensity distribution by numerical integration ($I_{H_V} = (E_S)_{H_V} \cdot (E_S)^*_{H_V}$) using the CDC 3600 computer at the University of Massachusetts Research Computing Center.

Fig. (2a) gives the theoretical patterns observed in the polar coordinates (U, μ). The patterns possess four-fold symmetry where the intensity along the 45° azimuthal direction [see Fig. (2b)] passes through several maxima whose position characterizes the size of the filler particles.

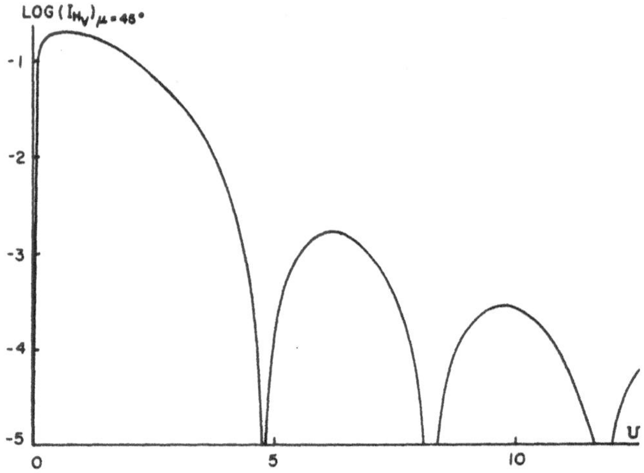

Figure 2. (b) Variation of I_{H_V} along the 45° azimuthal direction.

It is noted that the scattered intensity is zero at $\mu = 0°$ and 90° as a result of the sinμ cosμ term in Eqn. (5). With increasing size of the glass sphere, the scattering occurs at smaller angles.

FILM PREPARATION

Experiments were carried out on samples of synthetic cis-polyisoprene (kindly supplied by Dr. K. W. Scott of the Goodyear Tire and Rubber Company, Akron, Ohio). This rubber is called Natsyn-400.

Samples containing one to five parts of dicumyl peroxide (dicup) crosslinking agent per hundred parts of rubber were used to prepare two types of films; one containing a large number of small glass spheres and a second containing a smaller number of larger spheres.

For the first type, glass spheres having a size range of 5 - 10μ were used (Microbeads Division, Cataphote Corporation, Jackson, Mississippi). These were added to an approximately 3% solution of the rubber in benzene, mixed, and cast into films of about 0.5 mm thickness on a teflon pan. After evaporating the benzene at room temperature, these were dried for about 16 hours in an oven at 45°C, then transferred to a press where they were cured by heating at 90°C for 20 minutes and then heated to 140°C for 40 minutes under 3000 psi pressure.

A second type film was prepared which contained larger beads of about 29μ diameter identical with those used by Kotani and Sternstein[7] (3M Company Superbrite glass beads). These scattered sufficiently to permit observation of the scattering patterns arising from deformed regions about single beads. Some difficulty was experienced in binding these beads sufficiently well to the rubber so that they did not separate under the swelling stresses. Kotani and Sternstein used an epoxi adhesive for this purpose. We found a silane treatment more convenient.

The beads were first washed with acetone and then treated with a 10% aqueous NaOH solution at 50°C for 1 hour. After washing the beads were then treated with a freshly prepared 10% aqueous solution of Dow Corning Z-6020 silane coupling agent at room temperature for 30 minutes (Dow Corning Corporation, Midland, Michigan). This agent is N-β-aminoethyl-γ-aminopropyl trimethoxy silane. The beads were then removed from the silane and dried in an oven at 110°C for 3 hours.

Two procedures were used for dispersing these beads in the rubber. The first was a solution blending procedure similar to that previously described. The second, which was somewhat more satisfactory, involved dusting the glass spheres on the surface of the rubber film, folding the film, pressing at 60°C, releasing the pressure, refolding and repressing for about one hundred times to achieve uniformity of dispersion. In all cases the concentration of glass was less then 0.01 g/cm^3.

Films were swollen by immersing in xylene for 24 hours at room temperature to approach equilibrium swelling in the range of v_2 of about 0.35. They were then placed while wet between glass microscope slides for optical investigations.

LIGHT SCATTERING

Light scattering photographic patterns were obtained using a laser scattering apparatus as described by Rhodes and Stein[8].

The set of pictures [Fig. (3)] show some characteristic H_v patterns which have been obtained. Picture (3a) corresponds to unswollen filled rubber and is characterized by weak diffuse scattering, probably due to important depolarizing reflections at the particle (n = 1.486) and rubber (n = 1.519) interface. Figure (3b) shows that unfilled swollen rubber produces only a very weak anisotropic pattern which can be attributed to network

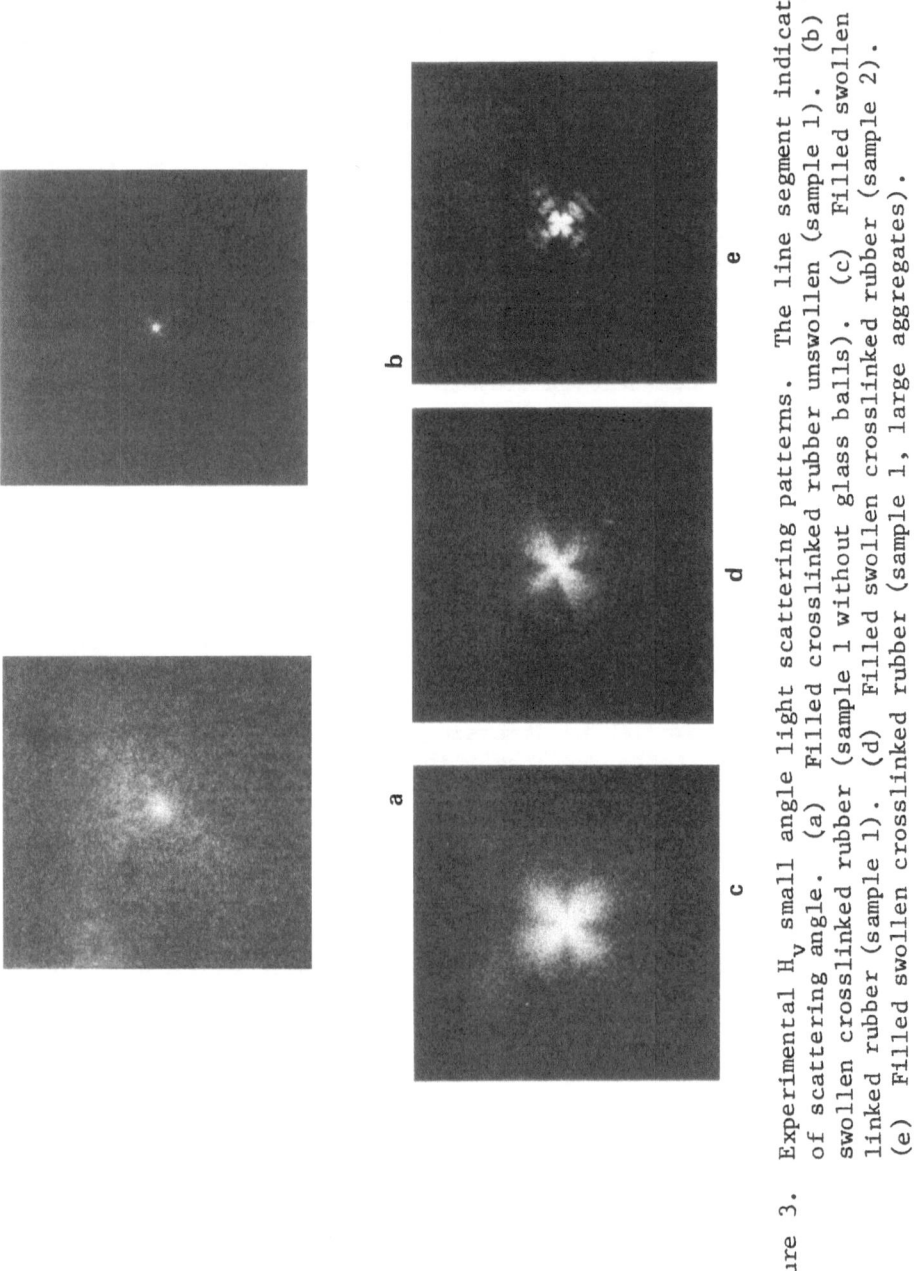

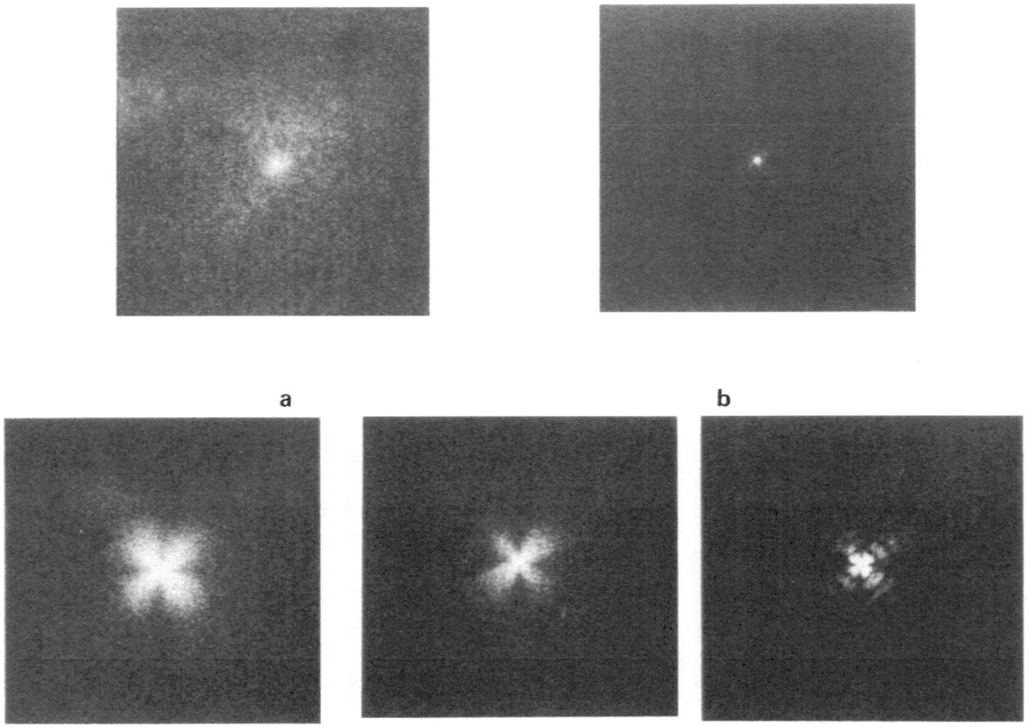

a b c d e

Figure 3. Experimental H_v small angle light scattering patterns. The line segment indicates 10° of scattering angle. (a) Filled crosslinked rubber unswollen (sample 1). (b) Unfilled swollen crosslinked rubber (sample 1 without glass balls). (c) Filled swollen crosslinked rubber (sample 1). (d) Filled swollen crosslinked rubber (sample 2). (e) Filled swollen crosslinked rubber (sample 1, large aggregates).

inhomogeneities giving rise to some local orientation fluctuation[9]. This contribution seems to be much smaller than the scattering arising from the anisotropic regions of a filled swollen rubber. This can be seen in Fig. (3c) which, for the same experimental conditions, exhibits a strong H_v scattering intensity.

Figure (3d) shows that by decreasing the crosslinking degree, the scattering intensity is lowered as a result of the decrease of the stress at the particle-rubber boundary. The comparison of experimental pictures to theoretical patterns of Fig. (2a) leads one to conclude that only the central part of the pattern can be observed experimentally. This hypothesis seems justified by the fact that in this condition the dimension of the pattern [Fig. (3c)] corresponds to particle sizes about 8μ which is in the range of sizes of the particles introduced in the rubber. On the other hand, the secondary maximum intensity is about a hundred times weaker than the first maximum intensity and as the distribution in size of the particles washes out the modulation of intensity, it appears difficult to record a complete scattering pattern.

Thus, it is seen that the presence of a filler within a swollen rubber leads to heterogeneity in anisotropy which gives rise to H_v scattering which is not seen with either the unfilled swollen rubber nor the filled unswollen rubber and characterizes the stress field in the vicinity of the heterogeneity.

The higher order part of the scattering pattern may be seen in Fig. (3e) which corresponds to a particular location of the sample where the glass balls aggregated, leading to an effective filler particle of about 25μ size which dominates the scattering. Such a sample effectively represents the scattering from this single large particle so that the higher order maxima are not averaged out because of heterogeneity of particle size.

This hypothesis may be tested by examining samples of the second type which are intentionally prepared with a low concentration of large glass spheres. In this case, the spheres are sufficiently far apart so that regions containing single or just a few spheres are localized in the area of the sample illuminated by the laser beam, and, because of the r_0^3 term in Eqn. (5), the intensity of light scattered is sufficient so that such particle scattering patterns may be conveniently observed.

In Fig. (4), a photomicrograph is shown for a xylene swollen sample between crossed polaroids of a crosslinked rubber containing 29μ average diameter silane coated glass beads. The four-leaf clover pattern which extends into the rubber over distances greatly exceeding the diameter of the beads is evident and is similar to

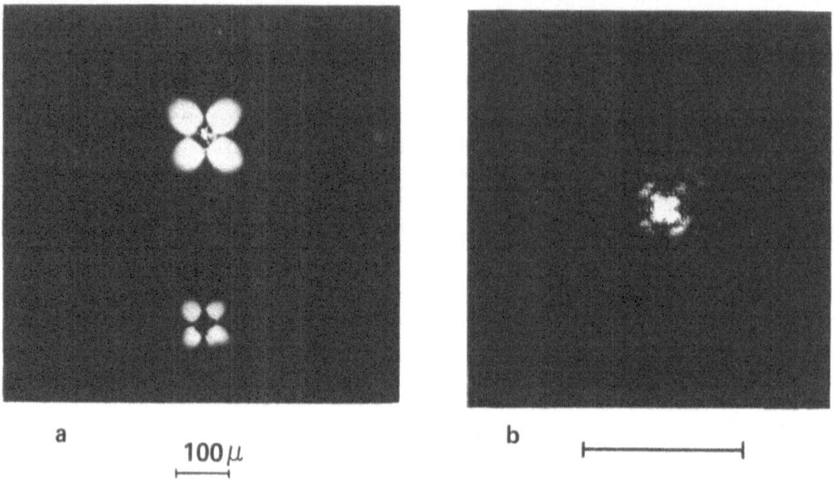

Figure 4. (a) A photomicrograph between crossed polaroids of a xylene swollen crosslinked rubber containing silane coated glass beads with an average particle size of 29μ and (b) an H_V small angle light scattering pattern from this sample. The line segment represents 5° of scattering angle.

the patterns first experimentally obtained and theoretically predicted by Kotani and Sternstein[7].

A scattering pattern is shown in Fig. (5), in which the higher order scattering maxima is clearly seen. It is noted that fine structure is seen within the pattern of a type previously seen for scattering from starch particles and polystyrene spherulites[10] which is believed to arise from interference between scattered rays from a small number of scattering regions. The patterns are qualitatively similar to those predicted in Fig. 2.

Scattering patterns similar to those observed here for model systems are observed for filled rubbers of commercial interest. For example, Fig. (5) shows a scattering pattern obtained for a swollen natural rubber sample obtained from Dr. P. Thirion of the Institut Francais du Caoutchouc, Paris, France, consisting of a natural rubber sample containing 20 parts/100 of Hi - Sil 233 (PPG Industries, Barberton, Ohio) hydrated silica filler. Similar

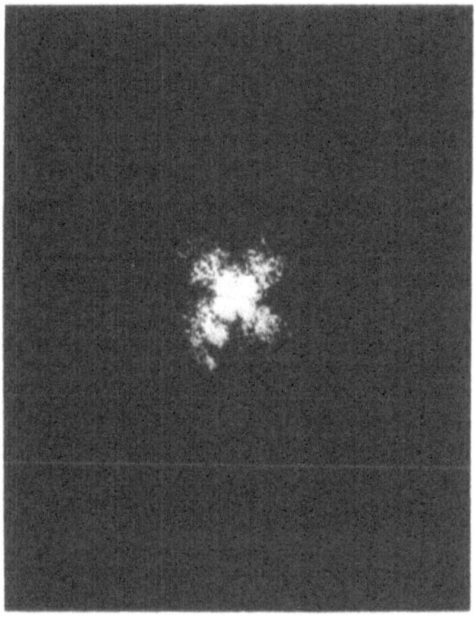

Figure 5. An H_v small angle light scattering pattern from a crosslinked swollen natural rubber sample containing 20 parts/100 rubber of Hi - Sil 233 silica filler.

patterns were obtained from crosslinked swollen films of Natsyn 400 containing 5 parts/100 of Hi - Sil 233 furnished through the courtesy of Dr. K. W. Scott of Goodyear Tire and Rubber Company. The patterns are characteristic of stress patterns surrounding heterogeneities several microns in size. The ultimate particle size of these filler particles is of the order of 200Å but histograms and electron micrographs furnished by Dr. M. P. Wagner of PPG Industries show the existance of aggregates ranging up to several microns in size. It is believed that the intense scattering arising from these larger aggregates completely dominates the scattering pattern.

An obvious extension of the present work (now in progress) is to observe the patterns arising from inhomogeneous deformation of the rubber in the vicinity of the filler particle. So long as no phase separation occurs at the particle-rubber interface, the resultant strains should predominately contribute to the H_v pattern. However, when phase separation does occur, the very pronounced density discontinuities resulting should give rise to an intense

V_V scattering pattern, the angular distribution of which should be characteristic of the size and shape of the phase separated region.

It is noted that the light scattering method complements the photoelastic investigation of strains in filled systems. Its advantages are that it permits extension of measurements to cases where the size of the heterogeneity is too small to be readily resolved in the microscope and also permits the ready determination of the average properties for systems containing a very large number of heterogeneities in the field of view. Furthermore, the light scattering technique may be readily adopted to the determination of the time-dependent changes of a system.

ACKNOWLEDGEMENTS

The authors are indebted to the assistance of several persons. In addition to Drs. K. W. Scott, P. Thirion and M. P. Wagner as referred to in the text, we appreciate the assistance and advice of Dr. S. Aggarwal of the General Tire and Rubber Company. We are indebted to Professor S. S. Sternstein and Dr. T. Kotani of Rensselaer Polytechnic Institute for their theoretical efforts in the calculation of the stress pattern in the vicinity of a spherical particle in the swollen rubber and their permission to use their results prior to publication. One of us (MF) appreciates the granting of an academic leave from his company, Denki Kagaku Kogyo of Tokyo, Japan.

REFERENCES

1. R. S. Stein and G. L. Wilkes, J. Polymer Sci. A-2, 7, 1696 (1969).
2. G. L. Wilkes and R. S. Stein, ONR Technical Report No. 122, Project NR 056-378, Contract: Nonr 3357(01), University of Massachusetts, Amherst, 1969.
3. S. S. Sternstein, J. Macromolecular Sci., B, in press.
4. L. R. G. Treloar, THE PHYSICS OF RUBBER ELASTICITY, Oxford, 1967.
5. A. N. Gent, Macromolecules 2, 262 (1969).
6. K. Nagai, J. Polymer Sci. A-2, 7, 1123 (1969); Polymer Journal (Japan) 1, 116 (1970).
7. T. Kotani and S. Sternstein, "Inhomogeneous Swelling and Resultant Birefringence in Filled Elastomers," A.C.S. Symposium on Highly Crosslinked Polymer Networks, Chicago, September, 1970, to be published in symposium proceedings by Plenum Press.

8. M. B. Rhodes and R. S. Stein, Symposium on Resinographic Methods, Special Tech. Publ. No. 348, ASTM (1963); M. B. Rhodes, D. A. Keedy and R. S. Stein, J. Polymer Sci. 62, S73 (1962).
9. R. S. Stein, J. Polymer Sci. B, 7, 657 (1969).
10. J. Borch, R. H. Marchessault, C. Picot and R. S. Stein, submitted for publication.

LIGHT SCATTERING AS A TOOL IN THE CHARACTERIZATION OF POLYMER MATERIALS

J. J. van Aartsen

AKZO Research Laboratories

Arnhem, The Netherlands

SUMMARY

The present contribution deals with light scattering as one of the techniques available to obtain information about the supermolecular order in polymeric materials. Existing theories are briefly reviewed, and it is shown that all theories have certain drawbacks. It is further shown that certain general assumptions enable a general theory to be derived. This newly derived theory is used to demonstrate

a) the kind of information one may expect to obtain from light scattering experiments
b) which experiments should be done in order to be able to evaluate the maximum number of structural parameters
c) the relation between the light scattering parameters and the supermolecular structure

The present theory is also compared with model calculations on a general model of a random collection of rods of finite length and radius.

For systems which are isotropic as a whole it is shown that the angular dependence of the scattered light intensity can give information about correlations in density and orientation. Together with absolute measurements it is possible to obtain a consistent picture in terms of such phenomena as phase separation and inhomogeneous crosslinking as well as the degree of local parallelization of chain segments. It is also shown that it is possible to distinguish between rod-like and disk-like types of local orientation.

The classical theory of rubber elasticity as used for calculating the mechanical properties of swollen and non-swollen, cross-linked, non-crystalline polymeric materials, is essentially based on a model of a random collection of macromolecular chains which are macroscopically homogenously cross-linked. Although this model has been rather successful, it has become quite evident in recent years, that such systems are not entirely disordered. In the literature (9,13) as well as at this symposium, proofs are given of the existence of a certain ordering in gels as well as amorphous (glassy) polymers. Although the exact type of ordering in these systems is far from understood, it is generally assumed that basically two types can be present.
In the first place there can be an inhomogeneity in the segment density, so that some volume elements of the sample contain more polymer segments than others. This is thought to be caused by such effects as inhomogeneous cross-linking and microsyneresis resulting from phase separations, which are due to the chemical effects of cross-linking in solution.
At the same time, however, it is conceivable that there occurs a local parallelization of chain segments. To what extent these effects are connected seems open to question and will probably depend upon the particular system chosen for study. For a better understanding of the kind of structure present in the material there are in principal two ways of investigation open, both based

Table I.

X-ray and light scattering used to obtain structural information

Technique	Dimension (Å)	Structural information
Wide Angle X-ray	1 - 10	Mainly for semi-crystalline material: Unit cell parameters Polymorphism Percentage crystallinity Mean crystallite dimensions Crystalline orientation
Small Angle X-ray	10 - 10^3	Super molecular structure: Electron-density fluctuation correlations Phase separations "Long period spacings" Deformation characteristics
(Small Angle) Light Scattering	10^2 - 10^5	Super molecular Structure: Polarizability fluctuation correlations Orientation fluctuation correlations Spherulite properties Deformation effects

on the scattering phenomenon. On the one hand are the image forming techniques as the optical microscope and electronmicroscope. These techniques have some drawbacks, because for many materials, especially gels, it is very difficult or maybe even impossible to use a sample-preparation method that does not influence the specimen. Furthermore, although the method as such is very direct, it is practically impossible to put such observations on a quantitative basis. Also there is always the possibility that part of the features observed are artefacts or a typical of the material as a whole. Nevertheless, if electron microscopy is applicable it can provide a firmer basis for structure models to be used for interpreting measurements by other techniques. On the other hand one can use techniques in which the desired information is derived from the diffracted intensity with of course loss of the phase information. Examples are X-ray and light scattering. Although the description of the scattering process is similar and even the same equations may be used in some cases there are two essential differences between the two techniques. The first difference is connected with the difference in wavelength of the radiation used. This results in a different dimension level for which these techniques can be used. In the second place the "refractive index" for X-ray is about one and does not depend on the polarization direction of the radiation, whereas (visible) light is influenced by local optical anisotropy. This causes light scattering to be more complex, but at the same time makes it possible to study also local optical anisotropy and spacial correlations of this property which in turn can be compared with structure models to be used for the calculation of mechanical properties. Table I gives a survey of the different techniques, the order of magnitude of the dimension level studied, and some examples of the kind of information that is obtained.
In this paper certain aspects of light scattering as a technique for polymer characterization will be discussed from the theoretical point of view. This does not mean that the experimental problems are regarded as easy. On the contrary, it is the authors opinion that only very careful experimentation will lead to results accurate enough to be fully used in the rather difficult theoretical formulas. A great help in this respect is a good and reliable apparatus. As an example the Bleeker Small Angle Scattering Photometer[15] may be mentioned here. Even so great care should be taken, especially in measurements on swollen gel systems- the scattering of which is usually rather low- to avoid surface scattering and to reduce the scattering of the container or cell as much as possible.

Light Scattering.

The intensity of scattered light (see Fig. 1) may be measured as a function of the scattering angle, θ, and the azimuthal angle, $(\mu+\phi)$. Since it is very difficult to lift the detector (photomultiplier) out of the horizontal plane, the angles ψ and φ are varied together in such a way that the difference between the two angles remains constant. In this way the (μ+φ) dependence can be obtained. In the figure the plane of the analyzer is perpendicular to the incident beam, but this is not a prerequisite.

In the general theory provision is also made for other arrangements of polarizer and analyzer but it should be stressed that a number of corrections have to be changed. The subject of correcting the measured intensities to values which are to be used in the equations developed is not treated here, because most of it can be found in the literature (11), (14). It should be emphasized, however, that a careful calculation of the so-called "vector $\tilde{O}$" should be made for every experimental setup in the way indicated by Keyzers et al. (11). The corrected intensities as a function of θ,ψ and φ contain information about different kinds of correlation. This can be demonstrated as follows:

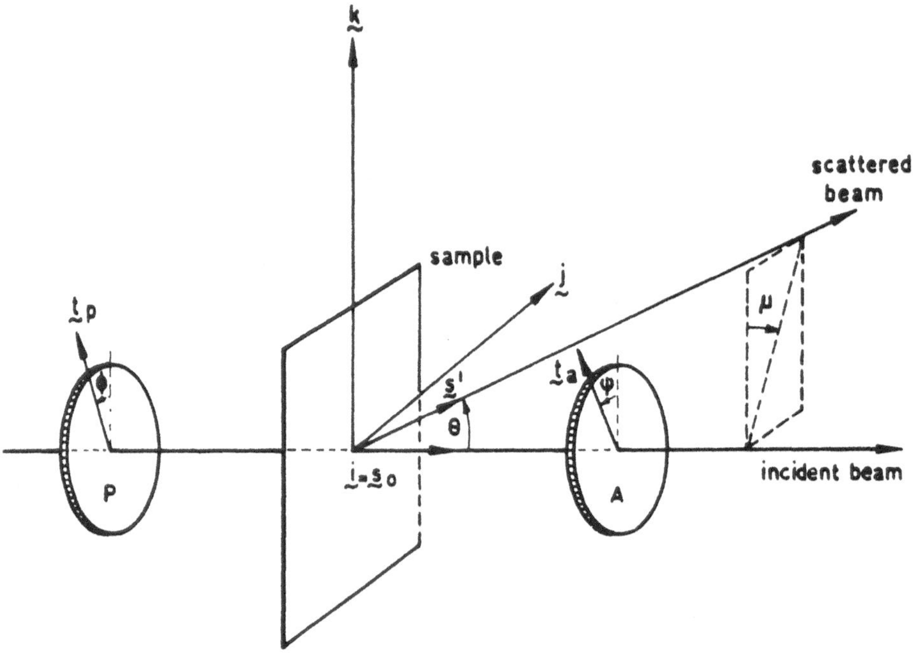

Fig. 1. Experimental arrangement for light scattering measurements. P and A are polarizer and analyzer polaroid plates, both perpendicular to the incident beam ($s_o = i$).
The vectors t_p and t_a denote the polarizer and analyzer transmission direction respectively.
The plane-parallel sample is assumed to be perpendicular to the incident beam direction too.

If the material is divided into small volume elements, v_i, then the optical properties for each of these elements may be specified by the direction of a unique optical axis $\underset{\sim}{a}_i$ (assuming cylindrical symmetry), an average polarizability

$$\alpha_i = \frac{1}{3}[(\alpha_{/\!/})_i + 2(\alpha_\perp)_i] \text{ and an anisotropy } \delta_i = (\alpha_{/\!/})_i - (\alpha_\perp)_i$$

In the simplest case of a homogeneous isotropic sample, where $\delta_i = o$ and $\alpha_i = \bar{\alpha}$, one does not get any light scattering at all. So only deviations of α_i from the mean $\bar{\alpha}$ give rise to scattering effects.
In view of this it is customary to speak of density fluctuations $\eta_i = \alpha_i - \bar{\alpha}$.

In the more general case where $\delta_i \neq o$ there can be correlations between

1. η_i and η_j : density fluctuation correlations
2. δ_i and δ_j : anisotropy fluctuation correlations
3. $\underset{\sim}{a}_i$ and $\underset{\sim}{a}_j$: orientation correlations
4. η_i and δ_j and/or η_j and δ_i : so called cross-correlations.

All these correlations are functions of the vector $\underset{\sim}{r}_{ij}$ connecting the ith and j^{th} volume elements and must be averaged over the whole sample.

One of the most general treatments of this problem is given by Goldstein and Michalik (1). This treatment has several drawbacks, however. First, it is not possible to introduce corrections for the fact that actual polarizers and analyzers do not work ideally and, secondly, this theory gives only certain special scattering components (V_v, H_v, V_h and H_h) Besides, the assumptions of Goldstein and Michalik are of such a general nature that an evaluation of experimental results in terms of their theory does not lead to more than certain parameter values. Hence this theory is not much used.

There are two ways out of this problem. In the first place one can develop a specific model, which is much easier to handle, and calculate the light scattering on this simplified basis. This was done by Stein and Wilson (2) and is called the random orientation correlation approach. The essential assumption of these authors is that all correlations are functions of the distance r_{ij} only and do not depend upon the direction of the vector $\underset{\sim}{r}_{ij}$. Cross correlations are assumed to be absent. This model was used successfully in a number of cases (4), (5), but proved to be too simple. The assumption mentioned above implies that the scattering will be cylinder-symmetrical around the incident beam and so will depend on θ only. Photographs of the scattering pattern have shown, however, that there are a number of cases where this evidently does not hold, see for instance (6), (7).

A second approach is to regard the sample as a collection of individual scattering entities with a certain specified optical structure. This kind of model calculations is used rather frequently. For example, Stein and Rhodes calculated the scattering from anisotropic spheres (6) as a model for spherulitic semicrystalline polymers, while the present author developed a model of a random collection of cylinders with finite length and width (8) for the interpretation of light scattering measurement of certain hydrogels (9), (10). These model calculations, however, do not take into account that most systems are densely packed and so will give rise to interparticle interference effects. Besides, one can never be sure that the model used for the scattering entity is the only reasonable one. Therefore it seemed of interest to develop a rather general theory based on the correlation function approach, which in the limiting case results in the random orientation correlation approach of Stein and Wilson (2), and on the other hand shows what kind of information is obtainable from light scattering measurements in the most favorable circumstances.

A first attempt to formulate such a theory was made by Stein et al (3), but this was a two-dimensional approach only. Moreover the resulting equations are rather cumbersome. A different approach was made by Keyzers et al (11), who interpreted their measurements on semi-crystalline polypropylene and polystyrene on the basis of a linear combination of the random orientation correlation approach and the spherulite model of Stein and Rhodes (6). Although this worked rather well in practice, the combination of two fundamentally different kinds of theory for one sample is not entirely satisfactory. The present author has therefore developed a new theory which will be published separately (12). The most important assumptions on which this theory is based are given below.

1. The volume elements have cylinder-symmetrical optical properties and so are defined by a unique optical axis $\underset{\sim}{a}_i$, an average polarizability α_i and an anisotropy δ_i as defined before.
2. Cross-correlations are absent.
3. The sample is macroscopically isotropic. This means that the theory cannot be used for stretched samples or samples possessing some kind of preferred orientation (such as caused by anisotropic swelling).
4. If the angle between two optical axes $\underset{\sim}{a}_i$ and $\underset{\sim}{a}_j$ is called θ_{ij}, then at constant θ_{ij} and constant vector separation $\underset{\sim}{r}_{ij}$ all orientations of $\underset{\sim}{a}_j$ around $\underset{\sim}{a}_i$ occur with equal probability.

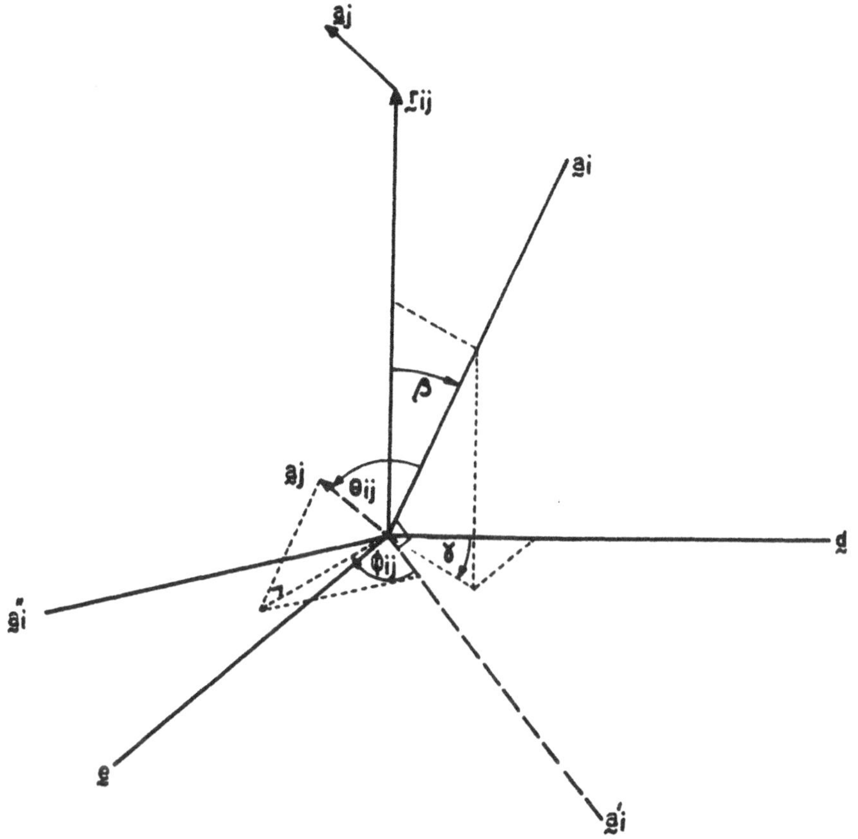

Fig. 2. The coordinate system used to describe the orientation of the optic axis a_j in volume-element j with respect to the optic axis a_i in volume-element i and the vector r_{ij} connecting both the elements.

In other words, the angle ϕ_{ij} (see Fig.2) takes on all values at random, so that averaging over ϕ_{ij} may be carried out separately.

5. It is assumed that the general correlation function $G = \langle\delta_i\ \delta_j\ \frac{1}{2}(3\cos^2\theta_{ij} - 1)\rangle$ can be developed in a series of spherical harmonics depending on $r_{ij} = r$, β and γ (see Fig. 2), so that

$$G = \langle\delta_i\delta_j\ \tfrac{1}{2}(3\cos^2\theta_{ij}-1)\rangle = \delta^2 \sum_{n=o}^{\infty}\ \sum_{m=o}^{n} A_{m,n}(r)P_n^m(\cos\beta)\cos m\gamma \quad \text{---(1)}$$

A unit vector in the direction of $s = s_o - s'$ is now called h so that (see Fig. 1)

$$s = 2\sin(\theta/_2)h \quad \text{---(2)}$$

and $\mathbf{h} = \sin(\theta/2)\mathbf{i} - \cos(\theta/2)\sin\mu\,\mathbf{j} - \cos(\theta/2)\cos\mu\,\mathbf{k}$ ---(3)

The incident light field is determined by the vector

$$\mathbf{E}' = E_o\,\mathbf{E} = E_o\,(-\sin\phi\,\mathbf{j} + \cos\phi\,\mathbf{k}) \qquad ---(4)$$

while the component of the scattered field which is seen through the analyzer is calculated with the aid of a vector $\mathbf{O}'$. A unit vector in this direction, inside the sample, is called $\mathbf{O}$ and for the case of Fig. 1 is given by (11)

$$\mathbf{O} = \left[\sin^2\psi + \cos^2\overset{*}{\theta}\overset{*}{p}\cos^2\psi\right]^{-\frac{1}{2}}\{-\sin\psi\,\mathbf{p}'' + \cos\overset{*}{\theta}\overset{*}{p}\cos\psi\ \ \mathbf{q}''\} \qquad ---(5)$$

where

$$\mathbf{p}'' = -\sin\theta\,\mathbf{i} + \cos\theta\sin\mu\,\mathbf{j} + \cos\theta\cos\mu\,\mathbf{k} \qquad ---(6)$$

and

$$\mathbf{q}'' = -\cos\mu\,\mathbf{j} + \sin\mu\,\mathbf{k} \qquad ---(7)$$

Using these assumptions and definitions, we find

$$I = CE_o^2\,O^2.$$

$$\{4\pi\int_0^\infty \delta^2\Big[(X-Y)\{\tfrac{1}{15}H_o^oP_o^o - \tfrac{2}{21}[H_2^oP_2^o + 3H_2^2P_2^2] + [\tfrac{1}{35}H_4^oP_4^o + \tfrac{4}{7}H_4^2P_4^2 + H_4^4P_4^4]\} +$$

$$+Y\{\tfrac{1}{15}H_o^oP_o^o + \tfrac{1}{21}[H_2^oP_2^o + 3H_2^2P_2^2] - 4[\tfrac{1}{35}H_4^oP_4^o + \tfrac{4}{7}H_4^2P_4^2 + H_4^4P_4^4]\} +$$

$$+35\,Z\,\{[\tfrac{1}{35}H_4^oP_4^o + \tfrac{4}{7}H_4^2P_4^2 + H_4^4P_4^4]\} +$$

$$-\tfrac{2}{3}(\mathbf{E}.\mathbf{O})\{(\mathbf{E}.\mathbf{O})(\tfrac{1}{3}H_o^oP_o^o - \tfrac{1}{3}[H_2^oP_2^o + 3H_2^2P_2^2]) + (\mathbf{h}.\mathbf{O})(\mathbf{h}.\mathbf{E})[H_2^oP_2^o + 3H_2^2P_2^2]\} +$$

$$+\tfrac{1}{9}(\mathbf{E}.\mathbf{O})^2H_o^oP_o^o\Big]r^2dr +$$

$$4\pi\,(\mathbf{E}.\mathbf{O})^2\int_0^\infty \Big[\tfrac{1}{4\pi}\int_0^\pi\int_0^{2\pi} \eta_i\eta_j\sin\beta\,d\gamma\,d\beta\Big]P_o^o r^2dr\} \qquad ---(8)$$

For the meaning of the different symbols used, see the appendix.

By putting

$x = \int_0^\infty \frac{1}{15}H_o^oP_o^o r^2dr,\ y = \int_0^\infty \frac{1}{21}[H_2^oP_2^o + 3H_2^2P_2^2]r^2dr$

and $z = \int_0^\infty [\frac{1}{35}H_4^oP_4^o + \frac{4}{7}H_4^2P_4^2 + H_4^4P_4^4]r^2dr$

the equation given above can be rearranged to

$$I = 4\pi CI_oO^2\{\delta^2\Big[x\{1 + \tfrac{1}{3}(\mathbf{E}.\mathbf{O})^2\} + y\{-2 + 3(\mathbf{h}.\mathbf{O})^2 + 3(\mathbf{h}.\mathbf{E})^2 - 2(\mathbf{h}.\mathbf{E})(\mathbf{h}.\mathbf{O})(\mathbf{E}.\mathbf{O}) + \tfrac{2}{3}(\mathbf{E}.\mathbf{O})^2\}\cdot$$

$$+z\{1 + 2(\mathbf{E}.\mathbf{O})^2 - 5(\mathbf{h}.\mathbf{O})^2 - 5(\mathbf{h}.\mathbf{E})^2 - 20(\mathbf{h}.\mathbf{E})(\mathbf{h}.\mathbf{O})(\mathbf{E}.\mathbf{O}) + 35(\mathbf{h}.\mathbf{E})^2(\mathbf{h}.\mathbf{O})^2\}\Big] +$$

$$+\overline{\eta^2}\,(\mathbf{E}.\mathbf{O})^2\int_0^\infty \gamma(r)\frac{\sin(hr)}{hr}r^2dr\} \qquad ---(9)$$

The (random) density fluctuation correlation function γ (r) is defined as

$$\gamma(r) = \frac{\langle \eta_i \; \eta_j \rangle_r}{\overline{\eta_i^2}} \qquad \text{---(10)}$$

where the product $\eta_i \; \eta_j$ is averaged over all volume elements i and j being a distance r apart.

It can be shown that for y=z=o the equation given (eq. 9) reduces to the equations for the random orientation correlation approach. So the more general theory presented here does indeed contain the Stein-Wilson theory (2) as a limiting case and at the same time shows that two other functions of the scattering angle, y(h) and z(h) respectively, are involved.

These functions are connected with the higher order terms of the orientational correlations. In order to see more clearly how the functions y(h) and z(h) can be interpreted it is necessary to make the following simplifying assumption, which is regarded as being rather reasonable, however.
If it is assumed that the general orientation correlation function G is a function of r and β only, and does not depend upon the angle γ (see Fig.2), all functions H_n^m (r) with m $\neq$ o can be shown to be zero. One then finds

$$x(h) = \frac{1}{15} \int_0^\infty A_{o,o}(r) \, \pi^{\frac{1}{2}} (2hr)^{-\frac{1}{2}} J_{1/2}(hr) \, r^2 dr$$

$$y(h) = \frac{-1}{105} \int_0^\infty A_{o,2}(r) \, \pi^{\frac{1}{2}} (2hr)^{-\frac{1}{2}} J_{5/2}(hr) \, r^2 dr$$

$$z(h) = \frac{1}{315} \int_0^\infty A_{o,4}(r) \, \pi^{\frac{1}{2}} (2hr)^{-\frac{1}{2}} J_{9/2}(hr) \, r^2 dr \qquad \text{---(11)}$$

where h = $(4\pi/\lambda)\sin(\theta/2)$ and $J_{1/2}(hr)$, $J_{5/2}(hr)$ and $J_{9/2}(hr)$ are Bessel functions of the first kind. So if the functions x(h), y(h) and z(h) can be obtained from experiments, it is evident from eq. 11 that these functions will give the coefficients $A_{o,o}(r)$, $A_{o,2}(r)$ and $A_{o,4}(r)$ via a Hankel transformation.

These functions give information about local orientation tendencies in the material. This can be seen more clearly if we write the general orientation correlation function G as:

$$G = \langle \delta_i \delta_j \, \tfrac{1}{2}(3\cos^2\theta_{ij} - 1) \rangle = \delta^2 \Big[A_{o,o}(r) + A_{o,2}(r) \tfrac{1}{2}(3\cos^2\beta - 1) + A_{o,4}(r) \left(\tfrac{35}{8}\cos^4\beta - \tfrac{30}{8}\cos^2\beta + \tfrac{3}{8}\right) + \quad \Big] \qquad \text{---(12)}$$

Although only the first three terms of this series can be obtained, it will be clear that if the values of the $A_{o,n}(r)$ functions are such as to yield a higher value for G at small β values and do so for much larger values of r, this points to a rod-like type of orientation. In the opposite case, where G is larger for β around 90°, we have a more disk-like type of orientation.

The conclusion is that, by making some simplifying assumptions, it is possible not only to prove the existence of some kind of ordering in the material but also to discriminate between the type of ordering, that is, between more rod-like and more disk-like configurations.

We now turn to the question whether the separate functions x(h), y(h) and z(h) can be obtained from experimental measurements. It then turns out that whatever theory is used to describe the sample structure, the intensity of scattered light can always be written in the form

$$I = K\left[A + B(\underset{\sim}{E}.\underset{\sim}{O})^2 + C\{(\underset{\sim}{h}.\underset{\sim}{O})^2 + (\underset{\sim}{h}.\underset{\sim}{E})^2\} - D(\underset{\sim}{h}.\underset{\sim}{E})(\underset{\sim}{h}.\underset{\sim}{O})(\underset{\sim}{E}.\underset{\sim}{O}) + E(\underset{\sim}{h}.\underset{\sim}{O})^2(\underset{\sim}{h}.\underset{\sim}{E})^2\right] \quad \text{---(13)}$$

where A,.....,E are five unknown functions of $h=(4\pi/\lambda)\sin(\theta/2)$.

With the same experimental geometry as used by Keyzers et al. (11) the corrected intensities of scattered light can be written as (see Fig. 1)

$$i = K\Big[\sin^2\psi\{A+C\cos^2(\theta/2)+\sin^2\phi[B\cos^2\theta+C\cos^2(\theta/2)-D\cos^2(\theta/2)\cos\theta+E\cos^4(\theta/2)]\}$$

$$+\cos^2\theta\overset{*}{p}\cos^2\psi\{A+C\cos^2(\theta/2)+\cos^2\phi[B-C\cos^2(\theta/2)]\}+$$

$$+\cos\theta\overset{*}{p}\sin\phi\cos\phi\sin\psi\cos\psi\{2B\cos\theta-D\cos^2(\theta/2)\}\Big] \quad \text{--(14)}$$

It is evident that only four relations between the five unknowns A,....,E are obtained, which is clearly caused by the fact that $H_v(\rightarrow\sin^2\psi=1,\ \sin^2\phi=o)=V_h(\rightarrow\cos^2\psi=1,\ \cos^2\phi=o)$.

Generally speaking these equations cannot be solved, unless some kind of assumption is made which leads to a fifth relation. The fifth relation can be derived by assuming that cross correlations do not occur, as was stated earlier. This yields an extra relation between C,D and E. Calculations on specific models implicitly assume a definite relationship between "cross-correlation terms" and the normal density and orientation fluctuation terms. So in that case there are only four unknowns and eq. 13 can be solved, too.

In the case of the general non-random orientation correlation theory as described before (see also (12)), the parameters A(h)... ..... E(h) are functions of x(h), y(h) and z(h) as well as of the density correlation, as can be seen by writing eq. 9 in the form of eq.13. It then follows that

$$
\begin{aligned}
A &= \delta^2 (x - 2y + z) \\
B &= \delta^2 (\tfrac{1}{3}x + \tfrac{2}{3}y + 2z) + \overline{\eta^2} \int_0^\infty \gamma(r)\frac{\sin (hr)}{hr} r^2 dr \\
C &= \delta^2 (3y - 5z) \\
D &= \delta^2 (2y + 20z) \\
E &= \delta^2\, 35z
\end{aligned}
\qquad ---(15)
$$

Furthermore it is now very instructive to compare the general theory given in eq. 9 with the expression derived by the present author for a model of a random assembly of rods of finite length and radius (8).
The scattered intensity in that case can be written as

$$
\begin{aligned}
I = {} & C\, V_o^2\, N\, E_o^2\, O^2 \\
& \{(\alpha_1-\alpha_2)^2 \tfrac{1}{8}\left[(X-Y)(D_o-2D_1+D_2)+Y(4D_1-4D_2)+Z(35\, D_2-30\, D_1+3D_o)\right]+ \\
& +\alpha_2(\alpha_1-\alpha_2)(\underset{\sim}{E}.\underset{\sim}{O})\left[(\underset{\sim}{E}.\underset{\sim}{O})(D_o-D_1)+(\underset{\sim}{h}.\underset{\sim}{O})(\underset{\sim}{h}.\underset{\sim}{E})(3D_1-D_o)\right]+ \\
& \alpha_2^2\, (\underset{\sim}{E}.\underset{\sim}{O})^2 D_o\}
\end{aligned}
\qquad ---(16)
$$

where the functions $D_n(h)$ are complicated functions of the scattering parameter h, the length 2L and the radius R of the rods (see Appendix). The polarizabilities α_1 and α_2 are chosen in the direction of the optic axis (2L) and perpendicular to it, respectively. The parameters X,Y and Z have the same meaning here as in eq. 8 (see Appendix). Writing eq.16 in the same form as eq. 13 it turns out that

$$
\begin{aligned}
A &= (\alpha_1-\alpha_2)^2\, \tfrac{1}{8}(D_o-2D_1+D_2) \\
B &= (\alpha_1-\alpha_2)^2\, \tfrac{2}{8}(D_o-2D_1+D_2)+\alpha_2^2 D_o+\alpha_2(\alpha_1-\alpha_2)(D_o-D_1) \\
C &= -(\alpha_1-\alpha_2)^2\, \tfrac{1}{8}(D_o-2D_1+D_2) + (\alpha_1-\alpha_2)^2 \tfrac{4}{8}(D_1-D_2) \\
D &= (\alpha_1-\alpha_2)^2\, \tfrac{4}{8}(D_o-2D_1+D_2)- \tfrac{16}{8}\, (\alpha_1-\alpha_2)^2 (D_1-D_2)-\alpha_2(\alpha_1-\alpha_2)(3D_1-D_o) \\
E &= (\alpha_1-\alpha_2)^2\, \tfrac{1}{8}(35D_2-30D_1+3D_o)
\end{aligned}
\qquad ---(17)
$$

In order to compare eq. 17 with eq. 15 it should be realized that model calculations implicitly assume the absence of density fluctuations inside the volume of a scattering entity, while at the same time these entities are thought to be embedded in a perfectly homogeneous medium. This means that cross correlations are implicitly taken into account:

writing $(\alpha_1-\alpha_2)= \delta$ and $\alpha_2=(\overline{\alpha}- \frac{1}{3}\delta)$ we find

$$A = \delta^2 \{ \tfrac{1}{8} (D_2- \tfrac{30}{35} D_1 + \tfrac{3}{35} D_o) - \tfrac{2}{42} (3 D_1 - D_o)+ \tfrac{1}{15} D_o\}$$

$$B = \delta^2 \{ \tfrac{2}{8} (D_2 - \tfrac{30}{35} D_1 + \tfrac{3}{35} D_o)+ \tfrac{2}{3}\cdot\tfrac{1}{42} (3D_1 - D_o)+ \tfrac{1}{3}\cdot\tfrac{1}{15} D_o\}+ + \overline{\alpha}^2 D_o - \tfrac{1}{3} \overline{\alpha} \delta (3D_1 - D_o)$$

$$C = \delta^2 \{ -\tfrac{5}{8} (D_2 - \tfrac{30}{35} D_1 + \tfrac{3}{35} D_o) + \tfrac{3}{42} (3D_1 - D_o)\}$$

$$D = \delta^2 \{ \tfrac{20}{8} (D_2 - \tfrac{30}{35} D_1 + \tfrac{3}{35} D_o) + \tfrac{2}{42} (3D_1 - D_o)\} -\overline{\alpha} \delta (3D_1 - D_o)$$

$$E = \delta^2\, 35\cdot\tfrac{1}{8} (D_2 - \tfrac{30}{35} D_1 + \tfrac{3}{35} D_o) \qquad ---(18)$$

A comparison between equations 18 and 15 reveals two differences. In the first place the term containing the density correlation function is replaced by a term $\overline{\alpha}^2 D_o$, which means that the model calculations automatically lead to the conclusion that in this case the correlation function for random orientation correlations and that for random density fluctuation correlations are identical. This seems to be a general result. The second difference is the occurrence of cross correlation terms in the equations for B and D. In the case of model calculations these extra terms are connected with the other functions of the system in the way given above. In the general case the cross correlation terms occur only in the B and D value too, but then there need not be a connection with the other functions of the system. Because of the fact that the assumption of a specific model automatically reduces the number of unknown parameters, the four relationships between A(h)......E(h), as obtained from experiments, can be used to calculate the functions $D_n(h)$ (for n=o, 1 and 2) as well as $\delta/\overline{\alpha}$. The latter is the fourth unknown quantity.

A last remark should be made about absolute measurements. It is shown that the angular dependence of the scattered light intensity can give information about correlations in density and orientation and at the same time permits conclusions about the type of orientational order present in the system under investigation. Absolute measurements, however, can add to these a quantitative estimation of the mean square of the polarizability fluctuation, which can be connected with phase separation and inhomogeneous cross-linking and the absolute value of the anisotropy which may be interpreted in terms of the kind of order and the relative parallelity of the chains on a microscale.

It can be shown in general that

$$\lim_{h \to o} A(h) = \frac{1}{15}\,\delta^2 \qquad \text{---(19)}$$

$$\lim_{h \to o} B(h) = \frac{1}{45}\,\delta^2 + \overline{\eta^2} \qquad \text{---(20)}$$

so that from absolute measurements the values of δ^2 and $\overline{\eta^2}$ may be calculated.

In conclusion it should again be stressed that interpretations of light scattering measurements on gels or any other transparent material as given above only apply to those materials which are isotropic as a whole. In that case it was shown that one may get information on anisotropy on a microscale.
For systems with macroscopic orientation, as may be caused by stretching or anisotropic swelling, the theories presented do not hold. Such cases call for an entirely different approach.

References.

1. M. Goldstein & E.R. Michalik, J.Appl.Phys. 26 1450(1955)
2. R.S. Stein & P.R. Wilson, J.Appl.Phys. 33, 1914(1962)
3. R.S. Stein et. al., in: Proc.Sec.Int.Conf. on Electromagnetic Scattering (R.L. Rowel & R.S. Stein ed.) New York (Gordon & Breach) 1967, 339.
4. R.S. Stein, in : Proc.Int.Conf. on Electromagnetic Scattering (M. Kerker ed.) New York (Pergamon Press) 1963, 439.
5. A.E.M. Keyzers et. al., J.Appl.Phys. 36, 2874(1965).
6. R.S. Stein & M.B. Rhodes, J.Appl.Phys. 31, 1873(1960).
7. M.B. Rhodes & R.S. Stein, in: ASTM Special Technical Publication No. 348 (1963).
8. J.J. van Aartsen, Eur. Polymer J., in press.
9. M.C.A. Donkersloot et.al., Rec.trav.chim. 86, 321(1967).
10. J.H. Gouda, Structural Characterization of non-crystalline hydrogels. Thesis Delft 1969.

11. A.E.M. Keyzers et.al., J.Am.Chem.Soc., 90, 3107 (1968).
12. J.J. van Aartsen, J.Appl.Physics, to be published.
13. J.H. Gouda et.al., J.Polymer Sci., part B, 8, 225(1970).
14. R.S. Stein & J.J. Keane, J. Polymer Sci. 17, 21(1955).
15. Described by Keyzers et al. (11) and commercially available from IMASS, Accord (Hingham) Mass 02018.

Appendix

The meaning of the abbreviations used in the text is as follows.

$$X = 1 + 2\,(\underset{\sim}{E}.\underset{\sim}{O})^2$$

$$Y = (\underset{\sim}{h}.\underset{\sim}{O})^2 + (\underset{\sim}{h}.\underset{\sim}{E})^2 + 4(\underset{\sim}{h}.\underset{\sim}{E})(\underset{\sim}{h}.\underset{\sim}{O})(\underset{\sim}{E}.\underset{\sim}{O})$$

$$Z = (\underset{\sim}{h}.\underset{\sim}{E})^2(\underset{\sim}{h}.\underset{\sim}{O})^2$$

$$H_0^0 = (4\pi\delta^2)^{-1}\int_0^{\pi}\int_0^{2\pi} G \sin\beta\, d\gamma\, d\beta \qquad = A_{o,o}(r)$$

$$H_2^0 = (4\pi\delta^2)^{-1}\int_0^{\pi}\int_0^{2\pi} G\, \tfrac{1}{2}(3\cos^2\beta - 1)\sin\beta\, d\gamma\, d\beta \qquad = \tfrac{1}{5}A_{o,2}(r)$$

$$H_2^2 = (4\pi\delta^2)^{-1}\int_0^{\pi}\int_0^{2\pi} G\, \tfrac{1}{2}\sin^3\beta\cos 2\gamma\, d\gamma\, d\beta \qquad = \tfrac{2}{5}A_{2,2}(r)$$

$$H_4^0 = (4\pi\delta^2)^{-1}\int_0^{\pi}\int_0^{2\pi} G\, (\tfrac{35}{8}\cos^4\beta - \tfrac{30}{8}\cos^2\beta + \tfrac{3}{8})\sin\beta\, d\gamma\, d\beta \qquad = \tfrac{1}{9}A_{o,4}(r)$$

$$H_4^2 = (4\pi\delta^2)^{-1}\int_0^{\pi}\int_0^{2\pi} G\, \tfrac{1}{8}\sin^3\beta(7\cos^2\beta - 1)\cos 2\gamma\, d\gamma\, d\beta \qquad = \tfrac{1}{3}A_{2,4}(r)$$

$$H_4^4 = (4\pi\delta^2)^{-1}\int_0^{\pi}\int_0^{2\pi} G\, \tfrac{1}{8}\sin^5\beta\cos 4\gamma\, d\gamma\, d\beta \qquad = \tfrac{8}{3}A_{4,4}(r)$$

$$P_0^0 = \int_0^{\pi/2} \cos(hr\cos\alpha)\sin\alpha\, d\alpha \qquad = (\pi/2hr)^{\frac{1}{2}}\, J_{1/2}(hr)$$

$$P_2^0 = \int_0^{\pi/2} \tfrac{1}{2}(3\cos^2\alpha - 1)\cos(hr\cos\alpha)\sin\alpha\, d\alpha \qquad =-(\pi/2hr)^{\frac{1}{2}}\, J_{5/2}(hr)$$

$$P_2^2 = \int_0^{\pi/2} \tfrac{1}{2}\sin^3\alpha\cos(hr\cos\alpha)\, d\alpha$$

$$P_4^0 = \int_0^{\pi/2} (\tfrac{35}{8}\cos^4\alpha - \tfrac{30}{8}\cos^2\alpha + \tfrac{3}{8})\cos(hr\cos\alpha)\sin\alpha\, d\alpha = (\pi/2hr)^{\frac{1}{2}}\, J_{9/2}(hr)$$

$$P_4^2 = \int_0^{\pi/2} \frac{1}{8} \sin^3\alpha \, (7\cos^2\alpha - 1)\cos(hr\cos\alpha)\, d\alpha$$

$$P_4^4 = \int_0^{\pi/2} \frac{1}{8} \sin^5\alpha \cos(hr\cos\alpha)\, d\alpha$$

$$D_n = \frac{2\pi}{4\pi.2\pi} \int_0^{\pi} \cos^{2n}\alpha \int_0^{2\pi} \left(\frac{1}{V_p} \int_V \cos k(r.s)\, dr \right)^2 d\phi \sin\alpha d\alpha$$

For cylindrical particles of length 2L and radius R it is found that

$$D_n = 4 \sum_{i=0}^{\infty} (-1)^i \frac{(2hL)^{2i}}{(2i+2)!} \sum_{k=0}^{\infty} (-1)^k \frac{(hR)^{2k}}{(k+1)!(k+2)!} \cdot \frac{\Gamma(k+3/2)\Gamma(i+n+1/2)}{\Gamma(1/2)\Gamma(i+n+k+3/2)}$$

LIGHT SCATTERING BY POLYMER NETWORKS

W. Prins

Department of Chemistry, Syracuse University

Syracuse, New York 13210

SUMMARY

Non-crystalline polymer networks, whether swollen or not, may exhibit interchain structuring effects due to incompatibility of various parts of the chains (e.g., backbone-side chain or blocks along a copolymer chain). Also, the network-formation process may lead to inhomogeneously crosslinked networks, possibly resulting in micro-phase separation. The complete polarized light scattering envelope provides structural information on the basis of the Rayleigh-Debye approximation in terms of either correlation functions or specific models for the structure. Small angle data can be utilized in conjunction with Gaussian correlation functions to obtain the scattered intensity at zero angle. This quantity yields the extent of polarizability fluctuations - sometimes relatable to fluctuations in crosslink density - as well as the mean anisotropy per volume element - sometimes relatable to regions of mesomorphic order. Wide angle data may reveal the existence of a more specific structure than that implied by Gaussian correlation functions. A comparison of the experimental data with calculated scattering patterns for specific models of the structure then becomes useful. Recent studies on various networks (poly saccharides, gelatin, poly vinylalcohol, hydrophilic poly methacrylates and poly urethanes) will be used to illustrate the above concepts.

Since about 1945, light scattering has become an important tool for the polymer scientist. Wide angle scattering (for instrumental reasons defined as scattering at angles $\theta > 20$) provides an excellent absolute method for the determination of particle size and shape in solution (1). The emergence of continuous, high intensity laser light sources of high monochromaticity in the 60's has added the possibility of observing particle dynamics through a frequency analysis of the Rayleigh line (2). Small angle scattering ($1 < \theta < 20$) has come of age as a tool for examining the submicroscopic morphology in crystalline polymers (3).

In the grey area between the dilute solution on the one hand and the well organized solid state on the other hand, lies a fruitful field for the application of light scattering techniques viz. the investigation of the structure and dynamics of concentrated polymer solutions, gels, glassy and rubbery polymers or, more generally, of the non-crystalline state, including the possible existence therein of mesomorphic phases of submicroscopic range.

This article will put in perspective some results obtained on swollen and unswollen crosslinked polymer networks by the author and his collaborators in recent years. Since, to the author's knowledge, no studies on the dynamics of non-crystalline polymer systems have as yet been undertaken, the emphasis will be on wide and small angle Rayleigh-Debye or (Rayleigh-Gans) scattering without frequency analysis, i.e., using conventional light sources such as mercury arcs. For small angle structural work the higher intensity provided by lasers is not of decisive importance.

Various reasons for supramolecular ordering in polymer systems, short of crystalline three-dimensional periodicity may be envisaged (4). Polymer chains in solution may form aggregates, a phenomenon often preceded by conformational transitions in the separate chains and followed by a sol $\rightarrow$ gel transition. Such a gelation process is usually sensitive to the temperature and the nature of the diluent including its pH and salt content. A typical example of reversible gelation is provided by degraded collagen (gelatin). Ordering effects of this type should be especially pronounced if the chains contain blocks of incompatible chemical structure or if the side chains are very different in chemical composition from the backbone of the chains. In analogy with the nomenclature for detergents, whose molecules always contain separated polar and non-polar segments, such polymer chains may be called amphiphilic. The gelation in a diluent then may lead to (micro) phase separations with the possibility that one phase becomes (micro) mesomorphic, i.e., of a one- or two-dimensional regularity extending over many molecules or parts thereof. Such phases could be nematic, smectic or cholesteric (5), but would be of limited extent in chemically crosslinked systems because of the disrupting influence of the crosslinks.

During crosslinking copolymerization an inhomogeneous network may be formed because of reactivity differences in the monomers, preferential crosslinking next to previously formed crosslinks and/or because the resulting polymer is incompatible with the monomers and/or the diluent present during copolymerization. The incompatibility may develop as a result of excessive local crosslinking (an elastic free energy effect) and/or as a result of unfavorable polymer-diluent interaction (a free energy of dilution effect). Both effects may lead to micro-syneresis and eventually to a micro phase separation resulting in heterogeneous polymer systems. In addition to these reasons for deviations from a random network structure, a supramolecular ordering may again be induced by a subsequently used diluent, as discussed earlier.

Finally, the possibility of some near-order, perhaps due to geometrical packing effects or to non-equilibrium states in the viscous polymer liquid which are trapped

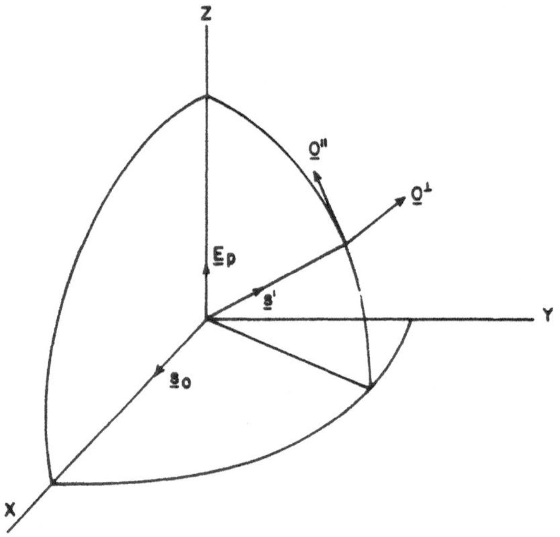

Figure 1. Definition of the Perpendicular and Parallel Component in Light Scattering. $\underline{E}_p$ - direction of polarization of primary light, $\underline{s}_o$ = propagation of primary light, $\underline{s}'$ = propagation of scattered light, $\underline{O}''$ = polarization for parallel component, $R_\theta^{\parallel}$; $\underline{O}^{\perp}$ = polarization for perpendicular component, $R_\theta^{\perp}$.

during the crosslinking process should be mentioned as a distinct possibility leading to some structure in polymer networks.

All structural information which the light scattering technique is able to furnish, is contained in the socalled parallel ($R_\theta^{\parallel}$) and perpendicular ($R_\theta^{\perp}$) Rayleigh ratios of the scattered light (see Figure 1). Most light scattering instruments scan only in the horizontal XY plane, but by using a polarizer and analyzer, the transmission directions of which can be set at any angle with the vertical but always perpendicular to the direction of light propagation in the XY plane it is possible to measure the complete spatial dependence of $R_\theta^{\parallel}$ and $R_\theta^{\perp}$ while keeping the detector arm in the horizontal plane (6,18). Because the scattering of non-crystalline materials is often low, it is advisable to place the analyzer in the detector arm so as to reduce the error due to its scattering. Surface or interface scattering of the sample and the cell can sometimes by reduced by immersion in silicon oil of matching refractive index (7) or - if unavoidable - it should be subtracted by employing a suitable blank. For reversible gels, one can subtract the scattering of the cell containing the solution prior to gelation. For crosslinked systems, the best one often can do is to subtract the scattering of the cell containing the diluent. At low levels of scattering which one encounters with $R_\theta^\perp$ this limits the accuracy of the determination. In addition, one has to apply the usual elaborate corrections for refraction, reflection and turbidity (8). The instrument used in the author's laboratory is an absolute photometer with chopper and lock-in amplifier, yielding absolute Rayleigh ratios (9). Photographic recording of light scattering patterns which has proven to be of value for the characterization of crystalline polymer films usually fails for networks - whether dry or swollen - because of the much lower levels of scattering.

The interpretation of scattering data can be done in terms of correlation functions. In these Proceedings J. J. van Aartsen discusses a very general theoretical approach to the determination of such correlation functions (10). If the correlation between volume elements with polarizabilities $\alpha^{\parallel}$ and $\alpha^{\perp}$ parallel and perpendicular to their optic axes depends only on the scalar distance of separation between the elements (random orientation correlation) there is no information in the $R_\theta^{\parallel}$ and $R_\theta^\perp$ envelopes that is not already contained in the so called H_v, V_v and H_h components. These quantities represent the Rayleigh ratios measured with vertically (horizontally) polarized incident light (small letters) and horizontally (vertically) polarized scattered light (capital letters), while the detector arm moves in the horizontal plane. One then finds (11)

$$H_V = (64\pi^5/15\lambda_o^4)\ \beta^2 \int_0^\infty f(r)\ \mu(r)\ \frac{\sin hr}{hr}\ r^2\ dr$$

-1-

$$V_V - 4H_V/3 = (64\pi^5/\lambda_o^4)\ \langle\eta^2\rangle \int_0^\infty \gamma(r)\ \frac{\sin hr}{hr}\ r^2\ dr$$

In these equations h is $(4\pi/\lambda)\sin\theta/2$ and β is the mean anisotropy $(\alpha^{\parallel} - \alpha^{\perp})$ per volume element; λ and λ_o are the wavelengths in the medium and vacuum respectively; $\gamma(r)$ and $f(r)$ are the polarizability and orientation correlation function respectively, defined by

$$\gamma(r) = \langle\eta_i\eta_j\rangle_r/\langle\eta^2\rangle$$

-2-

$$f(r) = (3\langle\cos^2\theta_{ij}\rangle_r - 1)/2$$

with η_i and η_j the polarizability fluctuations of the ith and j^{th} volume element separated by a distance r; θ_{ij} is the angle between the optic axes of these elements; $\mu(r) = 1 + \langle\eta^2\rangle\gamma(r)/\alpha^2$ with $\alpha (= (\alpha^{\parallel} + 2\alpha^{\perp})/3)$ the mean polarizability per volume element. One usually finds $\mu(r)$ to be essentially unity so that this function does not play a role in the analysis of the data (9). A test on the validity of the random orientation correlation assumption can be obtained from the relation

$$H_h = V_V \cos^2\theta + H_V \sin^2\theta \qquad -3-$$

which in practice often reduces to $H_h \simeq V_V \cos^2\theta$ because $H_V \ll V_V$.

In principle, the correlation functions appearing in equation (1) are obtainable by Fourier inversion; in practice one cannot determine the Rayleigh ratios down to zero scattering angle so that the limiting value has to be obtained by an approximate method. One can assume the correlation function to be Gaussian at least for the scattering over the last one or two degrees of scattering angle. In that case, log R_θ versus h^2 yields a straight line (see below), allowing extrapolation. Once the R_θ-data have been completed in this way, the Fourier inversion can be made (9). In several cases it has been profitable to assume that the scattering over the entire θ range can be described by a sum of Gaussian correlation functions:

$$\gamma(r) = \sum_{i=1}^{n} x_i \exp(-r^2/a_i^2) \; ; \qquad \sum_{i=1}^{n} x_i = 1 \qquad \text{-4a-}$$

$$f(r) = \sum_{j=1}^{m} y_j \exp(-r^2/b_j^2) \; ; \qquad \sum_{j=1}^{m} y_j = 1 \qquad \text{-4b-}$$

so that plots of log H_v and log $(V_v-4H_v/3)$ versus h^2 can be resolved in n and m straight lines respectively, provided the elements in the sets $\{a_i\}$ and $\{b_j\}$ are sufficiently spaced. An example of this procedure is given in Figure 2 for log H_v with m = 3, as taken from work on crosslinked, water swollen poly(2-hydroxy ethyl methacrylate) networks (PHEMA) (12). From such plots all parameters $\{a_i, x_i\}$, $\{b_j, y_j\}$ as well as β^2 and $\langle\eta^2\rangle$ can be evaluated. The set of size parameters $\{a_i\}$ and $\{b_j\}$ characterize the extent of the correlations, which is responsible for the angular dependence of the scattering. The square of the mean anisotropy, β^2, and the mean square polarizability fluctuation $\langle\eta^2\rangle$, govern the intensity of the scattering.

Light scattering studies on the network formation process were reported for the first time by Gallacher and Bettelheim (13). These authors analyzed their data on unsaturated polyester reacting with vinyl monomer in terms of density inhomogeneities or $\langle\eta^2\rangle$, neglecting possible β^2

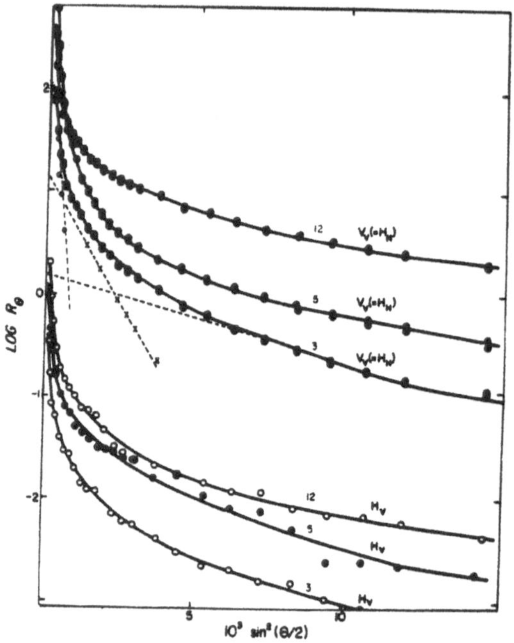

Figure 2. Small Angle Light Scattering for Poly(2-hydroxy ethyl methacrylate) (PHEMA) Networks Swollen in Water. Gel 12 has a 16 x higher crosslinking density than gel 5; gel 3 has the same crosslinking density as 3, but was polymerized in the presence of 40% water M. Ilavsky and W. Prins (12) .

contributions. More recently, Stein (14) has indicated that in a swollen network of randomly coiling Gaussian chains, $\langle\eta^2\rangle$ is relatable to the extent of inhomogeneous crosslinking. By employing the classical relation between crosslinking and swelling (4) Stein finds:

$$\langle\eta^2\rangle = (3n/10\pi)^2 \frac{(n_d - n_p)^2}{q^2} \frac{\langle\Delta M_c^2\rangle}{\langle M_c\rangle^2} \qquad -6-$$

In equation 6 n_d, n_p and n are the refractive indices of diluent, polymer and gel respectively, q is the volume degree of swelling and $\langle M_c\rangle$ is the number average molecular weight between crosslinks; $\langle\Delta M_c^2\rangle$ is the mean square fluctuation in this quantity and as such is a measure of the extent of inhomogeneous crosslinking. This procedure is illuminating but it should be remembered that its application in practice may be seriously impaired because (i) the corresponding polymer solution, the light scattering of which should be subtracted, may not be available, (ii) any supramolecular ordering in the gel and the corresponding solution may not be identical and (iii) any supramolecular ordering in the gel may invalidate the classical relation between swelling and crosslinking upon which equation (6) is founded. Almost simultaneously with Stein's communication, Bueche (14) published a similar analysis together with wide-angle experimental data. His data on swollen poly(methyl methacrylate) networks of moderate crosslinking densities support the general concept underlying equation (6). Bueche also observed angular dependencies in the range $30 < \theta < 120$, which indicate that in these swollen networks the polarizability correlation $\gamma(r)$ extends over many hundreds of angstroms. If the scattering can be exclusively attributed to regions of varying crosslinking density, then clearly the inhomogeneieties also extend over these distances. Small angle data may reveal even longer correlation distances.

Bueche neglects the possibility of H_v scattering altogether. This is not such a bad approximation since usually $H_v \ll V_v$ (compare Figure 2). Absolute photometry in small as well as wide angle instruments does, however, allow the determination of H_v (and even the complete $R_\theta^\perp$) so that the square of the mean anisotropy per volume element, β^2, as well as the extent of the orientation correlation, $\{b_j, y_i\}$, are experimentally accessible quantities. Within the framework of the Rayleigh-Debye theory the anisotropy does not depend on the refractive index of the surrounding, so that β^2 is a more reliable index of the structure existing in the

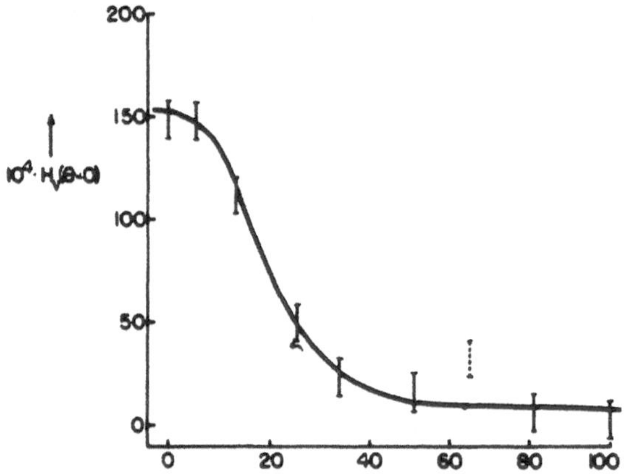

Figure 3. Extrapolated H_V Scattering of a PHEMA Gel as a Function of Diluent Composition. $(H_V)_{\theta=0}$ yields the square of the mean anisotropy, β^2 (Equation 1 or 5). The drastic reduction in β^2 indicates the breakdown of interchain ordering. J. H. Gouda, K. Povodator, T. C. Warren and W. Prins (16).

gel, than $\langle\eta^2\rangle$. An example of the usefulness of the parameter β^2 is shown in Figure 3 for PHEMA gels, swollen to equilibrium in a series of water/ethylene glycol mixtures. This polymer exhibits limited swelling in water because of the amphiphilic nature of its chains. The extrapolations of H_V to zero scattering angle were made as in Figure 2. The H_V scattering of the mixed diluent and the cell was subtracted rather than that of the corresponding polymer solution, so that β^2 reflects all anisotropy due to the polymer. The sudden drop in β^2 above a critical ethylene glycol concentration far exceeds the reduction in H_V scattering to be expected because of increased overall swelling (q ranges from 2 in water to 7 in pure ethylene glycol). This has been taken as an indication of a breakdown of cooperatively held supramolecular, anisotropic - perhaps mesomorphic - structural units, caused

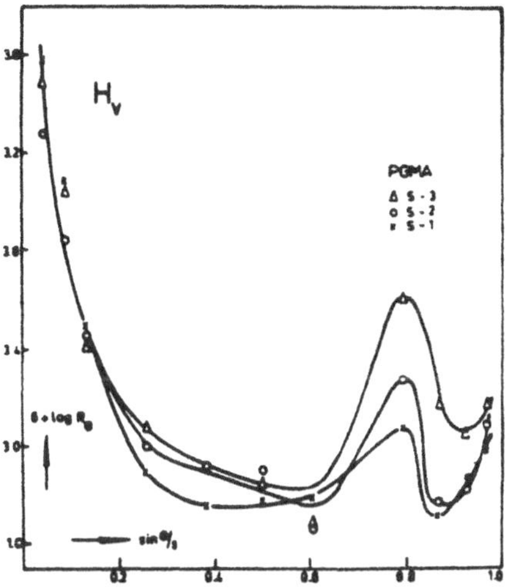

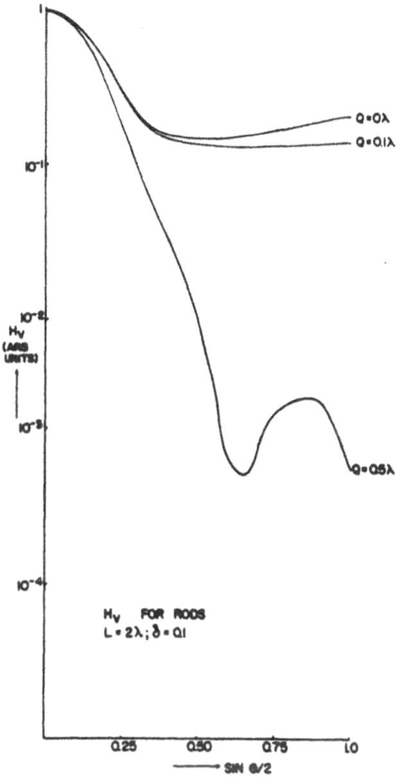

Figure 4. Wide Angle H_v Light Scattering of PHEMA Gels in Swelling Equilibrium with Water. All gels at low crosslinking density but polymerized at 41% (S-3), 20% (S-2) and 0% (S-1) water content respectively. M. C. A. Donkersloot, J. H. Gouda, J. J. van Aartsen and W. Prins (17) .

Figure 5. Theoretical H_v Scattering for a Random Collection of Anisotropic Rods Embedded in an Isotropic Matrix. L = length of the rods, Q = radius of the rods; δ in this Figure is $\alpha^{\parallel} - \alpha^{\perp}/(\alpha^{\parallel} + 2\alpha^{\perp})$. J. H. Gouda (18) .

by a reduction in the hydrophobic forces in the diluent mixture (16,12).

Wide angle H_v data on similar gels in water (17) show a maximum (Figure 4). Although there is some experimental

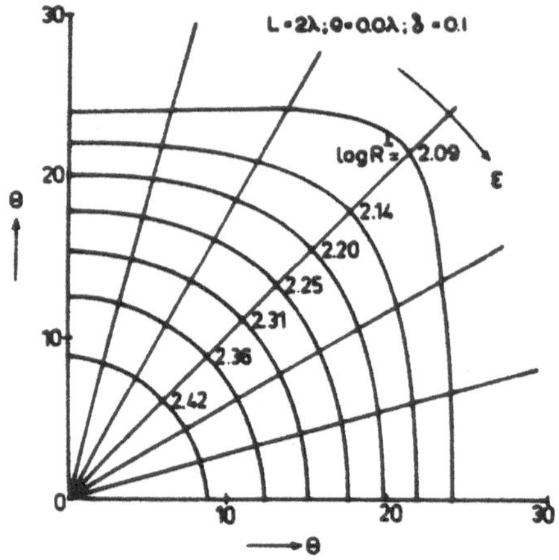

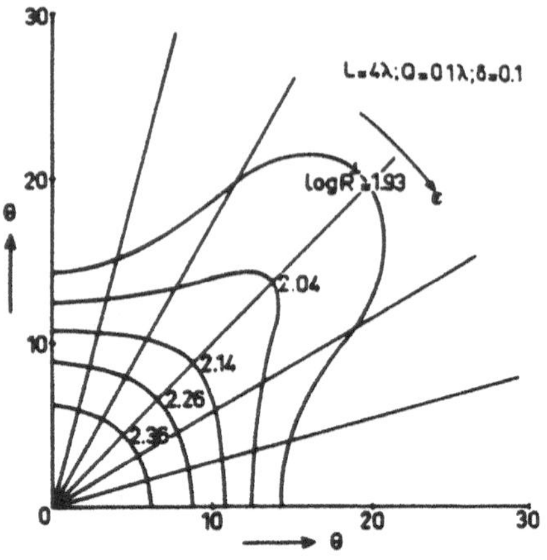

Figure 6. Theoretical $R_\theta^{\perp}$ Scattering for Infinitely Thin Rods of L = 2λ. ε = angle between the plane defined by $\underline{E}_p$ and $\underline{s}_o$ and the plane defined by $\underline{s}_o$ and $\underline{s}'$ (see Figure 1). J. H. Gouda (18) .

Figure 7. The same as Figure 6 for length 4λ and radius 0.1λ. J. H. Gouda (18) .

uncertainty as regards the reality of this maximum (17), it is an indication - if real - of a more pronounced structure than that implied by scalar correlation functions only. If one is confronted with such data, it becomes useful to develop more explicit models for the structure of the gel and to compare the theoretical with the experimental Rayleigh ratios. Figure 5 shows H_v for a model in which a random assembly of anisotropic rodlike aggregates of varying thickness embedded in an isotropic matrix is assumed, neglecting inter-rod interference effects. The optic axis is assumed to coincide with the rod axis (18,19). A comparison of Figures 4 and 5 points to the existence of rods with finite thickness in water-swollen PHEMA gels. The model also predicts that in this case there should be additional information in the com-

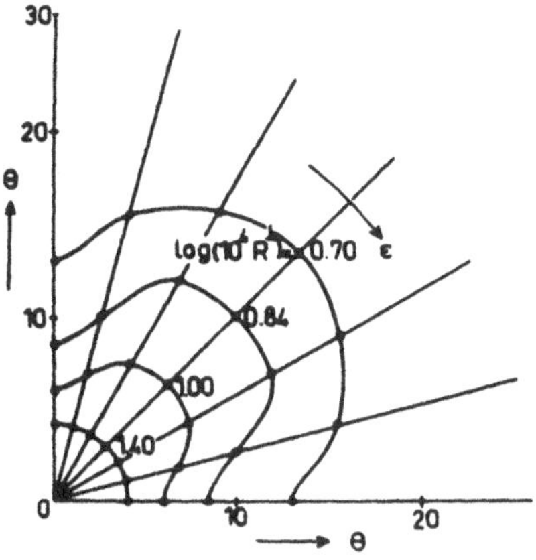

Figure 8. Experimental Small Angle $R_\theta^\perp$ Envelope for a PHEMA Gel, Swollen to Equilibrium in Water. Gel similar to S-2 in Figure 4, where the data do not extend to small angles. J. H. Gouda (18).

plete $R_\theta^{\parallel}$ and $R_\theta^\perp$ envelopes. Figures 6 and 7 show the $R_\theta^\perp$ envelopes as calculated for a rod thickness of 0.0λ and a length of 2λ as well for a thickness of 0.1λ and a length of 4λ. Figure 8 shows the $R_\theta^\perp$ envelope for one of the water-swollen PHEMA gels which at least qualitatively corresponds with the theoretical envelopes. On the basis of scalar correlation functions, $R_\theta^\perp$ would have been perfectly circular. It is worth while mentioning that the thinner the rods the more the $R_\theta^\perp$ envelopes are concentrated at 45°, at least as long as the rod axis and the optic axis coincide (18,19). This illustrates that the H_v scattering is not always very informative whereas $R_\theta^\perp$ is. This in turn exemplifies the usefulness of a rotatable polarizer and analyzer in light scattering instrumentation (see above and ref. 6).

In the limit of zero scattering angle the rod model and the random orientation correlation approach can formally be connected. At small angles H_v can be expanded as a series in h^2 (18,17,4), yielding

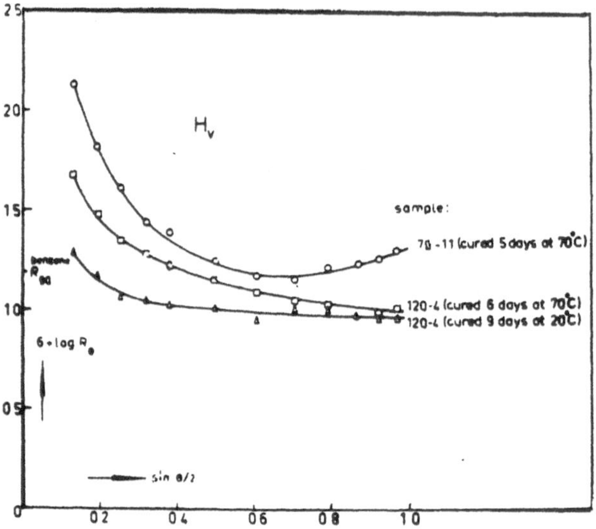

Figure 9. Wide Angle H_V Scattering of Polyurethane Elastomers. The first number in the code refers to the average number of oxypropylene units between crosslinks and the second to the average number of urethane links per chain. More urethane links and curing at higher temperature increase the amount of inter-chain ordering. R. Blokland and W. Prins (20) .

$$H_V = (16\pi^4/15\lambda_o^4)\ \delta_r^2 V_r \phi_r \left\{ 1 - (1/7)[(L/2)^2 + Q^2]h^2 + --- \right\} \qquad -7-$$

where δ_r is the anisotropy of the rod per unit volume, V_r its volume, L its length, Q its radius and ϕ_r the volume fraction of all rods together in the network. We can now, quite generally, consider the network as a collection of anisotropic elements of different types with anisotropies δ_ℓ, volumes V_ℓ and volume fractions ϕ_ℓ, such that $\sum_\ell \phi_\ell = 1$. For example, one might consider the network to be built up of chain segments and crosslinks, or of chain segments, crosslinks and supramolecular rod-like aggregates, or whatever. Upon comparison with equation (1) and (5) we find:

$$[H_v]_{\Theta=0} = (16\pi^4/15\lambda_0^4) \sum_\ell \delta_\ell^2 \, \emptyset_\ell \, V_\ell = (64\pi^5/15\lambda_0^4)\beta^2 \int_0^\infty f(\dot{r})r^2 dr$$

$$-8-$$

$$= (16\pi^5/15\lambda_0^4)\beta^2 \sum_{j=1}^m y_j b_j^3 \pi^{\frac{1}{2}}$$

Equation 8 can be useful if one wants to asses the relative contributions to the scattering (12).

Of course, other models than the ones mentioned here can be developed as the need arises. Work in progress in our laboratory for example shows that the light scattering of 1% aqueous agar-gels cannot be described by the rod model nor by scalar orientation correlation functions.

Light scattering of polymer networks is of importance because of the expectation that the structural make-up of a network will determine its time-dependent and equilibrium mechanical behavior. Evidence for this can be found in the work of Ilavsky on PHEMA gels (12) as well as in the work of Blokland (19) on chemically crosslinked poly-urethanes. Figure 9 shows an example of the latter. Equilibrium,wide-angle H_v data are plotted as they were obtained from a direct cure of linear propylene oxyde - extended diols, extended triols and toluene di-isocyanate in a cylindrical light scattering cell. Both the angular dependence (determined by f(r) in Eq. 1 and the intensity (determined by β^2 in Eq. 1) increase with increasing number of polar urethane links between crosslinks. Also, a slow, room temperature cure gives less scattering than a rapid, hot cure. These findings point toward the occurrence of a supramolecular ordering in the networks which is more pronounced the stronger the interaction between the network chains (more urethane links per chain). They also show the dependence of the interchain effects on the thermal history. Upon swelling in benzene the H_v scattering disappears. At the same time the markedly non-Gaussian (Mooney-Rivlin type) elasticity behavior is replaced by an ideal Gaussian behavior in the swollen state.

In conclusion, it seems that the relation between structure and (visco-)elasticity of non-crystalline polymer networks is well worth a further exploration (4). The light scattering technique should be useful tool in this endeavor. Studies on strained or dynamically loaded networks - not touched upon in this article - properly belong in such a program (21). A next step would be the investigation of the dynamics of non-crystalline polymer systems through a study of the spectral width of the Rayleigh scattering.

References

1. See e.g., D. McIntyre and F. Gornick "Light Scattering from Dilute Polymer Solutions". Gordon and Breach, N. Y., 1964, or M. Kerker, "The Scattering of Light", Academic Press, N. Y., 1969.

2. For theory, see e.g., R. Pecora and Y. Tagami, J. Chem. Phys., 51, 3298 (1969); R. Pecora, Macromolecules, 2, 31 (1969); for instrumentation, see e.g., N. C. Ford, G. B. Benedek, Phys. Rev. Letters, 15, 649 (1965); F. T. Arecchi, M. Giglio and U. Tartari, Phys. Rev., 163, 186 (1967).

3. See e.g., J. Polymer Sci. Part C, 5 (1964), 13 (1966) and R. S. Stein in "Electromagnetic Scattering", Proceedings of the Second Interdisciplinary Conference on Electromagnetic Scattering, R. L. Rowell and R. S. Stein eds., Gordon and Breach Publ., N. Y., 1967.

4. Reviewed in K. Dušek and W. Prins, Fortschr. Hochpolym. Forsch., 6, 1 (1969).

5. See e.g., R. S. Porter, E. M. Barrall, II, and J. F. Johnson, Accts. of Chem. Research, 2, 53 (1969).

6. J. H. Gouda and W. Prins, J. Polymer Sci. Part A2, in press. See also ref. 18.

7. E. V. Beebe, R. L. Coalson and R. H. Marchessault, J. Polymer Sci., Part C, 13, 103 (1966).

8. R. S. Stein and J. J. Keane, J. Polymer Sci., 17, 21 (1955).

9. This is the prototype of the instrument commercially available from Bleeker and Zonen, Zeist, The Netherlands, described in A. E. M. Keyzers, J. J. van Aartsen and W. Prins, J. Appl. Phys., 36, 2874 (1965).

10. J. J. van Aartsen, These proceedings.

11. R. S. Stein and P. R. Wilson, J. Appl. Phys., 33, 1914 (1962); also with corrected constant in ref. 9.

12. M. Ilavsky and W. Prins, Macromolecules, 1970, in press.

13. L. Gallacher and F. A. Bettelheim, J. Polymer Sci., 58, 697 (1962).

14. R. S. Stein, J. Polymer Sci., Part B, 7, 657 (1969).

15. F. Bueche, J. Coll. Interf. Sci., 33, 61 (1970).

16. J. H. Gouda, K. Povodator, T. C. Warren and W. Prins, J. Polymer Sci., Part B, 8, 225 (1970).

17. M. C. A. Donkersloot, J. H. Gouda, J. J. van Aartsen and W. Prins, Recl. Trav. Chim. Pays-Bas, 86, 321 (1967).

18. J. H. Gouda, Ph.D. Thesis, Technological University, Delft, 1969, The Netherlands, available upon request until supply is exhausted.

19. M. B. Rhodes and R. S. Stein, J. Polymer Sci., Part A2, 7, 1539 (1969).

20. R. Blokland and W. Prins, J. Polymer Sci., Part A2, 7, 1595 (1969). See also R. Blokland, "Elasticity and Structure of Polyurethane Networks", available through Gordon and Breach, N. Y., 1969.

21. W. Chu and R. S. Stein, J. Polymer Sci., Part A2, 8, 489 (1970). D. G. Legrand, J. Polymer Sci., Part A2, 7, 279 (1969) and other articles in press.

THE STRESS-STRAIN BEHAVIOR OF MECHANICALLY DEGRADABLE POLYMERS*

R. J. Farris

College of Engineering, University of Utah

Salt Lake City, Utah**

SUMMARY

Non-linear constitutive equations are developed for highly filled polymeric materials. These materials typically exhibit an irreversible stress softening called the "Mullins' Effect." The development stems from attempting to mathematically model the failing microstructure of these composite materials in terms of a linear cumulative damage model. It is demonstrated that p^{th} order Lebesgue norms of the deformation history can be used to describe the state of damage in these materials and can also be used in the constitutive equations to characterize their time dependent response to strain distrubances. This method of analysis produces time dependent constitutive equations, yet they need not contain any internal viscosity contributions. This theory is applied to experimental data and shown to yield accurate stress predictions for a variety of strain inputs. Included in the development are analysis methods for proportional stress boundary valued problems for special cases of the non-linear constitutive equation.

I. INTRODUCTION

This paper deals with non-linear homogeneous constitutive equations of degree one, a type of behavior that until recently [1] has not been mentioned in the field of mechanics. This type of

* The research reported herein was supported in part by the Air Force Office of Scientific Research THEMIS Contract F-44620-68-C-0022. The work was performed at the University of Utah, Department of Civil Engineering, and is part of the author's Ph.D. dissertation.
** Current address, Aerojet Solid Propulsion Company, Sacramento, Ca.

behavior satisfies one of the two requirements for linearity and in the author's opinion constitutes the simplest type of non-linear behavior. Because one of the linearity requirements is satisfied by these materials, they are often mistaken for linear materials [1,2], since the characterization procedures used by many laboratories do not differentiate between linear and homogeneous behavior [2]. The difficulty lies in differentiating between necessary conditions and sufficient conditions to guarantee linear behavior. If a material is linear, it will always have a homogeneous constitutive equation. However, if a material has a homogeneous constitutive, it need not be linear.

Composite solid propellants and other highly filled polymeric materials which exhibit "stress softening" [2-8] at strains below detectable dewetting [6,9,10,11] appear to have non-linear homogeneous constitutive equations. In the range of strain below detectable dewetting these materials have usually been treated and thought of as linear viscoelastic solids since they have relaxation moduli that are generally independent of strain magnitude [1,2,12]. Examination of the mathematical requirement for linearity indicates that the above criterion is simply a check on the homogeneity of the constitutive equation and does not guarantee linearity.

The purpose of this paper is, therefore, to (a) develop non-linear constitutive equations that are mathematically homogeneous for characterizing highly filled polymeric materials and (b) develop methods by which these constitutive equations can be used in solving boundary valued problems.

To accomplish these goals systematically, a brief discussion of constitutive linearity and non-linearity is introduced in Section 2. In Section 3 the "stress softening" or "Mullins' Effect" [2-8] is analyzed from a simple mechanical failure model of the composite material's microstructure and is seen to contain this type of non-linear behavior. Also the pth order Lebesgue norms [13,14] of the strain history are presented as being excellent memory measures of the strain history to use in the constitutive equations. In Section 4 the development is extended to include the general three-dimensional constitutive equation for isotropic materials.

Stress analysis procedures for materials having homogeneous constitutive equations are developed in Section 5. Here a correspondence principle is developed for proportional stress boundary valued problems demonstrating that for large classes of these constitutive equations linear analysis methods can be used to obtain the stress distribution, while the strain distribution must be obtained from the non-linear constitutive equation. In section 6 the theory is applied to experimental data and shown to yield accurate stress predictions for a variety of strain inputs.

II. LINEARITY REQUIREMENTS

There is apparently some confusion among many practicing engineers as to what exactly constitutes a linear constitutive equation and how much a relation is obtained. In the literature linear constitutive laws are usually simply given and the actual mathematical requirements for linearity are never stated [15-20]. The problem is greatly complicated when considering non-linear behavior, or more precisely what condition or conditions must be violated before the material is classified as non-linear. Most non-linear theories have had their origin in the addition of second order terms to a first order equation. Is this second order theory then the simplest non-linear equation or are there even simpler non-linear forms? These questions cannot be answered until the mathematical requirements for linearity are stated, for only by violating the linearity conditions can non-linearity be defined.

When solving boundary valued problems in the field of solid mechanics, non-linearities can arise in two ways, kinematic and material. Material non-linearities mean naturally a non-linear stress-strain constitutive law while kinematic non-linearities have to do with the strain-motion relationship.

2.1 Constitutive Linearity

In the field of mathematics, the requirements for linearity are the same whether they be applied to differential equations, functions, operators, transforms, functionals or other mathematical operations. When these linearity requirements are applied to constitutive theories they are applied best to functional equations [21] since in continuum mechanics a simple material is defined as material wherein the present state of stress can depend upon the history of the deformation gradients [18,19,20]. For the case of small displacements and rotations, the state of stress can be taken as being dependent upon the history of the strain tensor [21-26] and can be expressed functionally as

$$\sigma_{ij}(t) = F_{ij}\left[\varepsilon_{pq}\overset{t}{\underset{\tau=o}{(t,\tau)}}\right] . \tag{2.1}$$

For a constitutive equation to be linear it must satisfy the following functional equation [21].

$$F_{ij}\left[a\varepsilon_{pq}\overset{t}{\underset{\tau=o}{(t,\tau)}} + b\varepsilon'_{pq}\overset{t}{\underset{\tau=o}{(t,\tau)}}\right] = aF_{ij}\left[\varepsilon_{pq}\overset{t}{\underset{\tau=o}{(t,\tau)}}\right] + bF_{ij}\left[\varepsilon'_{pq}\overset{t}{\underset{\tau=o}{(t,\tau)}}\right] \tag{2.2}$$

The linearity conditions given in equation (2.2) can also be written as two separate rules instead of one as

$$F_{ij}\left[a\varepsilon_{pq}\left(t,\tau\right)_{\tau=o}^{t}\right] = aF_{ij}\left[\varepsilon_{pq}\left(t,\tau\right)_{\tau=o}^{t}\right] \text{ , and} \qquad (2.3a)$$

$$F_{ij}\left[\varepsilon_{pq}\left(t,\tau\right)_{\tau=o}^{t} + \varepsilon'_{pq}\left(t,\tau\right)_{\tau=o}^{t}\right] = F_{ij}\left[\varepsilon_{pq}\left(t,\tau\right)_{\tau=o}^{t}\right] + F_{ij}\left[\varepsilon'_{pq}\left(t,\tau\right)_{\tau=o}^{t}\right], \qquad (2.3b)$$

where $\varepsilon_{pq}(\tau)$, $\varepsilon'_{pq}(\tau)$, a, are arbitrary.

Careful examination of these two conditions indicates that the first rule of linearity, called scalar multiplication or homogeneity of degree one [21,27], is contained in the second rule of linearity, called additivity or Boltzmann superposition. This duplication can simply be shown for all scalars that are rational numbers [1]. Therefore only one mathematical requirement for linearity exists if a reasonable form of continuity requirement is enforced, and that is Boltzmann superposition. It can be shown also that scalar multiplication in no way implies superposition. In fact scalar multiplication is simply a homogeneity condition of degree one in the constitutive law, and many non-linear differential equations, functions, and functionals are homogeneous but not linear.

Non-linear ordinary differential equations that are homogeneous can always be separated [28] and solved quite simply by choosing a new variable that is the ratio of the two variables in the equation. Examples of functional equations that are homogeneous but not linear also can be constructed [21,27]. The main difficulty that arises from this observation is the commonly used and stated criteria for linearity of elastic and viscoelastic materials is, "doubling the strain input doubles the stress output". In light of the linearity conditions it is seen that this is only a check on the homogeneity of the materials constitutive law which is a necessary condition for linearity, but in itself cannot guarantee linearity. The homogeneity condition demands that the relaxation modulus for a linear viscoelastic material be independent of the magnitude of the applied strain, or that the first stretch behavior of an elastic solid have a constant moduli. These are necessary conditions for linearity but not sufficient conditions to guarantee linear response. Materials that possess this homogeneity property but are still non-linear perhaps are the simplest non-linear material since at least one of the conditions of linearity is satisfied. Because one of the linearity conditions has been satisfied, the material will possess some of the properties of linear materials. Unfortunately, the standard characterization methods used by many laboratories will cause all homogeneous materials, linear or non-linear, to be characterized as linear materials [1,2]. Examples of a non-linear viscoelastic material having a homogeneous constitutive law are solid propellants and most highly filled polymeric materials [1,2] such as asphalt concrete. Examples of non-linear elastic* materials having homogeneous constitutive laws are steel wire [29], rock [30], portland cement and masonry materials.

III. MODELING THE MULLINS' EFFECT IN FILLED POLYMERS

Viscoelastic materials have a "memory": that is, their present state depends upon their entire past history. Nearly all of the integral viscoelastic constitutive theories used to date [18-26, 31-41] are based on the concept of "fading memory". This means that a material is more sensitive to its immediate past than to its distant past. A physical interpretation of fading memory constitutive laws, both linear and nonlinear, indicates that such materials tend to forget the distant past. This theory implies no permanent change in microstructure, or damage caused by the deformation. A fading memory material can undergo no irreversible changes in structure and can be thought of as attributing the time effects, such as relaxation and creep, to internal viscosity.

Experience indicates that highly filled polymers do not fall into the category of fading memory materials even at small strains below detectable dewetting [1,2] or volume dilatation. These materials suffer from the "Mullins' Effect" [2-8], which is a stress-softening that occurs with deformation, and causes a permanent hysteresis on repeat loading.

There is considerable evidence that all the hysteresis effects observed in these materials and most of the viscoelastic behavior can be caused by the time dependent failure of the polymer on a molecular basis and are not due to internal viscosity [1,2]. At near equilibrium rates and small strains filled polymers exhibit the same type of hysteresis that many lowly filled, highly crosslinked rubbers demonstrate at large strains [1-8]. This phenomenon is called the "Mullins' Effect" and has been attributed to microstructural failure. Mullins postulated that a breakdown of particle-particle association and possibly also particle-polymer breakdown could account for the effect [3-5]. Later Bueche [7,8] proposed a molecular model for the Mullins' Effect based on the assumption that the centers of the filler particles are displaced in an affine manner during deformation of the composite. Such deformations would cause a highly non-uniform strain and stress gradient in the polymer

*Elastic is used in the classical sense which means complete recovery of geometry when the tractions are removed.

between particles, especially in the direction of stretch. He assumed that polymer chains attached themselves at both ends to neighboring filler particles and that these chains ruptured when the particles were separated enough to extend the chains to near their full elongation. He derived a model from which he could calculate the difference in stress levels at a given elongation for the first and second stretching cycles [7]. It is this type of model that is generally accepted as being representative of the molecular behavior which causes the "Mullins' Effect". Figure 3.1 illustrates this behavior for repetitive stretching to increasing strain levels. In highly cross-linked rubbers, the effect only depends upon strain and is generally irreversible [5,6]. However, if the prestressed composite is allowed to rest for long times in the relaxed state, a portion of the original stiffness might be regained [5]. This recovery or rehealing appears to be a complex function of the recovery temperature and time, nevertheless, it can and does greatly influence the materials behavior.

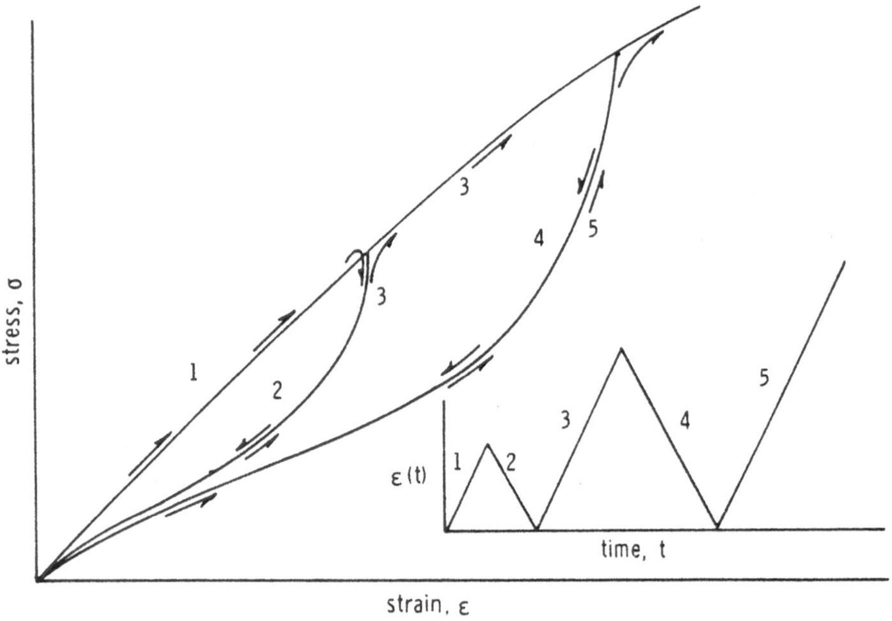

Figure 3.1. Typical Equilibrium Stress-Strain Behavior for a Highly Filled Polymer when the Direction of Strain is Reversed

All of the theoretical and nearly all of the experimental work done in studying this phenomenon has been on materials similar to the rubber found in automobile tires. These are highly cross-linked rubbers that are usually filled to about 20 volume percent with very fine carbon black. Composite solid propellants, on the other hand, are lowly crosslinked and highly filled with coarse particles. The relative particle spacing is consequently much more severe and the polymer chains are on the average hundreds of times longer in propellants than in tire rubber. The probability of finding a larger portion of the chains connecting particles would be greater in propellants and the effect therefore should be much stronger and occur at smaller strains [1,2,6], but the same basic mechanism proposed by Bueche still applies. This polymeric chain failure is therefore the step which precedes the vacuole formation process which causes the stress and dilatational non-linearities observed at larger strains [9,10]. Multiple stretch data on propellants at large strains with and without a superimposed pressure environment demonstrate that propellants also exhibit the Mullins' type hysteresis at large strains in the absence of measurable dilatation [6].

The time independent "Mullins' Effect" can account for the near equilibrium hysteresis observed in propellants at low strains, but cannot account for the nonlinear time effects [1,2]. There is considerable evidence however, that the "Mullins' Effect" in propellants is a very strong function of time [1,2]. Time dependent chain failure can be readily demonstrated by simply examining the influence of filler on viscoelasticity and some of the routine tests run on solid propellants.

One of the simplest ways of demonstrating a time dependent "Mullins' Effect" is through the strain endurance test [1,12]. In this test, a sample is strained to some level and held there for several days or longer. The only measurement taken is the time to failure, if the sample fails within the test period. The point of interest here is that samples fail while held at conditions of constant strain when the stress is slowly relaxing or at most constant. This type of failure is clear evidence of a time dependent "Mullins' Effect" and also demonstrates that some portion of the time dependent stress relaxation must be due to chain failure.

Another example of a time dependent "Mullins' Effect" is that of a cross-linked polymer, with little or no time dependency when unfilled, which becomes significantly time dependent when filled [42], as shown in Figure 3.2. The more filler incorporated into the system, the more marked the time effect. Many propellant polymers fall into this category and nearly all propellants show time dependence over such long times that true equilibrium data cannot be obtained. This time dependence in the composite material and no time dependence in the unfilled polymer cannot be

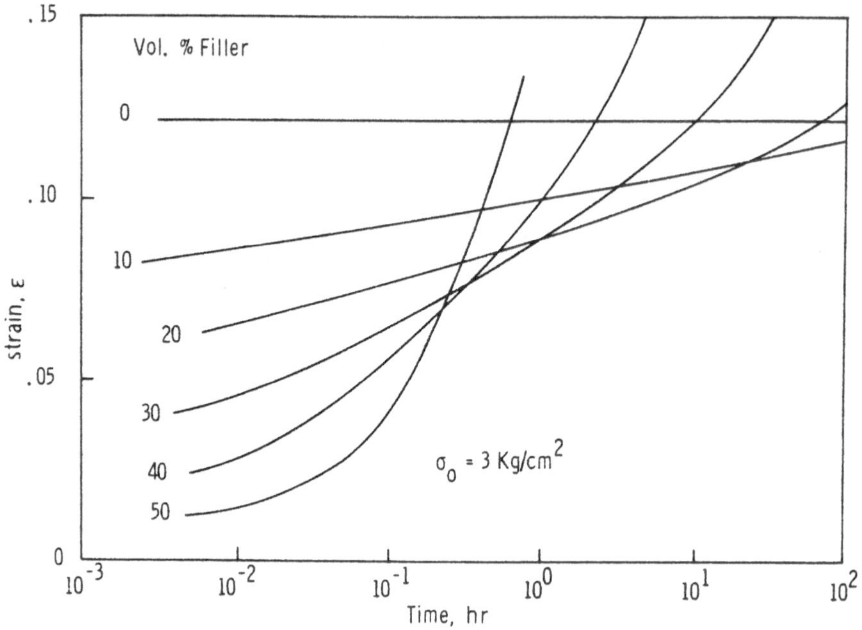

Figure 3.2. Creep of Sodium Chloride Filled Polyurethane Rubber

explained by the argument that the polymeric strain rates are higher in the composite than in the pure polymer since the time effects continue for such long times, and many propellant binders show no time dependence even at very short times.

Tests such as those described above indicate that the constitutive equations and degree of microstructural damage must be highly coupled effects. The commonly used assumption that the degree of damage and the constitutive equation are uncoupled can only lead to erroneous results for materials exhibiting the "Mullins' Effect". The model discussed below describes the "Mullins' Effect" and clearly demonstrates the coupling between the constitutive equation and the degree of damage. The model provides insight into the mechanism of behavior and indicates key variables or measures that should be used in the general multidimensional constitutive equation.

3.1 Modeling the Time Independent "Mullins' Effect"

There have been various models and mechanisms proposed for the "Mullins' Effect". Bueche proposed a model based on chains failing due to physically non-homogeneous local deformations [7,8]. His model was not sufficietly general and was designed to prove whether the chains were unbonding from the filler or actually failing. In this section a general one dimensional model will be developed for the "Mullins' Effect". Before proceeding, it would be wise to clarify the main difference between filled and unfilled polymers. The equilibrium constitutive equation for cross-linked amorphous polymers has been developed from the statistical theory of rubber elasticity assuming ideal rubber behavior [43,44]. There are essentially six basic assumptions made in the development of the statistical theory of ideal rubber behavior. They are:

1. There is no change in internal energy with isothermal deformations.

2. The end-end displacement of a polymer chain is small compared to its actual length.

3. The relative end-end displacements of all polymer chains in the system are equal for homogeneous motions.

4. The relative chain deformations occurring microscopically are the same as the relative deformation of the body for homogeneous motions.

5. There is no interaction between polymer chains.

6. A polymer chain never fails.

These assumptions dictate that the configurational entropy associated with a polymer chain be given by a Gaussian distribution and enable simple addition of the contributions of each chain. The Gaussian distribution is only valid for end-end chain displacements that are small compared to the actual chain length since they actually allow for end-end displacements from zero to infinity [43,44]. Corrected configurational statistics for large deformations provide what is called the Langevin Function [44]. The Langevin Function provides the correct configurational entropy since it limits the end-end separation of a chain to the chain's actual length [44]. The Gaussian distribution appears as the first term in the Langevin Function which is essentially a virial expansion. The main difference between these two distributions is the force-deformation relation they give for a polymer chain [44] which is illustrated in Figure 3.3. The great stiffening experienced when a chain is near fully extended can be simply observed by stretching a rubber band to failure.

The main difference between filled and unfilled systems is that even under equilibrium conditions the relative end-end

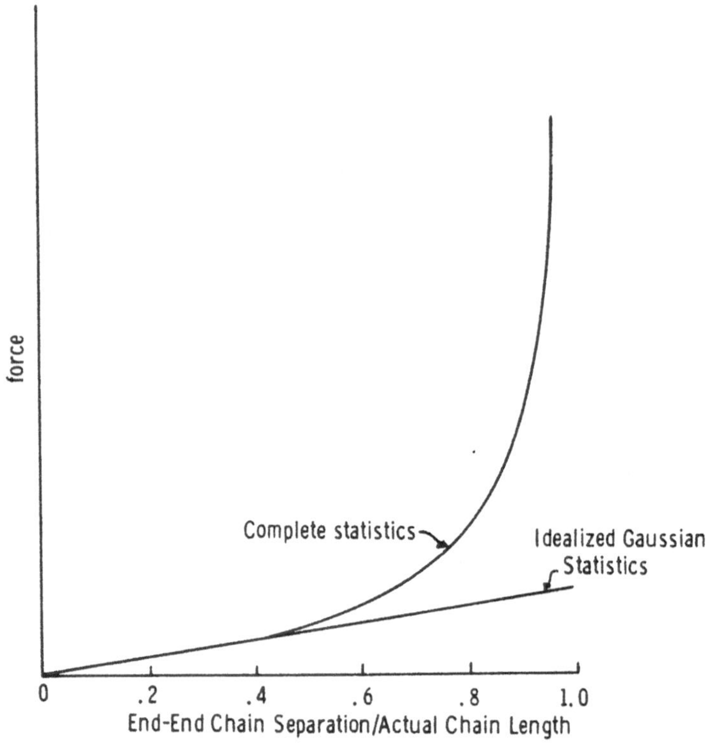

Figure 3.3. The Complete Force-Extension Relation for a Random Chain Compared to the Gaussian Prediction

displacement of all polymer chains in the system are not equal. Instead one finds, by any form of analysis, that the local strains in a filled system subjected to a physically homogeneous deformation are a very strong function of filler content, position, particle shape, and the distribution of chain lengths. The prime reason for the physically non-homogeneous local deformations of the polymer is that it is the centers of the filler particles that must undergo near affine or similar deformations since they are rigid and cannot occupy similar positions at the same time. The polymer being highly extensible and mobile is forced to undergo large variations in local strain. It is therefore not valid to assume the force contribution of each chain is similar, nor is it valid to assume the end-end displacements are small since very large strains can occur locally. Even small macroscopic strains can cause some fraction of the material to undergo very large local strains. For such conditions it is valid to assume that a chain will fail if some critical condition is exceeded. It is this type of localized failure that causes the "Mullins' Effect" in filled polymers. Such failure must preceed vacuole formation which is common in filled polymers. This behavior can be modeled for one dimensional behavior in a fairly general way by making the following assumptions [1]:

1. The relative axial deformation of any given polymer chain is proportional to the axial applied strain, the proportionality constant differing from chain to chain.

2. Each polymer chain has the same elastic, but not necessarily linear, stress-strain law.

3. Each polymer chain fails and remains failed if at any time in its history some failure criterion is exceeded.

In these assumptions physically non-homogeneous local deformations, a non-linear stress-strain law for each element, and the possibility of having some of the elements fail have all been taken into account. Since the desired end result of this work is accurate constitutive relations for materials exhibiting permanent memory phenomena that can be used in engineering analysis, emphasis will be placed on the behavior of elements, not necessarily on polymer chains. The resulting equations appear to be of value for describing many materials, not only amorphous polymers.

The first assumption can be expressed mathematically as

$$\varepsilon_i(x_j,t) = \varepsilon(t)\tau_i(x_j) \quad , \tag{3.1}$$

where ε_i = axial strain in the ith element

ε = applied axial strain

τ_i = strain intensity factor for the ith element

x_j = spatial coordinates

Assuming the elements are elastic and the problem strictly one dimensional and time independent, the failure criterion for an element can be expressed either as a maximum strain or maximum stress criterion. Assuming a chain fails when an extension ε_{max} or equivalently stress σ_{max} are exceeded, the following equation is obtained for the stress in the ith element.

$$\sigma_i = \begin{cases} f(\varepsilon_i) = f(\tau_i \varepsilon) & \text{if} \quad \tau_i ||\varepsilon|| \leq \varepsilon_{max} \\ 0 & \text{if} \quad \tau_i ||\varepsilon|| > \varepsilon_{max} \end{cases} \tag{3.2}$$

In equation (3.2) $f(\varepsilon_i)$ is an arbitrary function that is single valued, and $||\varepsilon||$ is the largest strain applied in the history of the deformation. The reason for using $||\varepsilon||$ is it assures that once an element fails, it remains failed.

The observed stress from such a model is simply the total force divided by the total area which is given by

$$\sigma = \frac{1}{A} \sum_{i=1}^{N} A_i \sigma_i = \frac{1}{A} \sum_{i=1}^{N} A_i f(\tau_i \varepsilon) \; . \tag{3.3}$$

The summation in equation (3.3) can be more conveniently expressed as an integral. Using distribution theory equation (3.3) becomes

$$\sigma = \int_0^{\varepsilon_{max}/||\varepsilon||} N(\tau) f(\tau\varepsilon) d\tau \; . \tag{3.4}$$

In this integral $N(\tau)d\tau$ is a weighting function that represents the fraction of elements in a unit cross-section having strain intensity factors between τ and $\tau + d\tau$. The lower limit of integration can be taken as zero since $N(\tau)$ can be zero until some lower limit of τ is reached. The upper limit of integration is a function of the deformation history. The resulting stress-strain equation is a function of two variables, the current strain ε, and the maximum strain in the deformation history, $||\varepsilon||$. Before proceeding further, it should be pointed out that this simple stress-strain law can contain reversible as well as irreversible elastic responses which can be both linear and non-linear, up to and including failure by proper selection of the function $N(\tau)$. For example if all elements had the same intensity factor (e.g., rubber elasticity where $\tau_i = 1$) we obtain $N(\tau) = (\delta - 1)$, where δ is the Dirac delta function. The resulting integration yields

$$\sigma = \begin{cases} f(\varepsilon) & \text{if} \quad ||\varepsilon|| \leq \varepsilon_{max} \\ 0 & \text{if} \quad ||\varepsilon|| > \varepsilon_{max} \end{cases} \tag{3.5}$$

Similarly, reversible behavior for only some small region of strain can be obtained by having $N(\tau)$ non-zero only for the same range of τ. An example of this case would be

$$N(\tau) = \begin{cases} g(\tau) & \tau \leq a \\ 0 & \tau > a \end{cases} \quad (3.6)$$

The resulting integration yields

$$\sigma = \int_0^{\varepsilon_{max}/||\varepsilon||} N(\tau)f(\tau\varepsilon)d\tau = \int_0^{a} g(\tau)f(\tau\varepsilon)d\tau = G[\varepsilon,a]$$

$$\text{when} \quad a||\varepsilon|| < \varepsilon_{max} . \quad (3.7)$$

The problem at hand is not reversible behavior, but instead is irreversible phenomenon such as the "Mullins' Effect". Consider that $N(\tau)$ and $f(\tau\varepsilon)$ are arbitrary but non-zero, and the material was subjected to the strain history given below, it can be shown that the Mullins' type hysteresis is contained in equation (3.4).

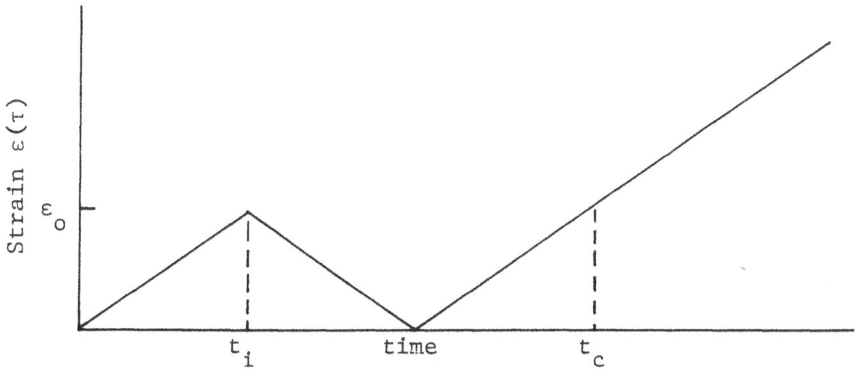

Since $||\varepsilon||$ is by definition the maximum strain experienced in the deformation history, this input yields

$$||\varepsilon|| = \begin{cases} \varepsilon(t) & \text{if} \quad 0 \leq t \leq t_1 \\ \varepsilon_o & \text{if} \quad t_1 \leq t \leq t_2 \\ \varepsilon(t) & \text{if} \quad t_2 \leq t \end{cases} . \quad (3.8)$$

The resulting stress output for this case, when the values of range from zero to infinity would be

$$\sigma(t) = \int_0^{\varepsilon_{max}/\varepsilon(t)} N(\tau)f(\tau\varepsilon)d\tau = G[\varepsilon(t),\varepsilon(t)] \text{ if } 0 \leq t \leq t_1, \quad (3.8a)$$

$$\sigma(t) = \int_0^{\varepsilon_{max}/\varepsilon_o} N(\tau)f(\tau\varepsilon)d\tau = G[\varepsilon(t),\varepsilon_o] \quad \text{if } t_1 \leq t \leq t_2, \quad (3.8b)$$

$$\sigma(t) = \int_0^{\varepsilon_{max}/\varepsilon(t)} N(\tau)f(\tau\varepsilon)d\tau = G[\varepsilon(t),\varepsilon(t)] \text{ if } t_2 \leq t \quad . \quad (3.8c)$$

This type of behavior is illustrated in the sketch below.

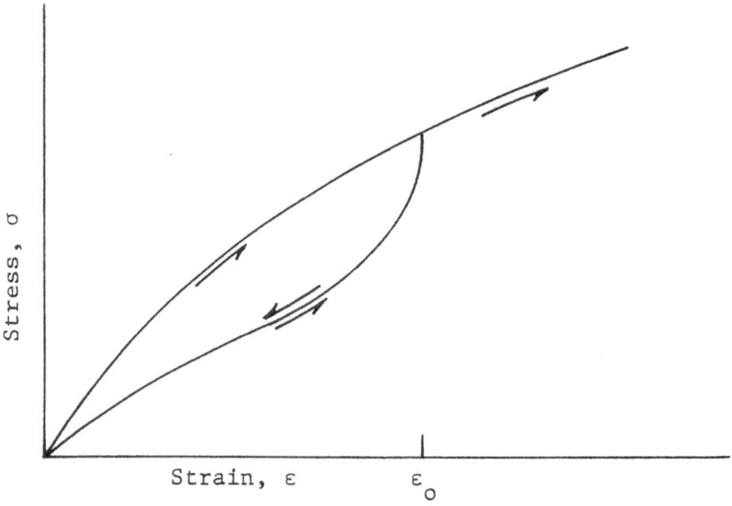

Clearly equation (3.4) contains the Mullins' type hysteresis discussed earlier. Data from tests like the one illustrated above to increasing values of ε_o can be used to determine the distribution function $N(\tau)$ independent of the local stress-strain function $f(\tau\varepsilon)$. Differentiating equation (3.8a) and 3.8b) with respect to strain, substracting one from the other, and evaluating each at $\varepsilon = \varepsilon_o$, produces

$$N(\tau = \varepsilon_{max}/\varepsilon_o) = \frac{C}{\tau^2}\left[G'[\varepsilon,\varepsilon_o] - G'[\varepsilon,\varepsilon]\right]_{\varepsilon=\varepsilon_o} \tag{3.9}$$

$$\text{where } C = \frac{\varepsilon_{max}}{f(\varepsilon_{max})} = \text{constant}$$

If for example the difference between these derivatives was found to be independent of ε_o, then $N(\tau)$ would be proportional to $1/\tau^2$. For the general case the difference between these derivatives could be expanded in a polynomial in ε_o to obtain

$$\left[G'[\varepsilon,\varepsilon_o] - G'[\varepsilon,\varepsilon]\right]_{\varepsilon=\varepsilon_o} = \sum_{k=o}^{N} a_k \varepsilon_o^k \tag{3.10}$$

$$\text{where } a_k = \text{constants}$$

The distribution function $N(\tau)$ could be determined as a polynominal in $1/\tau$ since in equation (3.9) the variable τ is evaluated at $\tau = \varepsilon_{max}/\varepsilon_o$. Substituting a polynomial for $N(\tau)$ and $f(\tau\varepsilon)$ and performing the integration, the resulting equation for the stress output can be expressed as

$$\sigma = A_1 P_1(\varepsilon/||\varepsilon||) + A_2\varepsilon^2 P_2(\varepsilon/||\varepsilon||) + A_3\varepsilon^3 P_3(\varepsilon/||\varepsilon||) + \ldots,$$

where A_i = constants

and P_i = polynomials in the variable $(\varepsilon/||\varepsilon||)$

$P_i(1) = 1$ (3.11)

There are only two possibilities for the variable $||\varepsilon||$ in equation (3.11) or its predecessors, either $||\varepsilon|| = \varepsilon$ or $||\varepsilon||$ = constant. In the first case the unique stress-strain behavior for the first stretch is given. In the second case a hysteresis behavior dependent upon $||\varepsilon||$ as well as ε results.

3.2 Modeling the Irreversible Time Dependent "Mullins' Effect"

In the irreversible elastic case above a constitutive equation was derived that was dependent upon two variables, the current strain ε, and the maximum strain in the history of the deformation, $||\varepsilon||$.

At first, one might think that by using the time dependent equation derived above as a multiplier to a hereditary type fading memory viscoelastic constitutive equation, an equation describing all the irregularities and non-linearities a filled polymer demonstrates could be obtained. This is not the case, and can be clearly demonstrated in a number of ways. The simplest proof is that for monotonically increasing strains $\varepsilon = ||\varepsilon||$, and the equation would contribute nothing new and no permanent memory phenomenon. Yet, filled polymers exhibit non-linearities of the type that cannot be handled by fading memory equations for such cases. Another feature that must be contained in an accurate coupled constitutive equation is the concept of time dependent failure of elements and ultimately the material. These two possibilities clearly negate the possibility of so simple an equation of state.

There are two likely possibliities for incorporating time dependent failures. The first is to apply a kinetic reaction rate theory to the elastic elements [1,45], and the second is to apply a cumulative damage law [37,46] to the elastic elements. The first of these approaches is inconsistent with the time independent case and leads to very cumbersome mathematics. The cumulative damage concept on the other hand appears to be a natural extension of the time dependent case since it contains equation (3.4) as a special case. It is this type of model that will be pursued here.

Linear cumulative damage theory based on Miners' law [1,37,46] requires that

$$D(t) = \sum_{k=1}^{M} (t_k/t_{fk}) \quad . \tag{3.12}$$

In the above equation t_k is the time the material is held in the kth state of stress or strain, t_{fk} is the time to failure for the material if this k^{th} state of stress or strain were acting alone, and $D(t)$ is the measure of damage. For such a theory, failure occurs when $D(t) = 1$. Since the elements of our model are elastic, it makes no difference if a stress or strain damage formulation is assumed. It simplifies the mathematics however if a strain cumulative damage criterion is assumed since our model expresses stress in terms of the strain history.

Using a power law strain-time to failure relation [46] for the singly applied strain gives for the ith element

$$t_{fk} = C|\varepsilon_i|^{-P} \quad , \tag{3.13}$$

where $| \quad |$ indicates absolute value or magnitude,

and P = material property,

C = constant,

the damage equation becomes

$$CD_i(t) = \sum_{k=1}^{M} t_k \left| \varepsilon_i \right|^P . \tag{3.14}$$

Equation (3.14) like equation (3.3) can be more conveniently expressed as the integral

$$CD_i(t) = \int_0^t |\varepsilon_i(\xi)|^P d\xi, \text{ where } \xi = \text{dummy time} \tag{3.15}$$

Recalling the ε_i is the local strain in the ith element and that this strain is related to the applied strain by $\tau\varepsilon$, the equation for the stress becomes

$$\sigma(t) = \int_0^{\tau_c(t)} N(\tau) f(\tau\varepsilon) d\tau , \tag{3.16}$$

where τ_c = maximum permissible value of τ, and from equation (3.15) we obtain

$$\tau(t) = \left[CD_i(t) \bigg/ \int_0^t |\varepsilon(\xi)|^P d\xi \right]^{1/P} \tag{3.17}$$

The maximum permissible value of τ, can be obtained by maximizing the numerator and minimizing the denominator in equation (3.17), or equivalently answering the question, what must the intensity factor τ on an element be such that its time to failure is the current time t? Upon setting $D_i(t) = 1$, equation (3.17) becomes

$$\tau_c(t) = C' \bigg/ \left(\int_0^t |\varepsilon(\xi)|^P d\xi \right)^{1/P} = C'/||\varepsilon||_P \tag{3.18}$$

where C' = constant.

Mathematically the quantity $||\varepsilon||_p$ is called the pth order. Lebesgue norm [13,14], L_p. The L_p norm has properties that are worth noting and these are listed below.

$$||f||_p = \left(\int_0^t |f(\xi)|^p d\xi\right)^{1/p}$$

(a) $||af||_p = |a|\,||f||_p$

(b) $||f+g||_p \leq ||f||_p + ||g||_p$

(c) $||fg||_1 \leq ||f||_p ||g||_p$

(d) $||f-h||_p \leq ||f-g||_p + ||g-h||_p$

(e) $||f||_\infty = \lim_{p\to\infty}\left(\int_0^t |f(\xi)|^p d\xi\right)^{1/p} = \underset{0}{\overset{t}{\text{Maximum}}}\ |f(\xi)|$.

In the above equations f, g, and h are time functions, p and a are scalars, and ξ is a dummy time.

Thus for the time dependent case with permanent hysteresis our stress-strain equation becomes

$$\sigma(t) = \int_0^{c'/||\varepsilon||_p} N(\tau) f(\tau\varepsilon) d\tau . \tag{3.19}$$

If $N(\tau)$ is non-zero in the range $0 \leq \tau < \infty$, then it can be expanded as suggested by equation (3.9) and (3.10). If $f(\tau\varepsilon)$ is similarly expanded in a polynomial a form similar to equation (3.11) is obtained, the only difference being the replacement of $||\varepsilon||$, which was defined as what is now known to be $||\varepsilon||_\infty$, by $||\varepsilon||_p$. The resulting stress-strain equation is

$$\sigma(t) = A_1\varepsilon P_1(\varepsilon/||\varepsilon||_p) + A_2\varepsilon^2 P_2(\varepsilon/||\varepsilon||_p) + A_3\varepsilon^3 P_3(\varepsilon/||\varepsilon||_p) + \ldots, \tag{3.20}$$

where A_i = constants

P_i = polynomials in the variables $(\varepsilon/||\varepsilon||_p)$

The use of pth order Lebesgue norms in the constitutive equation is not original to this paper. Fitzgerald [47] has proposed a constitutive equation wherein the stress is a functional of the present value of the deformation gradient and its pth order Lebesgue norm. Coleman, Noll and Mizel have also proposed using these norms as approximations to the constitutive functionals [14,48]. Certain restricted forms of the constitutive equations developed in this study can be shown to be contained in these earlier works. The development herein however was not motivated by the earlier works which were developed from a pure mathematical continuum approach. Instead the development in this paper stems from attempting to mathematically model the microstructural behavior of highly filled polymeric materials. Key variables, which were measures of microstructural damage, that were obtained from these models so happened to have the exact same mathematical definition as L^p norms and were brought to the author's attention by Fitzgerald. This work may therefore in some way physically justify the use of norms in constitutive theory.

It is clear that equation (3.19) and (3.20) contain the time independent behavior given by equations (3.4) and (3.11) as special cases by letting $p = \infty$.

In order to obtain equation (3.18) and therefore equations (3.19) and (3.21), it was assumed that t_{fk} in the cumulative damage relations was given by a simple power law. Equivalently this meant that the damage relation, $D(t)$, for this special case could be expressed in terms of the L^p norm as

$$[D_i(t)]^{1/p} = a||\varepsilon_i||_p = D_i'(t) \quad , \tag{3.21}$$

where a is a constant and $D_i'(t)$ is some new measure of damage. Since failure was defined to occur when $D_i(t) = 1$, failure also occurs when $D_i'(t) = 1$ for all p. Although $D_i'(t)$ is a non-linear damage measure whenever $D(t)$ is linear, it has more useful properties than $D(t)$. An example of this is that by using equation (3.21) instead of equation (3.12) for this simple power law case, the strain cumulative damage criterion contains the maximum strain failure criterion simply by letting $p = \infty$. Also equation (3.21) and equation (3.12) predict precisely the same time to failure for all arbitrary strain histories providing t_{fk} is given by a simple power law. Since any monotonically increasing function can be approximated in terms of L^p norms, cases when t_{fk} is not a simple power law can also be handled. Consider the case when $D_i'(t)$ can be given by

$$D_i'(t) = a_1||\varepsilon_i||_1 + a_2||\varepsilon_i||_2 + \cdots + a_p||\varepsilon_i||_p . \tag{3.22}$$

For a simple step strain of magnitude ε_{io}, the time to failure, t_f, is given by the equation

$$\varepsilon_{io}^{-1} = a_1 t_f + a_2 t_f^{1/2} + \cdot\cdot + a_p t_f^{1/p} , \qquad (3.23)$$

Letting $\varepsilon_i = \tau\varepsilon$ as before the critical value of τ at any time t becomes

$$\tau_c(t)^{-1} = a_1||\varepsilon||_2 + a_2||\varepsilon||_2 + \cdot\cdot + a_p||\varepsilon||_p . \qquad (3.24)$$

Equation (3.24) could be used to give the upper limit of integration in equation (3.16) and equations similar to equation (3.20) could be generated. This approach is not necessary since in the next section the approach given here is generalized to obtain three dimensional constitutive equations with permanent memory phenomenon. The importance of the model approach selected is to shed light on key variables or measures to use in the constitutive equations.

Before proceeding further it would be wise to point out some of the behaviors possible using equation (3.19) or (3.20). Earlier in this paper existing integral viscoelastic consititutive theories were criticized as being of limited value since they did not contain the special case of a homogeneous non-linear constitutive equation of degree one. Equation (3.19) clearly satisfies the concept of a simple material since if the history of the strain is known, the stress can be computed. To see if the homogeneity condition can be satisfied, the history $a\varepsilon(t)$ is substituted for the history $\varepsilon(t)$ and the two equations can be compared. Doing so we find after using the properties of L^P norms given above that if $a^2N(a\tau) = N(\tau)$ and if the indefinite integral involves only odd powers of τ, then the homogeneity condition will be satisfied. Note that $N(\tau) \simeq 1/\tau^2$ meets the first requirement and the second requirement is satisfied simply by making the function f odd. Also it has been shown that $N(\tau) \simeq 1/\tau^2$ is representative of filled systems. This conclusion can be readily demonstrated by examining the distribution of effective gage lengths between two particles in a system filled with spherical particles [1]. This method of analysis also predicts that if in filled polymers exhibiting the "Mullins' Effect" the stress-strain behavior of an element is very non-linear as indicated by the Langevin statistic, then nearly all of the stress is being supported only by a small fraction of the polymer chains. This observation is consistent with experimental data and agrees with the observations of Mullins [3,5].

Before proceeding further to the development of three dimensional constitutive equations, considerable insight into the problems of mechanical characterization of materials can be obtained by analyzing the simple one dimensional equation given in equation (3.19) and (3.20). To restrict this equation to homogeneous functionals of degree one a sufficient condition is to require that equation (3.19) have the form

$$\sigma(t) = A_1 \varepsilon P((\varepsilon/||\varepsilon||_p)^2) . \tag{3.25}$$

In the above equation P is a polynomial in the variable $(\varepsilon/||\varepsilon||_p)^2$. This condition is necessary if scalar multiplication is to hold for all scalars, positive or negative. A special case of this equation is

$$\sigma(t) = 100\varepsilon[1 + (\varepsilon/||\varepsilon||_p)^n], \text{ where n is an even integer.} \tag{3.26}$$

Equation (3.26) gives a relaxation modulus that is independent of the applied strain magnitude, and will obey the homogeneity requirement of linearity for all scalars, with any arbitrary strain input. Equation (3.26) is not linear however as norms are not superposable except in the most trivial examples. The hypothetical material represented by Equation (3.26) is therefore non-linear, but for many types of tests used for material characterization it could not be distinguished from a linear viscoelastic material. In fact, the parameters n and P appearing in this constitutive equation can be adjusted so that the time derivative of the stress for a constant strain rate input is proportional to the relaxation modulus; a commonly cited property of a linear viscoelastic material [12,37]. Careful examination of the stress output to various strain inputs confirms the non-linear nature of this equation and indicate it is within the range of this simple equation to describe the one-dimensional response of solid propellants at small strains. To demonstrate this ability, the stress output for a variety of strain inputs have been determined for different values of n and p.

These data are illustrated in Figure 3.4 through 3.9. In these calculations the ratio n/P has been kept constant; therefore all of the equations would exhibit the same relaxation modulus. However as clearly indicated by these figures, the behavior to other inputs is different for the different values of n and P. This feature of giving the same output for one test, yet a different output for other tests, is characteristic of non-linear systems. If the material were linear, this feature would be impossible since one test dictates the results of all other tests for linear systems. Characterization of non-linear materials is therefore a difficult task as many tests must be used and one never knows if the chosen representation is complete. Individuals familiar with the behavior of linear viscoelastic materials and the non-linear behavior of

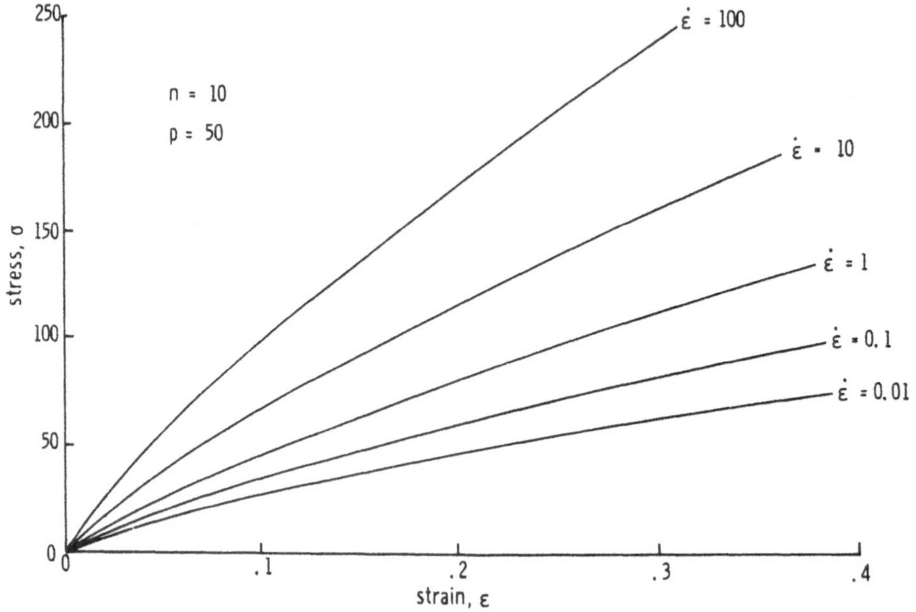

Figure 3.4. Calculated Constant Rate Stress-Strain Behavior of a Permanent Memory Material

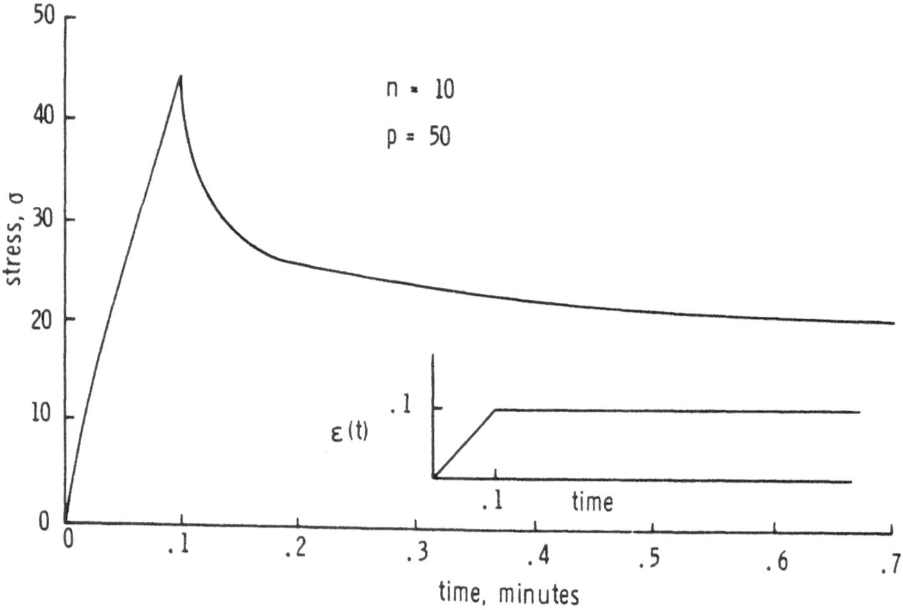

Figure 3.5. Calculated Ramp Strain Stress Relaxation Behavior of a Permanent Memory Material

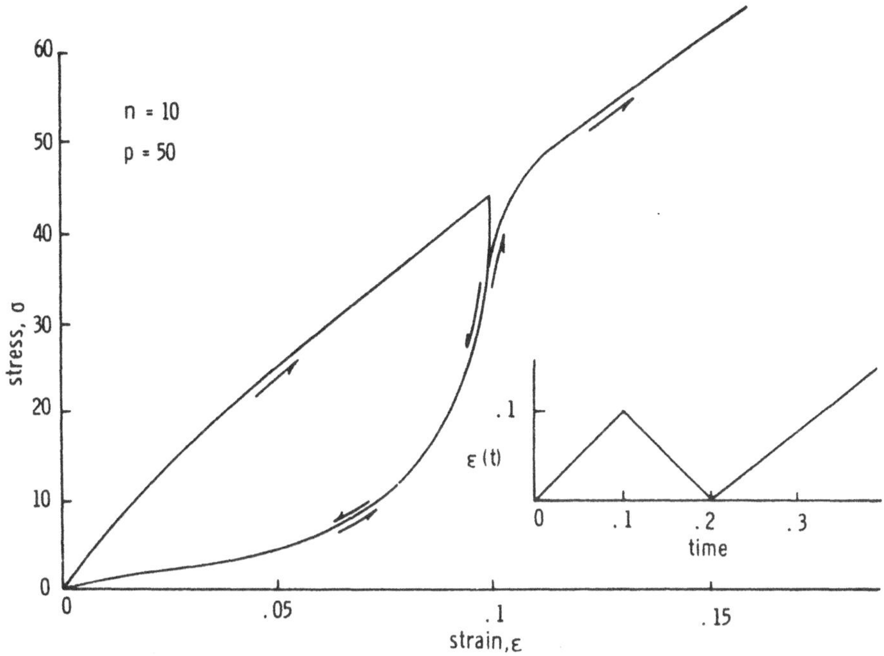

Figure 3.6. Calculated Permanent Memory Hysteresis Response to a Reversed Ramp Strain Input

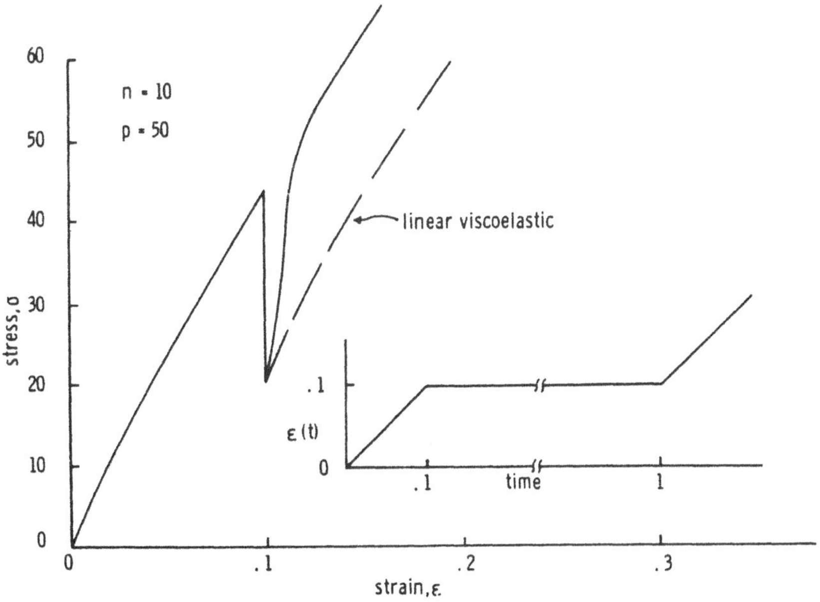

Figure 3.7. Calculated Permanent Memory Stress-Strain Response to an Interrupted Ramp Input

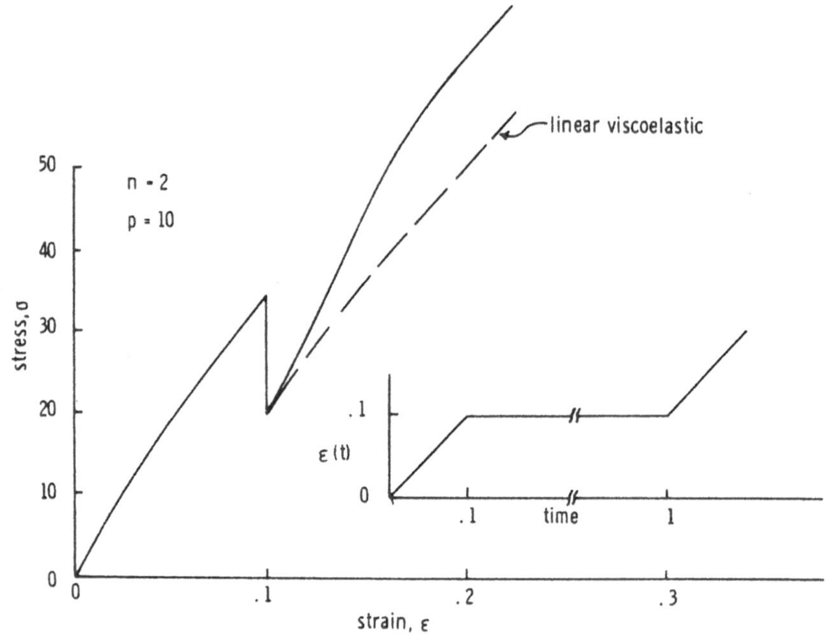

Figure 3.8. Calculated Permanent Memory Stress-Strain Response to an Interrupted Ramp Input

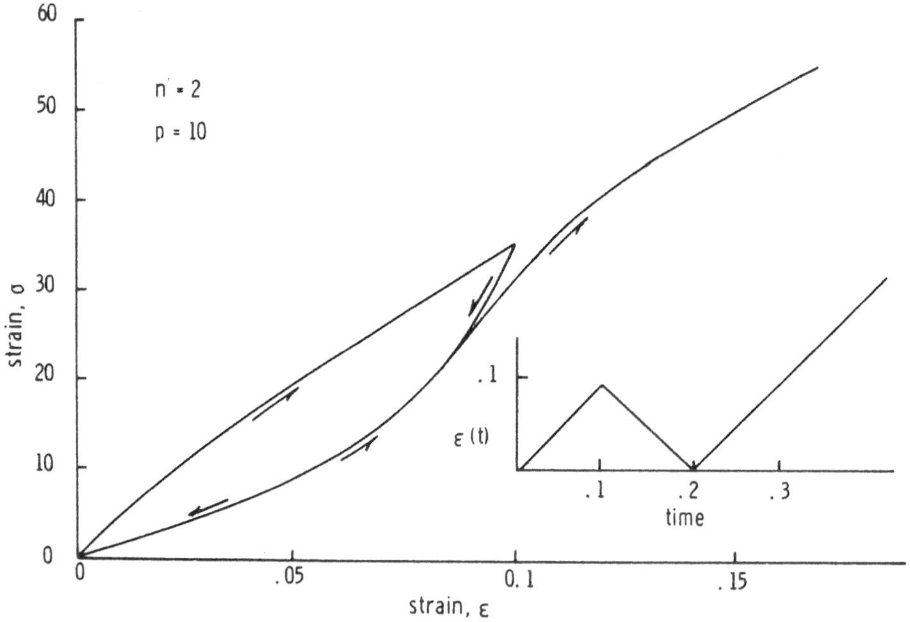

Figure 3.9. Calculated Permanent Memory Hysteresis Response to a Reversed Ramp-Strain Input

composite solid propellants and other highly filled polymers will observe great similarity between the data illustrated in Figures 3.4 through 3.9 and the behavior of these materials.

IV. THREE DIMENSIONAL CONSTITUTIVE EQUATIONS

This section will primarily deal with constitutive equations for non-linear materials having homogeneous constitutive equations of degree one that can describe the mechanical response of highly filled polymers and are amenable to three dimensional stress analysis. The simplest way to proceed in the development of constitutive equations homogeneous to degree one is to begin with the stress-strain equation for isotropic materials given by Green and Rivlin [22] was

$$\underline{\sigma}(t) = K_o(t)\underline{I} + \int_o^t K_1(t,\tau_1)\underline{\dot{\varepsilon}}(\tau_1)d\tau_1 +$$

$$+ \int_o^t \int_o^t K_1(t,\tau_1,\tau_2)\underline{\dot{\varepsilon}}(\tau_1)\underline{\dot{\varepsilon}}(\tau_1)d\tau_1 d\tau_2 +$$

$$+ \cdots + \int_o^t \cdots\cdots \int_o^t K_n(t,\tau_1,\cdots,\tau_n)\underline{\dot{\varepsilon}}(\tau_1)\cdots\underline{\dot{\varepsilon}}(\tau_n)d\tau_1\cdots d\tau_n + \cdots,$$

$$\text{where } K_n(t,\tau_1,\tau_2,\cdots\tau_n) = K_n\left[\overset{t}{\underset{o}{I_1}}(\xi),\overset{t}{\underset{o}{I_2}}(\xi),\overset{t}{\underset{o}{I_3}}(\xi),t,\tau_1,\cdots\tau_n\right], \qquad (4.1)$$

ξ = dummy time*,
$\underline{\sigma}$ = stress tensor,
and $\underline{\varepsilon}$ = strain tensor

In this theory the kernels were functionals of the history of the scalar invariants of the Cauchy strain tensor as well as the generic and current values of time. If the form of these kernel functionals was given again as a Fréchet multiple integral expansion, then the class of materials having non-linear homogeneous constitutive equations was shown not to be contained by the theory [1]. There is great similarity between equation (4.1) and the one dimensional irreversible equation obtained from the models in section 3, which was given as

*The dummy time ξ is introduced so that no confusion can arise as to what variables enter into the integration process.

$$\sigma = A_1\varepsilon P_1(\varepsilon/||\varepsilon||_p) + A_2\varepsilon^2 P_2(\varepsilon/||\varepsilon||_p) + A_3\varepsilon^3 P_3(\varepsilon/||\varepsilon||_p) + \cdots + A_n\varepsilon^n P_n(\varepsilon/||\varepsilon||_p) + \cdots \quad (4.2)$$

Equation (4.2) allows for no fading memory viscoelasticity, only permanent strain-time memory. If the kernel functionals of equation (4.1) were allowed to take on terms like $(||I_1||_p/||I_1||_q)$, then the equation could contain two types of memory phenomenon; the fading memory viscoelasticity contained in the hereditary integral representation and the permanent memory behavior registered in the L^p norms.

Before proceeding in the development of homogeneous constitutive equations of degree one, a special case of equation (4.1) is worth mentioning. Note that when the kernel functionals are independent of the variables τ_i, $i = 1,\cdots,n$, then the integrations can be performed and the result is

$$\underline{\sigma}(t) = K_o(t)\underline{I} + K_1(t)\underline{\varepsilon}(t) + K_2(t)\underline{\varepsilon}^2(t) + \cdots + K_n(t)\underline{\varepsilon}^n(t) + \cdots ,$$

$$\text{where} \quad K_i(t) = K_i\left[\mathop{I_1(\xi)}_{o}^{t}, \mathop{I_2(\xi)}_{o}^{t}, \mathop{I_3(\xi)}_{o}^{t}, t\right] . \quad (4.3)$$

Applying the Cayley-Hamilton theorem [18,20,28], equation (4.3) can be reduced to the form

$$\underline{\sigma}(t) = \psi_o\underline{I} + \psi_1\underline{\varepsilon}(t) + \psi_2\underline{\varepsilon}^2(t) ,$$

$$\text{where } \psi_i = \psi_i\left[\mathop{I_1(\xi)}_{o}^{t}, \mathop{I_2(\xi)}_{o}^{t}, \mathop{I_3(\xi)}_{o}^{t}, I_1(t), I_2(t), I_3(t), t\right] . \quad (4.4)$$

Equation (4.4) represents a viscoelastic material where all the time effects come from the history of the scalar invariants of strain and also from aging effects, which can be eliminated by removing the variable t. Depending on the form of the functionals ψ_i, this particular constitutive equation can describe both permanent and fading memory viscoelasticity with strain coupling. When the history dependence is eliminated from the functionals ψ_i, then equation (4.4) reduces to the standard non-linear elastic equation for isotropic materials [18,19,20].

The development of constitutive equations which are homogeneous to degree one is quite simple and can be done by simply imposing restrictions or constraints on equation (4.1). Recall homogeneity of degree one simply means that scalar multiplication is valid for all scalars. Recall also that the strain invariants are given by

$$I_1 = \varepsilon_{11} + \varepsilon_{22} + \varepsilon_{33}$$

$$I_2 = \varepsilon_{11}\varepsilon_{22} + \varepsilon_{11}\varepsilon_{33} + \varepsilon_{22}\varepsilon_{33} - \varepsilon_{12}^2 - \varepsilon_{13}^2 - \varepsilon_{23}^2$$

$$I_3 = \varepsilon_{11}\varepsilon_{22}\varepsilon_{33} + 2\varepsilon_{12}\varepsilon_{13}\varepsilon_{23} - \varepsilon_{11}\varepsilon_{23}^2 - \varepsilon_{22}\varepsilon_{13}^2 - \varepsilon_{33}\varepsilon_{12}^2 \qquad (4.5)$$

Mathematically the homogeneous constitutive equation of degree one has the property that the equation

$$F_{ij}\left[a\varepsilon_{pq}\overset{t}{\underset{\tau=o}{(t,\tau)}}\right] = aF_{ij}\left[\varepsilon_{pq}\overset{t}{\underset{\tau=o}{(t,\tau)}}\right] \qquad (4.6)$$

holds for all real scalars, all strain inputs and all time t. Since the first strain invariant is homogeneous to degree one, the second to degree two, and the third to degree three, non-linear homogeneous constitutive functionals can be constructed within the framework of the Green-Rivlin theory.

From physical reasoning kernel functionals homogeneous to degree less than zero cannot be admitted since they can yield unbounded stresses or singularities which are not real. With this added restriction the most general constitutive equation homogeneous to degree one within the range of applicability of equation (4.1) is

$$\sigma_{ij}(t) = \delta_{ij}\int_o^t K_o\left[I_1\overset{t}{\underset{o}{(\xi)}}, I_2\overset{t}{\underset{o}{(\xi)}}, I_3\overset{t}{\underset{o}{(\xi)}}, t, \tau\right]\dot{\varepsilon}_{kk}(\tau)d\tau$$
$$+ \int_o^t K_1\left[I_1\overset{t}{\underset{o}{(\xi)}}, I_2\overset{t}{\underset{o}{(\xi)}}, I_3\overset{t}{\underset{o}{(\xi)}}, t, \tau\right]\dot{\varepsilon}_{ij}(\xi)d\tau \quad , \qquad (4.7)$$

where the kernels have the property that

$$K_i\left[aI_1\overset{t}{\underset{o}{(\xi)}}, a^2I_2\overset{t}{\underset{o}{(\xi)}}, a^3I_3\overset{t}{\underset{o}{(\xi)}}, t, \tau\right] = K_i\left[I_1\overset{t}{\underset{o}{(\xi)}}, I_2\overset{t}{\underset{o}{(\xi)}}, I_3\overset{t}{\underset{o}{(\xi)}}, t, \tau\right]. \qquad (4.8)$$

In equation (4.7) the kernel functionals are homogeneous to degree zero. If the kernels are independent of the history of the invariants then the equation reduces to that of linear visco-elasticity. If the material is non-aging, equation (4.7) must reduce to [21,22]

$$\sigma_{ij}(t) = \delta_{ij} \int_0^t K_o\left[\overset{t}{\underset{o}{I_1}}(\xi), \overset{t}{\underset{o}{I_2}}(\xi), \overset{t}{\underset{o}{I_3}}(\xi), t-\tau\right] \dot{\varepsilon}_{kk}(\tau)d\tau$$

$$+ \int_0^t K_1\left[\overset{t}{\underset{o}{I_1}}(\xi), \overset{t}{\underset{o}{I_2}}(\xi), \overset{t}{\underset{o}{I_3}}(\xi), t-\tau\right] \dot{\varepsilon}_{ij}(\tau)d\tau \ . \tag{4.9}$$

In a similar manner the state of strain can be expressed in terms of the history of the stresses for homogeneous equations of degree one as

$$\varepsilon_{ij}(t) = \delta_{ij} \int_0^t L_o\left[\overset{t}{\underset{o}{J_1}}(\xi), \overset{t}{\underset{o}{J_2}}(\xi), \overset{t}{\underset{o}{J_3}}(\xi), t-\tau\right] \dot{\sigma}_{kk}(\tau)d\tau$$

$$+ \int_0^t L_1\left[\overset{t}{\underset{o}{J_1}}(\xi), \overset{t}{\underset{o}{J_2}}(\xi), \overset{t}{\underset{o}{J_3}}(\xi), t-\tau\right] \dot{\sigma}_{ij}(\tau)d\tau \quad , \tag{4.10}$$

In equation (4.10) the J_i are the principle stress invariants, and the kernels have the property that

$$L_i\left[a\overset{t}{\underset{o}{J_1}}(\xi), a^2\overset{t}{\underset{o}{J_2}}(\xi), a^3\overset{t}{\underset{o}{J_3}}(\xi), t-\tau\right] = L_i\left[\overset{t}{\underset{o}{J_1}}(\xi), \overset{t}{\underset{o}{J_2}}(\xi), \overset{t}{\underset{o}{J_3}}(\xi), t-\tau\right] \ . \tag{4.11}$$

Except for the case when the equations reduce to linear equations, direct inversion from equation (4.9) to equation (4.10) appears to be vertually impossible. Unlike the linear constitutive equations the power of Laplace transforms cannot be applied since these transforms can be applied only to linear hereditary functionals. This difficulty does not mean that the inversion does not exist. In fact the homogeneity condition alone intuitively suggests that the inversion does exist since scalar multiplication must hold for all scalars and this type of one-to one behavior is characteristic of invertible systems. For the purpose of this paper however, equation (4.10) is given as the inverse form of equation (4.9) when such an inverse exists.

V. SOLUTION OF BOUNDARY VALUED PROBLEMS

The equations developed thus far are non-linear but homogeneous constitutive equations of degree one that describe a large class of memory phenomena. The purpose of this section is to demonstrate the applicability of this type of constitutive equation in the solution stress boundary valued problems. The types of problems considered will be proportional boundary valued problems with constant body forces and no inertial effects. By a proportional boundary valued problem it is meant that the conditions at the boundary surface are given as a single product term involving a spatial function and a time function. Since no inertial effects are being accounted for, the time variance of the boundary conditions must be reasonably slow or quasi-static to justify having no inertia terms in the equations of equilibrium. Proportional boundary valued problems encompass a majority of the engineering problems encountered since it can allow for the boundary values to change with time. For linear elasticity or linear viscoelasticity, the procedure for developing a solution is straightforward since all the equations that must be solved are linear and superposition is applicable. For non-linear materials however, one is quite fortunate if a large class of problems can be contained in a solution scheme. Such is the case for the homogeneous constitutive equation of degree one.

5.1 Proportional Stress Boundary Valued Problems

In the introduction of this section it was indicated that the solution to proportional boundary valued problems could be obtained for non-linear constitutive equations homogeneous to degree one. For plane strain or plane stress problems which have proportional stress boundary values it is found that a linear elastic solution for the stresses is a solution for the stress-time distribution whenever the kernel functionals of the constitutive equation can be decomposed into a product form. The strain-time distribution for this case will be given by substituting the linear stress solution into the non-linear constitutive equation which is homogeneous to degree one. That such a solution is applicable is demonstrated in the following discussion.

By a proportional stress boundary valued problem it is meant that the boundary conditions are space and time separable.

$$\sigma_{ij}(x_k,\tau)\nu_j(x_k) = \overset{\circ}{\sigma}_{ij}(x_k)\nu_j(x_k)f(\tau), \quad \text{all } x_k \varepsilon \text{ boundary.}$$

where ν_j = direction consines of a unit vector normal to the boundary

$\sigma^{\circ}_{ij}(x_k)$ = stresses prescribed at some reference time

$f(\tau)$ = time function (5.1)

In the above description it was assumed that the boundary position does not change significantly with time which naturally restricts this discussion to infinitesimal strain theory. Hence the strain tensor in the constitutive equation will be given as the Cauchy strain tensor ε_{ij} where

$$\varepsilon_{ij} = \frac{1}{2}\left(\partial u_i/\partial x_j + \partial u_j/\partial x_i\right) . \tag{5.2}$$

With this definition of strain, the constitutive equation becomes

$$\varepsilon_{ij}(x_k,t) = \delta_{ij} \int_0^t L_o\left[J_1(x_k,\overset{t}{\underset{o}{\xi}}),J_2(x_k,\overset{t}{\underset{o}{\xi}}),J_3(x_k,\overset{t}{\underset{o}{\xi}}),t-\tau\right]\dot{\sigma}_{ii}(x_k,\tau)d\tau$$

$$+ \int_0^t L_1\left[J_1(x_k,\overset{t}{\underset{o}{\xi}}),J_2(x_k,\overset{t}{\underset{o}{\xi}}),J_3(x_k,\overset{t}{\underset{o}{\xi}}),t-\tau\right]\dot{\sigma}_{ij}(x_k,\tau)d\tau . \tag{5.3}$$

The kernel functionals in equation (5.3) were specified as being homogeneous to degree zero. Note that the kernel functionals contain spatial measures since the invariants are simply combinations of the stress which except in trivial cases, are functions of the spatial coordinates x_k. For the purpose of clarity, suppose that a linear elastic solution for the stresses within the body is valid. For the proportional boundary valued problem this linear elastic solution can be represented as

$$\sigma_{ij}(x_k,\tau) = \sigma^{\circ}_{ij}(x_k)f(\tau) \quad ,$$

$$\text{all } x_k \varepsilon \text{ volume, and } 0 \le \tau \le t . \tag{5.4}$$

In equation (5.4) $\sigma^{\circ}_{ij}(x_k)$ is the solution to the boundary valued problem when $f(\tau) = 1$. This proportionality of the solution is a direct consequence of the first linearity rule. If the linear elastic solution for the stresses is valid, then the stress invariants $J_i(x_k,\xi)$ are given as

$$J_1(x_k,\xi) = J_1^\circ(x_k)f(\xi),$$

$$J_2(x_k,\xi) = J_2^\circ(x_k)f^2(\xi), \text{ and}$$

$$J_3(x_k,\xi) = J_3^\circ(x_k)f^3(\xi), \text{ for all } x_k \varepsilon \text{ volume and } 0 \leq \xi \leq t. \tag{5.5}$$

In equation (5.5) the invariants $J_i^\circ(x_k)$ are simply the values of the invariants when the time function $f(\xi)$ is equal to unity. Observe that at some particular location within the body $x_k = (a_1,a_2,a_3)$ the invariants are given as

$$J_i(x_k,\xi) = J_i^\circ(a_1,a_2,a_3)f^i(\xi),$$

$$\text{where } J_i^\circ(a_1,a_2,a_3) = \text{constant function} \tag{5.6}$$

Assuming now that the kernel functions can be separated into product functions yields

$$L_i\left[\overset{t}{\underset{o}{J_1(x_k,\xi)}},\overset{t}{\underset{o}{J_2(x_k,\xi)}},\overset{t}{\underset{o}{J_3(x_k,\xi)}},t-\tau\right] =$$

$$= \sum_{r=1}^{N} M_{ir}\left[\overset{t}{\underset{o}{J_1(x_k,\xi)}},t-\tau\right] N_{ir}\left[\overset{t}{\underset{o}{J_2(x_k,\xi)}},t-\tau\right] P_{ir}\left[\overset{t}{\underset{o}{J_3(x_k,\xi)}},t-\tau\right]. \tag{5.7}$$

Recall that these kernel functions L_o and L_1 were homogeneous to degree zero which means

$$L_i\left[a\overset{t}{\underset{o}{J_1(x_k,\xi)}},a^2\overset{t}{\underset{o}{J_2(x_k,\xi)}},a^3\overset{t}{\underset{o}{J_3(x_k,\xi)}},t-\tau\right] =$$

$$L_i\left[\overset{t}{\underset{o}{J_1(x_k,\xi)}},\overset{t}{\underset{o}{J_2(x_k,\xi)}},\overset{t}{\underset{o}{J_3(x_k,\xi)}},t-\tau\right], \tag{5.8}$$

where a is an arbitrary constant and $\sigma_{pq}(\tau)$ are arbitrary.

Therefore, it is required that each of the components of the kernel decomposition be likewise homogeneous to degree zero. Substituting the invariants given by equation (5.6) into equation (5.7) produces

$$M_{ir}\left[J_1(x_k,\overset{t}{\underset{o}{\xi}}),t-\tau\right] = M_{ir}\left[J_1^\circ(x_k)f(\overset{t}{\underset{o}{\xi}}),t-\tau\right] ,$$

$$N_{ir}\left[J_2(x_k,\overset{t}{\underset{o}{\xi}}),t-\tau\right] = N_{ir}\left[J_2^\circ(x_k)f^2(\overset{t}{\underset{o}{\xi}}),t-\tau\right] , \quad \text{and}$$

$$P_{ir}\left[J_3(x_k,\overset{t}{\underset{o}{\xi}}),t-\tau\right] = P_{ir}\left[J_3^\circ(x_k)f^3(\overset{t}{\underset{o}{\xi}}),t-\tau\right] . \tag{5.9}$$

The constitutive equation gives the strain at some arbitrary, but fixed, point within the body. At some fixed point in space, however, the functions $J_1^\circ(x_k)$, $J_2^\circ(x_k)$, and $J_3^\circ(x_k)$ are simply constants. Since the kernel functions M_{ir}, P_{ir}, and N_{ir} were specified as being homogeneous to degree zero, these constants have no effect on the values these kernels take on. For the special case when a linear elastic solution is valid and the kernels can be decomposed, the conclusion is as follows

$$M_{ir}\left[J_1(x_k,\overset{t}{\underset{o}{\xi}}),t-\tau\right] = M_{ir}\left[J_1^\circ(x_k)f(\overset{t}{\underset{o}{\xi}}),t-\tau\right] = M'_{ir}\left[f(\overset{t}{\underset{o}{\xi}}),t-\tau\right] ,$$

$$N_{ir}\left[J_2(x_k,\overset{t}{\underset{o}{\xi}}),t-\tau\right] = N_{ir}\left[J_2^\circ(x_k)f^2(\overset{t}{\underset{o}{\xi}}),t-\tau\right] = N_{ir}\left[f^2(\overset{t}{\underset{o}{\xi}}),t-\tau\right] ,$$

$$P_{ir}\left[J_3(x_k,\overset{t}{\underset{o}{\xi}}),t-\tau\right] = P_{ir}\left[J_3^\circ(x_k)f^3(\overset{t}{\underset{o}{\xi}}),t-\tau\right] = P_{ir}\left[f^3(\overset{t}{\underset{o}{\xi}}),t-\tau\right] . \tag{5.10}$$

Note that the kernel functions now contain no spatial variables and are only functions of the history of $f(\xi)$ and the variable $t-\tau$. Since the forms of these kernels are still arbitrary, there is no loss in generality by assuming that a functional of $f^2(\xi)$ or $f^3(\xi)$ is contained in a general functional of $f(\xi)$. Therefore the assumptions of a linear elastic solution for the stresses and separable kernels results in

$$L_i\left[J_1(x_k,\underset{o}{\overset{t}{\xi}}),\ J_2(x_k,\underset{o}{\overset{t}{\xi}}), J_3(x_k,\underset{o}{\overset{t}{\xi}}), t-\tau \right] =$$

$$= L_i\left[f(\underset{o}{\overset{t}{\xi}}), f^2(\underset{o}{\overset{t}{\xi}}), f^3(\underset{o}{\overset{t}{\xi}}), t-\tau \right] = L_i' \left[f(\underset{o}{\overset{t}{\xi}}), t-\tau \right] . \tag{5.11}$$

Dropping the prime notation in the last of equation (5.11) the constitutive equation now becomes

$$\varepsilon_{ij}(x_k,t) = \delta_{ij} \int_o^t L_1 \left[f(\underset{o}{\overset{t}{\xi}}), t-\tau \right] \dot{\sigma}_{ii}(x_k,\tau)\, d\tau \; +$$

$$+ \int_o^t L_2 \left[f(\underset{o}{\overset{t}{\xi}}), t-\tau \right] \dot{\sigma}_{ij}(x_k,\tau)\, d\tau \; . \tag{5.12}$$

Equation (5.12) would be identical to the linear viscoelastic constitutive equation if the kernels L_o and L_1 were independent of $f(\xi)$. Because $f(\xi)$ is present in the equation, linear transforms cannot be utilized on the constitutive equation to demonstrate that the assumed linear elastic solution for the stresses was valid. If it can be shown that equation (5.12) will satisfy the equations of equilibrium and compatibility whenever the linear elastic solution is valid, a type of correspondence principle will have been developed similar to what has already been done in the theory of linear viscoelasticity. The validity of the elastic solution can be demonstrated by substituting the constitutive equation directly into the equations of equilibrium and compatibility. It should be immediately obvious that no complications can arise in such a procedure since the only spatially dependent quantities in the constitutive equation are the stresses σ_{ij} and the strains ε_{ij}. For the two dimensional problem only one equation of compatibility* is present,

$$\frac{\partial^2 \varepsilon_{11}(x_k,t)}{\partial x_2^{\,2}} + \frac{\partial^2 \varepsilon_{22}(x_k,t)}{\partial x_1^{\,2}} = \frac{2\partial^2 \varepsilon_{12}(x_k,t)}{\partial x_1 \partial x_2} \tag{5.13}$$

---

*There are other compatibility conditions for plane stress that are sometimes not satisfied by this method. See Timoshenko and Goodier [15], page 25.

Substituting the constitutive equation directly into the compatibility equation and interchanging the roles of integration and differentiation equation (5.13) becomes

$$\int_0^t \left(L_o \left[\overset{t}{\underset{o}{f(\xi)}}, t-\tau \right] \nabla^2 \dot{\sigma}_{ii}(x_k,\tau) + L_1 \left[\overset{t}{\underset{o}{f(\xi)}}, t-\tau \right] \left(\frac{\partial^2 \dot{\sigma}_{11}(x_k,\tau)}{\partial x_2^{\,2}} + \frac{\partial^2 \dot{\sigma}_{22}(x_k,\tau)}{\partial x_1^{\,2}} \right) \right.$$

$$\left. - 2L_1 \left[\overset{t}{\underset{o}{f(\xi)}}, t-\tau \right] \frac{\partial^2 \dot{\sigma}_{12}(x_k,\tau)}{\partial x_1 \partial x_2} \right) d\tau = 0 \tag{5.14}$$

The equilibrium equations for a two dimensional problem are

$$\frac{\partial \sigma_{11}(x_k,\tau)}{\partial x_1} + \frac{\partial \sigma_{12}(x_k,\tau)}{\partial x_2} + \bar{X}_1 = 0 , \quad \text{and}$$

$$\frac{\partial \sigma_{12}(x_k, \ )}{\partial x_1} + \frac{\partial \sigma_{22}(x_k, \ )}{\partial x_2} + \bar{X}_2 = 0 , \tag{5.15}$$

where $\bar{X}_i$ are body forces and were specified as being constant.

Differentiating the first equilibrium equation with respect to x_1, the second with respect to x_2 and adding the following is obtained

$$2 \frac{\partial^2 \sigma_{12}(x_k,\tau)}{\partial x_1 \partial x_2} = - \left(\frac{\partial^2 \sigma_{11}(x_k,\tau)}{\partial x_1^{\,2}} + \frac{\partial^2 \sigma_{22}(x_k,\tau)}{\partial x_2^{\,2}} \right) \tag{5.16}$$

Substituting this result into equation (5.14) yields

$$\int_o^t \left(L_o\left[\overset{t}{\underset{o}{f(\xi)}}, t-\tau\right] \nabla^2 \dot{\sigma}_{ii}(x_k,\tau) + L_1\left[\overset{t}{\underset{o}{f(\xi)}}, t-\tau\right] \nabla^2 \left(\dot{\sigma}_{11}(x_k,\tau) + \dot{\sigma}_{22}(x_k,\tau) \right) \right) d\tau = 0,$$

where $\nabla^2 = \frac{\partial^2}{\partial x_1^2} + \frac{\partial^2}{\partial x_2^2}$. (5.17)

5.2 Specialization to Plane Stress

For the condition of plane stress $\sigma_{33}(x_k,\tau) = 0$, with this restriction equation (5.17) reduces to

$$\int_o^t \left(L_o\left[\overset{t}{\underset{o}{f(\xi)}}, t-\tau\right] + L_1\left[\overset{t}{\underset{o}{f(\xi)}}, t-\tau\right] \right) \frac{\partial}{\partial\tau} \nabla^2 \sigma_{ii}(x_k,\tau) \quad d\tau = 0 \tag{5.18}$$

A sufficient condition to make the integral in equation (5.18) vanish is to require that

$$\nabla^2 \sigma^\circ_{ii}(x_k) = 0, \text{ since } \sigma_{ii}(x_k,\tau) = f(\tau)\sigma^\circ_{ii}(x_k) \tag{5.19}$$

Equation (5.19) is precisely the identical condition placed on the stress distribution for a linear elastic body. The assumption made in equation (5.4) therefore in no way violates the equations of equilibrium or compatibility.

5.3 Specialization to Plane Strain

For the condition of plane strain $\varepsilon_{33}(x_k,t) = 0$, therefore taking the Laplacian of the constitutive equation produces

$$0 = \nabla^2 \varepsilon_{33}(x_k,t) = \int_o^t \left(L_o\left[\overset{t}{\underset{o}{f(\xi)}}, t-\tau\right] \nabla^2 \dot{\sigma}_{ii}(x_k,) + L_1\left[\overset{t}{\underset{o}{f(\xi)}}, t-\tau\right] \nabla^2 \dot{\sigma}_{33}(x_k \tau) \right) d\tau. \tag{5.20}$$

Substituting this restriction into (5.17) gives

$$\int_o^t L_i\left[\overset{t}{\underset{o}{f(\xi)}}, t-\tau\right] \frac{\partial}{\partial\tau} \nabla^2 \sigma_{ii}(x_k,\tau) d\tau = 0 \tag{5.21}$$

In order to require this integral to vanish for all times and all space a sufficient condition is to require

$$\nabla^2\sigma^\circ_{ii}(x_k) = 0, \text{ since } \sigma_{ii}(x_k,\tau) = f(\tau)\sigma^\circ_{ii}(x_k) \qquad (5.22)$$

Equation (5.22) is again the exact condition placed on a plane strain solution for a linear elastic solid.

For the case of plane strain and plane stress a type of correspondence principle between the linear elastic solution and the solution for a homogeneous but non-linear material has been established. This correspondence principle can be stated as follows:

Correspondence Principle - Given a plane strain or plane stress proportional boundary valued problem of the form

(i) $\sigma_{ij}(x_k,\tau)\nu_j(x_k) = \sigma^\circ_{ij}(x_k)\nu_j(x_k)f(\tau)$ $x_k \varepsilon$ Boundary and $0 \leq \tau \leq t$

for a material having a non-linear but homogeneous constitutive equation of degree one of the form

(ii) $$\varepsilon_{ij}(x_k,t) = \delta_{ij}\int_0^t L_o\left[J_1(x_k,\overset{t}{\underset{o}{\xi}}),J_2(x_k,\overset{t}{\underset{o}{\xi}}),J_3(x_k,\overset{t}{\underset{o}{\xi}}),t-\tau\right]\dot{\sigma}_{ii}(x_k,\tau\, d\tau$$

$$+\int_0^t L_1\left[J_1(x_k,\overset{t}{\underset{o}{\xi}}),J_2(x_k,\overset{t}{\underset{o}{\xi}}),J_3(x_k,\overset{t}{\underset{o}{\xi}}),t-\tau\right]\dot{\sigma}_{ij}(x_k,\tau)d\tau\ .$$

Then a plane strain or plane stress linear elastic solution for the stress distribution is valid and when substituted into the non-linear constitutive equation, the equations of equilibrium and compatibility used in two dimensional elasticity will be identically satisfied provided

(1) the body forces are constant

(2) the kernel functions can be decomposed into a product form, each term of which only involves one invariant history $J_1(x_k,\overset{t}{\underset{o}{\xi}})$ and the variable $t-\tau$.

The strains can then be obtained by substituting the time dependent elastic solution into the non-linear constitutive equation which can be reduced to the form

$$\text{(iii)} \qquad \varepsilon_{ij}(x_k,t) = \delta_{ij}\sigma^{\circ}_{ii}(x_k) \int_{0}^{t} L_0\left[\overset{t}{\underset{o}{f(\xi)}}, t-\tau \right] \dot{f}(\tau)d\tau$$

$$+ \sigma^{\circ}_{ij}(x_k) \int_{0}^{t} L_1\left[\overset{t}{\underset{o}{f(\xi)}}, t-\tau \right] \dot{f}(\tau)d\tau \ .$$

Other correspondence principles have been developed by the same method [1] by first assuming a solution of a particular form exists and then showing what conditions are necessary to satisfy the equilibrium and compatibility equations. In this manner correspondence principles have been developed for certain stress and displacement boundary valued problems [1].

VI. REALISTIC CHARACTERIZATION OF COMPOSITE PROPELLANTS

For nearly a decade composite solid propellant materials have for the most part been treated as linear viscoelastic materials. Today the propellant industry has the ability to perform complex thermoviscoelastic stress analysis using thermorheologically simple linear viscoelastic constitutive theory. Careful examination of propellant data however indicates the materials are not linear viscoelastic even for small strain, isothermal conditions. Researchers in the propellant industry have been applying incorrect criteria of linearity to their materials [1,2,12]. Assumptions have been made that if the material has a relaxation modulus that is independent of strain, then the material is linear. This assumption is not correct. Having a relaxation modulus that is strain independent is but a single check of the homogeneity condition and in no way checks the validity of the superposition principle, which is the real test for linearity. Hence for over a decade complex computer analyses have been performed using linear viscoelastic theory yielding highly questionable results.

To demonstrate that propellants are non-linear materials even at small strains, one need only check the superposition principle experimentally. In the range of small strains below detectable dewetting or volumetric dilatation [6,9-11], most propellants have a relaxation that is independent of strain and in general closely obey the scalar multiplication homogeneity rule. Yet this relaxation modulus cannot be used to accurately predict the response due to other isothermal, low rate, small strain inputs. To demonstrate the inadequacies of linear viscoelastic predictions on solid propellants, laboratory tests where superposition is applicable can be performed. Figure 6.1 illustrates the stress-strain-dilatational behavior of a typical composite propellant. The dilatation-strain behavior is caused by vacuole formation within the microstructure

[9-11] and causes a stress-softening; an obvious type of nonlinearity. Below significant dilatation the material appears to have a relaxation modulus that is independent of strain as illustrated in figure 6.2. To determine if superposition is applicable, the interrupted ramp-strain stress relaxation test can be employed. Linear viscoelasticity theory would predict the stress output for the second loading and would be simply the superposition of the initial response with the continuation of the original stress-relaxation response. Figure 6.3 illustrates the linear viscoelastic prediction and the actual experimental results for this interrupted ramp strain test. From this and other tests it is apparent errors of over plus or minus one hundred percent are typical when linear viscoelastic theory is used to predict the response of propellant materials. To clarify the point, the data in figure 6.3 are plotted stress against strain in figure 6.4. Here it is apparent that when the straining is again commenced, the response rapidly rejoins the original constant rate response, whereas the linear theory would indicate it should parallel the original response. Figure 6.5 illustrates similar test results for the doubly interrupted ramp test plotted stress vs strain. Again the same behavior of rejoining the original constant rate response is shown and also that the errors of linear theory grow with each cycle.

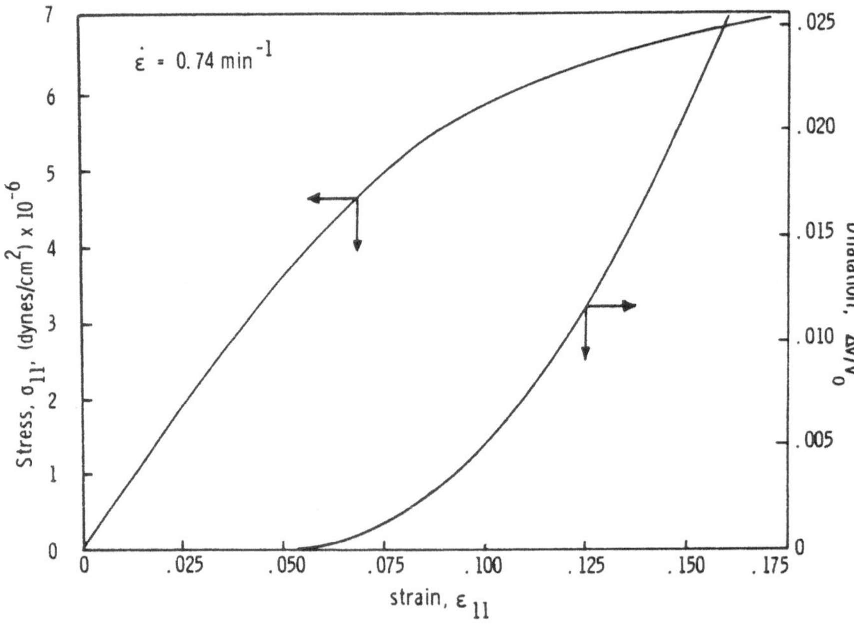

Figure 6.1. Uniaxial Stress-Strain and Dilatation-Strain Behavior of a Highly Filled Polymer

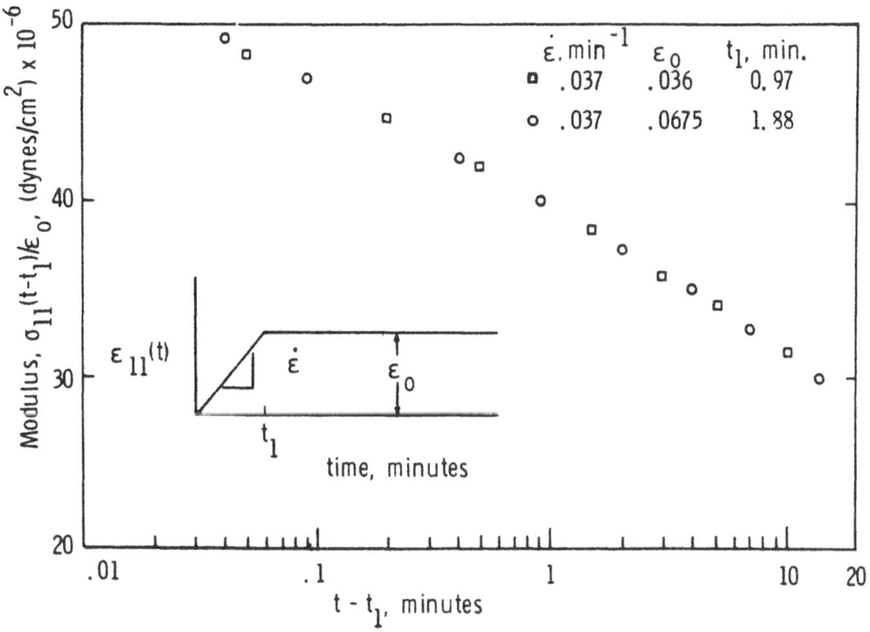

Figure 6.2. Ramp Relaxation Modulus for Two Samples Tested at Different Strain Levels

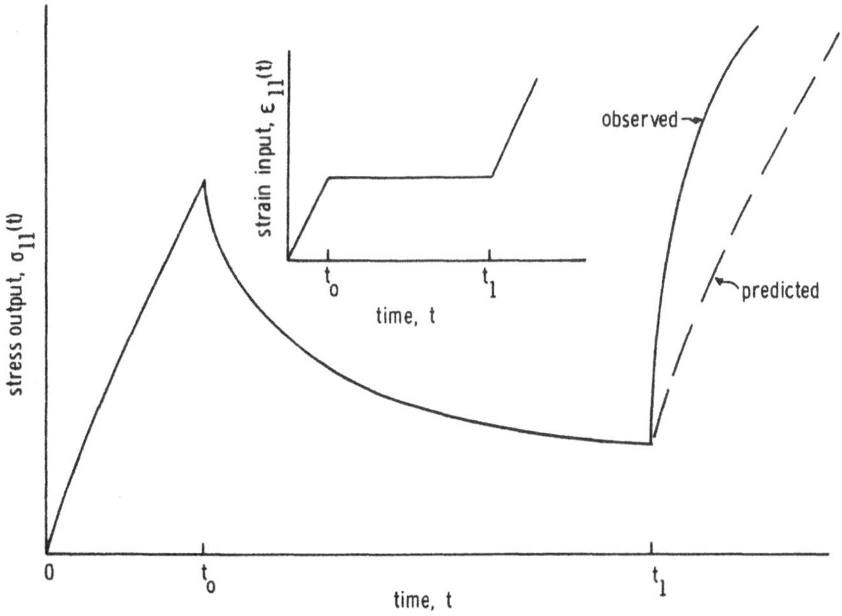

Figure 6.3. Linear Viscoelastic Stress-Time Predictions and Experimental Data for an Interrupted Ramp Strain Input on a Typical Highly Filled Polymer

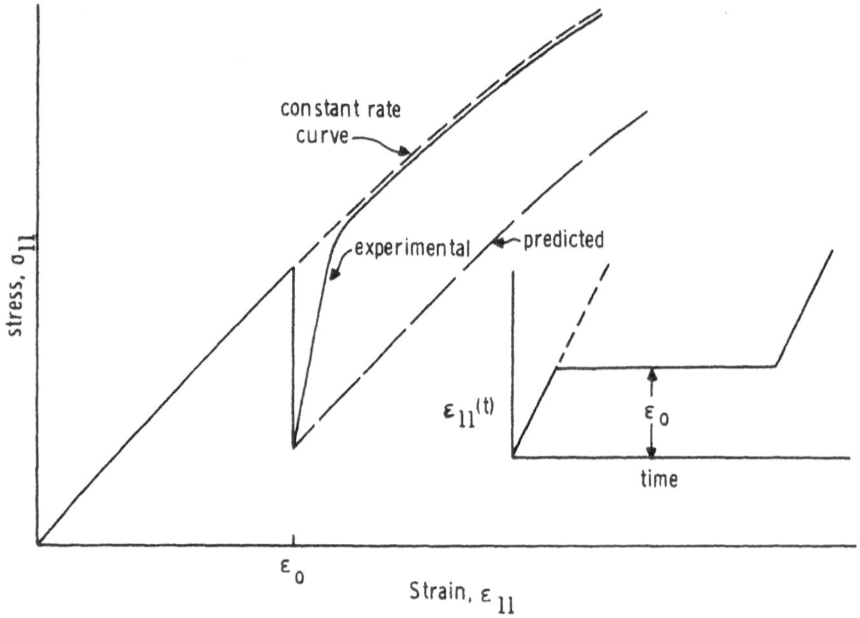

Figure 6.4. Linear Viscoelastic Stress-Strain Prediction and Experimental Data for an Interrupted Ramp Strain Input on a Typical Highly Filled Polymer

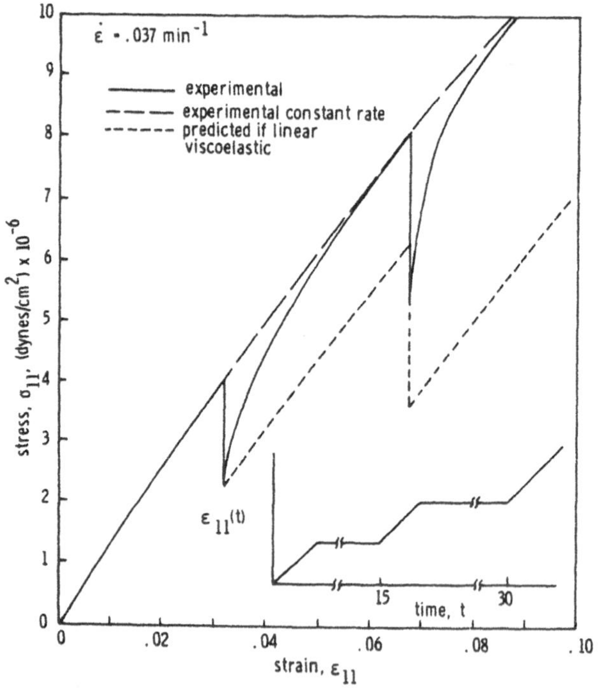

Figure 6.5. Stress Output for Interrupted Constant Strain Rate Test

Plotting the relaxation data from each portion of this test in figure 6.6 further demonstrates that the relaxation response for the first and second straining are identical when time is measured from the beginning of each relaxation. Interestingly enough this means that part of the memory of its past has somehow been completely annihilated and that the previous relaxation history has no influence on the second relaxation. This is not fading memory response since all of the past has not been forgotten. It does however indicate that the fading memory portion of the viscoelastic constitutive equation is for all practical purposes zero as long as the strain is increasing.

In figure 6.7 further verification of the homogeneity principle for these materials is presented. In this figure the stress output is compared for two cyclic inputs that differ in amplitude only. As dictated by the homogeneity principle, these data indicate that the ratio of the stress outputs is equal to the ratio of amplitudes of the cyclic strain inputs. The data in figure 6.8 compares the linear-viscoelastic prediction for the cyclic data presented in figure 6.7. The agreement between the linear predictions and the experimental data are quite good for this test whereas it was found to be poor for the tests discussed earlier in this section. Good agreement between experimental data and linear predictions might be expected for some tests since the material has been shown to satisfy one of the conditions of linearity. Furthermore the dilatation data on propellants in this range of small strains prior to dewetting indicate near incompressible elastic behavior [9-11]. Incorporating these features into the constitutive equation indicate a valid form would be

$$\sigma_{ij}(t) = \delta_{ij}P = G\left[\overset{t}{\underset{o}{f(\xi)}}\right]\varepsilon_{ij}(t) \quad , \tag{6.1}$$

where p is an arbitrary pressure.

The relaxation data for most propellants obeys a simple power law expression as indicated by these data when plotted logarithmically in figure 6.9. From Farris' discussion dealing with material characterization [1], a logical choice for the functional is

$$G\left[\overset{t}{\underset{o}{f(\xi)}}\right] = \sum_{i=o}^{N} A_i \left(\frac{||f||_{p_i}}{||f||_{q_i}}\right)^{r_i} , \text{ where } r_i\left(\frac{1}{p_i} - \frac{1}{q_i}\right) = -n \; . \tag{6.2}$$

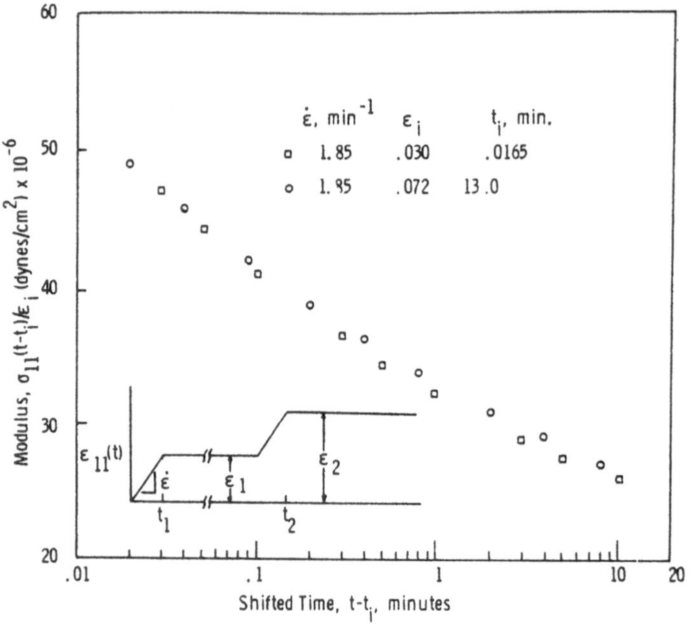

Figure 6.6. Ramp Relaxation Modulus for One Sample Tested at Two Different Strain Levels

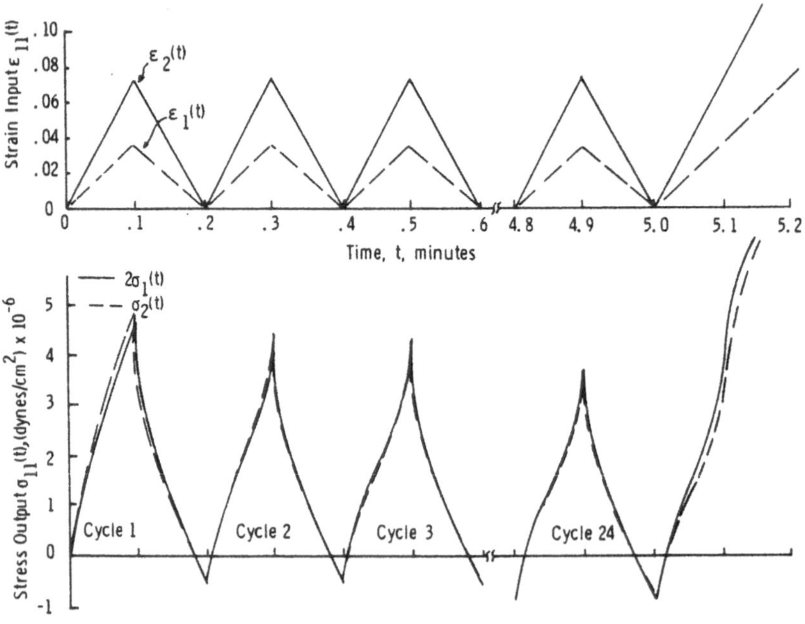

Figure 6.7. Verification of Homogeneity Principle for a Cyclic Stress Input

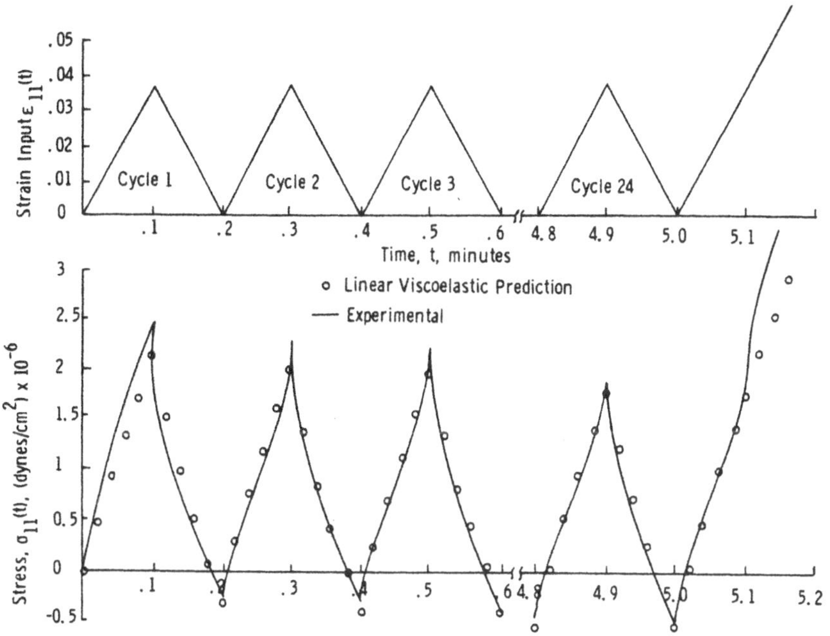

Figure 6.8. Comparison of Linear Viscoelastic Prediction and Experimental Data for a Cyclic Strain Input

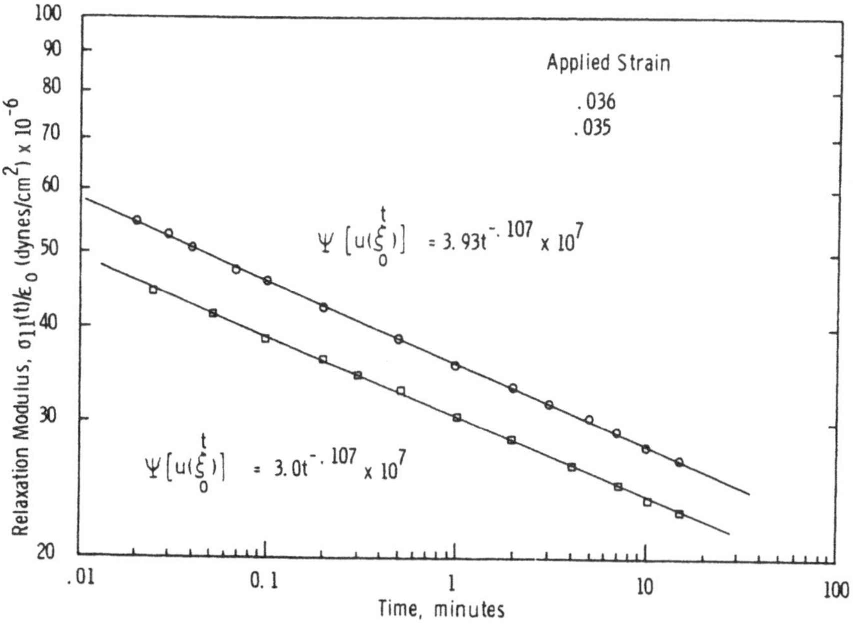

Figure 6.9. Stress Output to a Rapidly Applied Constant Strain Input

For a jump strain equation (6.2) then reduces to

$$G\left[\underset{o}{\overset{t}{u(\xi)}}\right] = E_r(t) = kt^{-n} = t^{-n}\sum_{i=o}^{N} A_i \tag{6.3}$$

To complete the characterization process we need only determine A_i, p_i, q_i, and r_i. The complex characterization procedure discussed in the previous section was not used for this purpose. Instead only one term was taken choosing $p_1 = \infty$, $A_1 = k$ and graphically determining r_1 and q_1 to fit a few tests. The results of this very simple analysis is demonstrated in figures 6.10 through 6.12 which compare calculated and observed response for several tests plotted both stress-time and stress-strain. The close agreement between experiment and theory for this single term representation would appear to indicate this is a powerful method of characterization and valid for propellant materials. However comparing the predictions with the actual data for the cyclic tests, figure 6.13, shows the agreement is not as good as demonstrated in the previous figures. The reason for this disagreement between prediction and observation lies in the need to have some fading memory viscoelasticity present, since compressive stresses for the state of positive tensile strain cannot come from the permanent memory portion of the constitutive equation. Proper characterization procedures will bring out such defects in the chosen representation.

In an attempt to improve the characterization process a three term expansion was chosen that contained a fading memory term, a permanent memory term, and an interaction term. The constitutive equation chosen has the form

$$\sigma_{ij}(t) = \delta_{ij}P + A_1\left(\frac{|f|}{\||f|\|_{q_1}}\right)^{r_1}\varepsilon_{ij} + A_2\int_o^t (t-\tau)^{-n_2}\,\dot{\varepsilon}_{ij}(\tau)\,d\tau$$

$$+ A_3\left(\frac{|f|}{\||f|\|_{q_3}}\right)^{r_3}\int_o^t (t-\tau)^{-n_3}\,\dot{\varepsilon}_{ij}(\tau)\,d\tau\ ,$$

where P = arbitrary pressure, and

$|\cdot|$ = absolute value. (6.4)

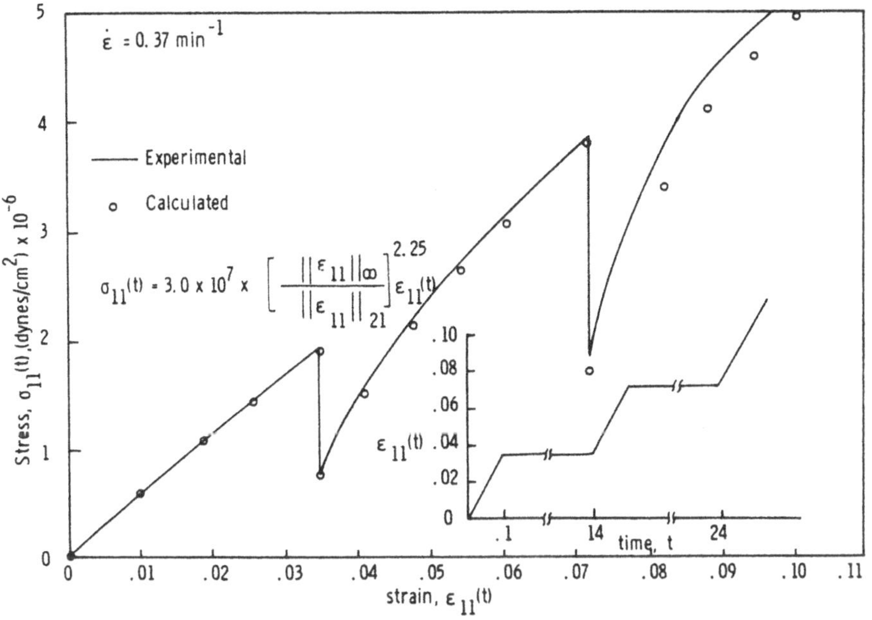

Figure 6.10. Comparison of Calculated and Observed Stress-Strain Output for an Interrupted Ramp Strain Input

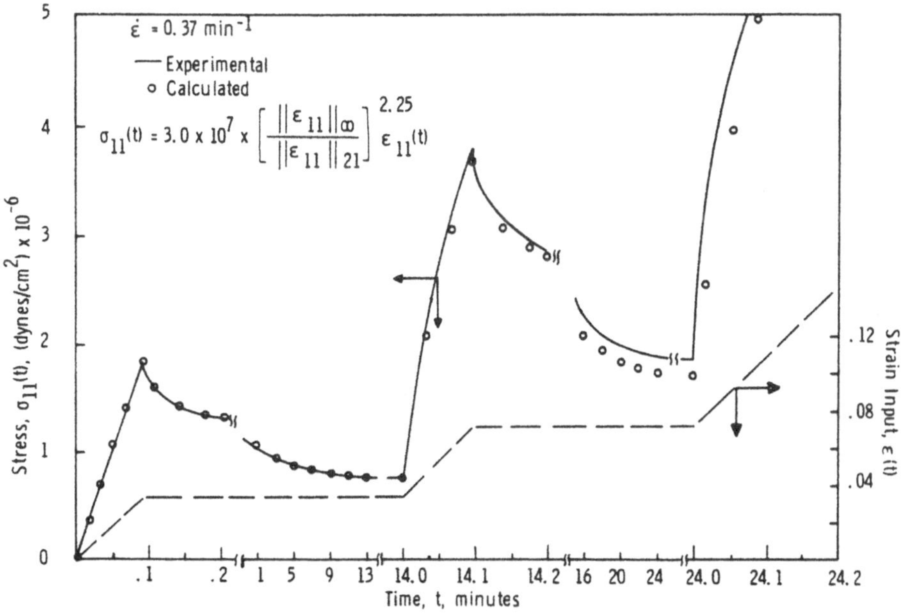

Figure 6.11. Comparison of Calculated and Observed Stress-Time Output for an Interrupted Ramp Strain Input

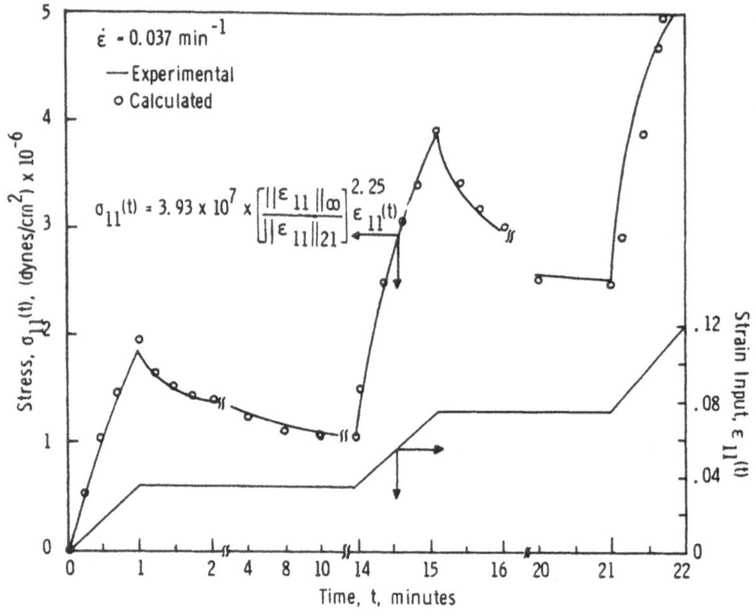

Figure 6.12. Comparison of Calculated and Observed Stress-Time Output to an Interrupted Ramp Strain Input

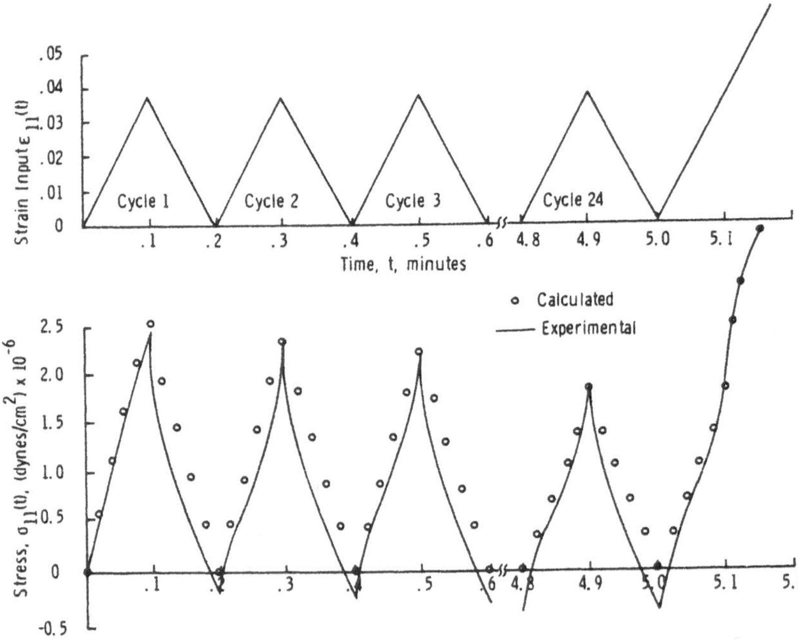

Figure 6.13. Comparison of Experimental and Calculated Stress Output for a Cyclic Strain Input

A simple analysis of all the test data indicate that the material would be well characterized if the parameters in the equation took on the following values

$$
\begin{aligned}
n_2 = n_3 &= 0.1 \\
r_1 &= 2.25 \\
r_3 &= 1.0 \\
q_1 &= 21 \\
q_3 &= \infty \\
A_1 = A_2 = -A_3 &= 3.93 \times 10^7
\end{aligned}
$$

Substituting these values into equation (6.4) and rearranging terms yields

$$
\sigma_{ij}(t) = \delta_{ij}P + 3.93 \times 10^7 \left(\frac{|f|}{||f||_{21}} \right)^{2.25} \varepsilon_{ij}(t) +
$$

$$
+ 3.93 \times 10^7 \left(1 - \frac{|f|}{||f||_{\infty}} \right) \int_0^t (t-\tau)^{-.1} \dot{\varepsilon}_{ij}(\tau)\,d\tau \,. \qquad (6.5)
$$

For tests where the strain is never decreasing (or never increasing), such as those illustrated in figures 6.10 through 6.12, the last term in equation (6.5) contributes nothing since for these tests $|f|$ equals $||f||_{\infty}$. The data in figures 6.10 through 6.12 calculated by equation (6.5) and that calculated by equation (6.1) is therefore identical. Comparisons between the calculated and observed data for the cyclic strain inputs is illustrated in figure 6.14. As seen by the data illustrated in this figure, the calculated and experimental data agree quite well. Perhaps other inputs exist, where again the agreement of even this modified constitutive equation will predict poorly. The only way to be assured a non-linear constitutive equation will predict accurately is to perform all possible tests and compare. Naturally this is impossible, however the least one should do is use a large number of greatly different tests. Inputs of a similar type to those the material in question will be subjected to in its lifetime, performed at the same temperatures and over the same time scales, should be used in realistic characterization procedures.

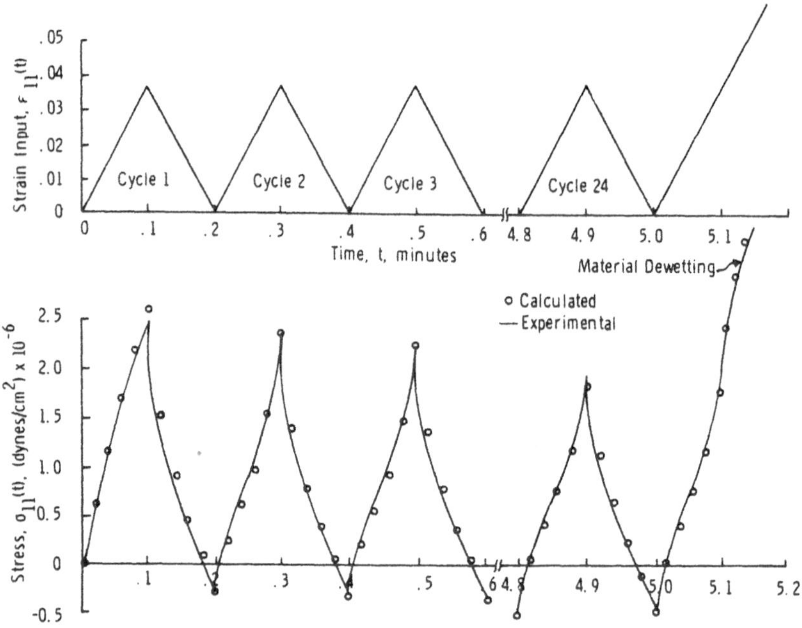

Figure 6.14. Comparison of Experimental and Calculated Stress Output for a Cyclic Strain Input

In particular there are two things a rheologist attempting to mathematically describe the behavior of materials should always remember

(1) Simply because the chosen representation accurately curve fits the tests used in the characterization procedure does not guarantee accurate predictions for other different tests.

(2) If a constitutive equation cannot predict the output to an arbitrary input, it is of little value in general stress analysis.

REFERENCES

1. Farris, R. J., "Homogeneous Constitutive Equations for Materials with Permanent Memory," Ph.D Thesis, University of Utah, Department of Civil Engineering (June 1970).

2. Farris, R. J., "Applications of Viscoelasticity to Filled Materials," Master's Thesis, University of Utah, Department of Civil Engineering (June 1969).

3. Mullins, L. J., "Effect of Stretching on the Properties of Rubber," J. Rubber Res., 16, 275-289, (1947).

4. Mullins, L. J., "Permanent Set in Vulcanized Rubber," Ind. Rubber World, 63-69 (1949).

5. Mullins, L. J., "Studies in the Absorption of Energy by Rubber," J. Rubber Res., 16, 180-185 (1947).

6. Oberth, A. E., "Principle of Strength Reinforcement in Filled Polymers," Rubber Chem. Tech., 40, 1337-1362 (1967).

7. Bueche, F., "Molecular Basis for the Mullins Effect," J. Appl. Polymer Sci., 4, 107-114 (1960).

8. Bueche, F., "Mullins Effect and Rubber-Filler Interaction," J. Appl. Polymer Sci., 5, 271-281 (1961).

9. Farris, R. J., "Dilatation of Granular Filled Elastomers Under High Rates of Strain," J. Appl. Polymer Sci., 8, 25-25 (1964).

10. Farris, R. J., "The Character of the Stress-Strain Function for Solid Propellants," Trans. Soc. Rheol., 12, 281-301 (1968).

11. Farris, R. J., "The Influence of Vacuole Formation on the Response and Failure of Highly Filled Polymers," Trans. Soc. Rheo., 12, 315-334 (1968).

12. Williams, M. L., Blatz, P. J., and Schapery, R. A., "Fundamental Studies Relations to Systems Analysis of Solid Propellants," GALCIT SM 61-S, California Institute of Technology, Pasadena, California (Feb 1961). Also published in Interagency Chemical Rocket Propulsion Group, Solid Propellant Mechanical Behavior Manual, Chemical Propulsion Information Agency Pub. No. 21, Section 2.3 (1963).

13. Royden, H. L., Real Analysis, III, 2nd Edition, The Macmillan Company, New York (1968).

14. Coleman, B. D., and Noll, W., "An Approximation Theorem for Functionals with Applications in Continuum Mechanics," Arch. Rat. Mech. Anal., 6, 355-370 (1960).

15. Timoshenko, S., and Goodier, J. N., Theory of Elasticity, McGraw-Hill Book Co., New York (1951).

16. Fung, Y. C., Foundations of Solid Mechanics, Prentice-Hall Inc., Englewood Cliffs, New Jersey (1965).

17. Sokolnikoff, I. S., Mathematical Theory of Elasticity, McGraw-Hill Book Co., New York (1956).

18. Eringen, A. C., Mechanics of Continua, John Wiley and Sons, Inc., New York (1967).

19. Eringen, A. C., Non-Linear Theory of Continuous Media, McGraw-Hill Book Co., New York (1962).

20. Malvern, L. E., Introduction to the Mechanics of a Continuous Media, Prentice-Hall, Inc., New Jersey (1969).

21. Volterra, V., Theory of Functionals and of Integral and Integro-Differential Equations, Dover Publications, Inc., New York (1959).

22. Green, A. E., and Rivlin, R. S., "The Mechanics of Non-Linear Materials with Memory, Part One," Arch. Rat. Mech. Anal., 1, 1-21 (1959).

23. Green, A. E., Rivlin, R. S., and Spencer, A. J. M., "The Mechanics of Non-Linear Materials with Memory, Part Two," Arch. Rat. Mech. Anal., 3, 82-90 (1959).

24. Green, A. E., and Rivlin, R. S., "The Mechanics of Non-Linear Materials with Memory, Part Three," Arch. Rat. Mech. Anal., 4, 387-404 (1959).

25. Rivlin, R. S., "Non-Linear Viscoelastic Solids," SIAM Review, 7, 323-340 (1965).

26. Pipkin, A. C., and Rivlin, R. S., "Small Deformations Superposed on Large Deformations in Materials with Fading Memory," Arch. Rat. Mech. Anal., 8, 297-308 (1961).

27. Freda, E., "Il Teorema di Eulers per le Funzioni di Linea Omogenee," R. Acc. du LincEi, Rend., XXIV, 5, (1915).

28. Wylie, C. R., Jr., Advanced Engineering Mathematics, McGraw-Hill Book Co., New York (1960).

29. Love, A. E. H., The Mathematical Theory of Elasticity, 4th Ed., Dover Publications, Inc., New York (1944).

30. Swanson, S. R., "Development of Constiutive Equations for Rocks," Ph.D. Dissertation, Dept. of Mechanical Eng., Univ. of Utah (1969).

31. Noll, W., "Mathematical Theory of the Mechanical Behavior of Continuous Media," Arch. Rat. Mech. Anal., 2, 197-226, (1958).

32. Pipkin, A. C., "Small Finite Deformations of Viscoelastic Solids," Rev. Mod. Phys., 36, 1034-1041 (1964).

33. Pipkin, A. C., and Rogers, T. G., "A Non-Linear Integral Representation for Viscoelastic Behavior," J. Mech. Phys. Solids, 16, 59-72, (1968).

34. Fréchet, M., "Sur Les Fonctionnelles Continues," Ann. de L'Ecole Normale Sup., 27, 3rd Series (1910).

35. Herrmann, L. R., "On a General Theory of Viscoelasticity," J. Franklin Inst., 280, 244-255 (1965).

36. Onaran, K., and Findley, W. N., "Combined Stress-Creep Experiments on a Non-Linear Viscoelastic Material to Determine the Kernel Functions for a Multiple Integral Representation of Creep," Trans. Soc. Rheol. 1, 299-327, (1965).

37. Williams, M. L., "Structural Analysis of Viscoelastic Materials," AIAA J., 2, 785-808 (1964).

38. Tobolsky, A. V., Properties and Structure of Polymers, 160, John Wiley and Sons, Inc., New York (1960).

39. Schapery, R. A., "A Theory of Non-Linear Thermoviscoelasticity Based on Irreversible Thermodynamics," Proc. Fifth U.S. Nat. Cong. of Appl. Mech., ASME, 511-530 (1966).

40. Schapery, R. A., "On the Characterization of Non-Linear Viscoelastic Materials," Polymer Eng. Sci., 9, 295-310 (1969).

41. McGuirt, C. W., and Lianis, G., "Experimental Investigation of Non-Linear Non-Isothermal Viscoelasticity," Int. J. Eng. Sci., 7, 579-599 (1969).

42. Frudenthal, A. M., "Strain Sensitive Response of Filled Elastomers," Technical Report No. 24, Department of Civil Engineering and Engineering Mechanics, Columbia Univ., etc.

43. Flory, P. J., Principles of Polymer Chemistry, Cornell University Press, Ithaca, New York (1953).

44. Treloar, L. R. G., The Physics of Rubber Elasticity, Oxford, Clarendon Press, (1949).

45. Graham, P. H., and Robinson, C. N., "Analysis of Cumulative Damage in Solid Rocket Propellants by Application of Reaction Rate Methods to the Binder-Filler Separation Process," Minutes of the Fourth Cumulative Damage Technical Coordination Meeting, North American Rockwell Corporation, McGregor, Texas (May 1968).

46. Bills, K. W., Jr., Svob, G. J., Planck, R. W., and Erickson, T. L., "A Cumulative-Damage Concept for Propellant-Liner Bonds in Solid Rocket Motors," J. of Spacecraft, 3, 408-412, (1966).

47. Fitzgerald, J. E., "Thermomechanical Coupling in Viscoelastic Materials," Presentation at the International Conference on Structure, Solid Mechanics and Engineering Design in Civil Engineering Materials, Southampton, England (1969). Discussion to be published by John Wiley and Sons, Inc., New York.

48. Coleman, B. D., and Mizel, V. J., "Norms and Semi-Groups in the Field of Fading Memory," Arch. Rat. Mech. Anal., 23, 87-123 (1966).

DISCUSSION

A. J. Chompff (Ford Motor Company):

To decide upon the amount of mechanical degradation in a certain experiment, shouldn't you heat the sample to a sufficiently high temperature and measure the amount of recoverable strain?

R.J.Farris:

No. The models developed in my paper were for amorphous elastic polymers It was assumed that they always recovered their geometry (given sufficient time) whenever the surface tractions were removed, regardless of the amount of mechanical degradation. To assess the degree of damage, one should perform tests wherein the first and second responses to some deformation history are compared. Naturally for such tests it is desirable to allow the material to recover sufficiently between tests and temperature would surely speed up this process, however it could also influence the data in other ways. Another excellent way to assess damage would be to compare the equilibrium stress states for two different relaxation tests where sample 1 was strained to, say, 5% strain and sample 2 was strained to, say, 10% and then returned to 5% strain. The difference in the equilibrium stress states at 5% strain for these two tests would be a measure of the damage that results from going from 5% strain to 10% strain. If these stress states were different, then the material would not fall into the class of a fading memory material and a permanent memory representation such as I have described in my paper is needed to characterize the material's response.

M. Gordon (Essex University):

Your model contemplates bond breakage under local strain conditions. Bueche had a model of tensile failure where breakage of one bond led to distribution of its load among neighbor-bonds; hence a cascade of breakage. How is this picture related to yours?

R.J.Farris:

The assumptions going into Bueche's model and mine are nearly identical. Both models allow for nonuniform local strains, localized failure, and nonlinear response of the elements. The models differ in that (1) Bueche attempted to calculate the distribution of local strain; (2) he used the inverse Langevin function for the local stress-strain behavior; (3) he assumed a critical strain failure criterion; and (4) his representation was one-dimensional. In my model, I proposed a simple representation wherein the local stress-strain function and the distribution of local strains were arbitrary and for the case of a critical strain failure criteria, I have described how these two functions can be assessed from experimental tests. This portion of my work can be looked upon as an extension of Bueche's research. The main difference between my work and Bueche's is that I also handle the case when the local failure is time and strain history dependent and I also show how these constitutive assumptions can be contained in a general three dimensional constitutive equation. When one is working with the models proposed by Bueche or myself, it is relatively simple to calculate the state of stress for a given strain input, however for both of these models it is extremely difficult to calculate the strain output for a given stress history. In this later case, one must determine how each bond rupture redistributes its load among neighbors to still satisfy the equation of equilibrium in the direction of stretch. To the best of my knowledge, Bueche did not handle this case.

A CONSTITUTIVE REPRESENTATION OF INHOMOGENEOUS POLYMERIC SYSTEMS

C. C. Hsiao and W. Chen, Department of Aerospace
Engineering and Mechanics, University of Minnesota
Minneapolis, Minnesota 55455

SUMMARY

Highly crosslinked network polymers appear to be inhomogeneous in both crosslink density and local orientations in their microstructure. To obtain some understanding of these systems a constitutive representation of such inhomogeneous polymeric systems is presented. The state of stress is calculated in terms of the contributions resulted in from deformations of microstructural systems. The approach is kept fairly general to maintain its validity for a wide variety of somewhat similar systems. The mathematical model used is composed of a large number of star-shaped basic units which are simple enough to reduce the complexity of the problem. However, attempts are made particularly to bring out the inhomogeneous nature of the crosslink density as well as the local orientations in the individual microstructural units. In order to assure a proper distribution for all the units in the network system the strain energy is optimized through the use of variational principles with respect to basic quantities in any deformational processes. A simple constrained condition is introduced in the variational process to guide the mode of deformation of the network system. In the case that the network system deforms elastically, all quantities introduced in the analysis are properly interpreted.

The crosslinking phenomenon in polymer growth is a common occurring fact. The mechanical behavior of the polymeric networks resulted from crosslinking is not yet fully considered especially with regard to deformation and strength problems. In order to gain a better understanding of network systems of this type, it is felt that the effects of crosslink density and local molecular orientations on macroscopic behavior must be treated. The physical motive underlying the subsequent treatment of the problem is that in a given highly crosslinked network system somewhat similar microscopic units forming the system can be isolated for analysis. Each individual unit is inhomogeneous in its crosslink density and local orientations. Under load the links deform and withstand applied forces according to their orientations. The interest of this report is to obtain a constitutive equation describing the mechanical behavior of a medium through analyzing the total contributions of all linking forces in the system.

In the following analysis, it is considered that the intermolecular forces responsible for crosslinking together with other bonding forces in a polymeric system are randomly distributed as individual decoupled units. They may be represented by a system of forcing elements radiating out from a central position forming a number of star-shaped units. Assuming that the configuration of each of the units be consisted of one principal link P of length l and modulus function K and q groups of crosslinks with each group consisted of R_q identical links which are uniformly distributed around the principal link. A typical crosslink C_{qr} of the qth group has length $a_q l$ and modulus $b_q K$ and makes an angle α_q with P. This covers a finite number of crosslinks by allowing q to vary from 1, 2, ... to Q, and r from 1, 2, ... to R_q. The number q can be permitted to reach infinity. However, to a first approximation, since our interest lies in an inhomogeneous crosslink density and local orientations in the molecular structure, it is likely that a limited finite number will be adequate.

Using this basic idea, now let us consider a system of crosslinked network in a rectangular coordinate frame of reference OX_i (i = 1, 2, 3). With the assumption that these individual units are basically similar and are decoupled from their surroundings. For any arbitrary unit we can expect to obtain the components of the stress tensor σ_{ij} (ij = 1, 2, 3) in the vicinity of the central position of that unit. If the density of the probability distribution function of orientation of all units in the network system is known together with the number of units η per unit volume, the expected stress tensor can be calculated.

In the past it had been accomplished that if only a single link is present in each of the basic units, the components of the stress tensor in the vicinity of a point is expressible [1] as follows:

$$\sigma_{ij} = \int \kappa l^2 \eta \, \varepsilon_{mn} e_m e_n e_i e_j \rho \, d\omega \tag{1}$$

where κ, l, and η as mentioned earlier are constants describing the characteristics of the link, ε_{mn} represents the components of the strain tensor. e_i, e_j, e_m, e_n are unit vectors and $d\omega$ is the infinitesimal solid angle. The integration is to be carried out with respect to a solid angle directed toward all possible directions. Here $\rho d\omega$ is interpreted as the number of all links lying in the small solid angle $d\omega$. The strain tensor ε_{mn} can be regarded as nonlinear [2] if large finite deformations are considered instead of the small deformations [1] intended when the formulation was first derived.

Following the above general considerations, if the principal directions of both stresses and strains are assumed coincident, there will be nine integrals when (1) is expanded.

They are:

$$I_{ij} = \int \kappa l^2 \eta \, e_i^2 e_j^2 \rho \, d\omega \tag{2}$$

Now taking into consideration of the contributions from all the links (2) will become [3] :

$$\begin{aligned} \Sigma I_{ij} &= (I_{ij})_P + \sum_{q=1}^{Q} \sum_{r=1}^{R_q} (I_{ij})_{C_{qr}} \\ &= 2\pi \int_\omega \kappa l^2 \eta \, e_i^2 e_j^2 \rho \, d\omega \\ &\quad + 2\pi \sum_{q=1}^{Q} R_q a_q^2 b_q \int_\omega \kappa l^2 \eta \, e_{qi}^2 e_{qj}^2 \rho \, d\omega \end{aligned} \tag{3}$$

where

$$e_i = \{\sin\theta\cos\phi,\ \sin\theta\sin\phi,\ \cos\theta\}, \qquad (4)$$

$$\begin{aligned} e_{qi} = \{&[(\cos\alpha_q \sin\theta + \cos(\beta+\gamma)\cos\theta\sin\alpha_q)\cos\phi \\ &-\sin\alpha_q \sin(\beta+\gamma)\sin\phi], \\ &[(\cos\alpha_q \sin\theta + \cos(\beta+\gamma)\cos\theta\sin\alpha_q)\sin\phi \\ &+\sin\alpha_q \sin(\beta+\gamma)\cos\phi], \\ &[\cos\alpha_q \cos\theta - \sin\alpha_q \sin(\beta+\gamma)\sin\theta]\} \end{aligned} \qquad (5)$$

with $(\beta+\gamma)$ together as a reference angle equivalent to the angle β between the plane formed by the principal link p and its projection on $Ox_1 - Ox_2$ and the plane formed by p and the qth link plus a convenient constant angle $\gamma = (r-1)2\pi/R_q$.

As an illustration, if an applied stress σ_{33} is present in the X_3 direction, there will be an associated strain ε_{33} in the same direction and $\varepsilon_{11} = \varepsilon_{22}$ with all $\sigma_{ij} = 0$ except σ_{33} . Assuming that for highly crosslinked polymer networks an elastic stress strain relation can be obtained in the following form:

$$\sigma_{33} = \left(\frac{2\Sigma I_{12}\Sigma I_{13}\Sigma I_{23} - \Sigma I_{23}(\Sigma I_{13})^2 - \Sigma I_{11}(\Sigma I_{23})^2}{\Sigma I_{11}\Sigma I_{22} - (\Sigma I_{12})^2} + \Sigma I_{33}\right)\varepsilon_{33} \qquad (6)$$

When large deformations occur, nonlinear behavior will be present and the analysis becomes more complex and involved. However, the general deformational behavior under a one dimensional behavior is expected to be somewhat similar to the result shown in [2] after the distribution function ρ of orientation is properly introduced.

To obtain an optimum ρ it may be desirable to employ the variational theory for optimizing the total energy of the system with respect to any deformation under certain constrained conditions. Since highly crosslinked network systems are fairly "rigid," without losing the generality, they may be considered to deform linearly elastic. This is particularly true if deformation is small.

Now consider that the total energy of the system be composed of internal energy and strain energy for all the links, then

$$U = \int_\omega (I_p + \int \psi_p \, d\mathcal{E}_p)\rho \, d\omega + \sum_{q=1}^{Q} R_q a_q^2 b_q \int_\omega (I_q + \int \psi_q \, d\mathcal{E}_q)\rho \, d\omega \tag{7}$$

where I_p and I_q are respectively the internal energy of the principal link p and that of any of the q crosslinks, ψ_p and ψ_q are stresses associated with the principal link p and the qth crosslink respectively, similarly $d\mathcal{E}_p$ and $d\mathcal{E}_q$ are respectively the stretches for the links. If the stresses are derivable from a potential W or an elastic strain energy as it is true for an elastic system, then the internal and strain energy for the system can be expressed in the following manner:

$$U = \int_\omega (I_p + \sum_{q=1}^{Q} R_q a_q^2 b_q I_q)\rho \, d\omega + (1 + \sum_{q=1}^{Q} R_q a_q^2 b_q) \int_\omega W \rho \, d\omega \tag{8}$$

where I_p and I_q are internal energy quantities,

$$\psi_p = \partial W / \partial \mathcal{E}_p,$$

and $$\psi_q = \partial W / \partial \mathcal{E}_q \quad (q = 1, 2, 3, \ldots\ldots, Q) \tag{9}$$

To properly restrict the strain energy we allow our system to deform macroscopically from spherical bodies into ellipsoids. Utilizing the invariant quantities associated with the strain tensor $\mathcal{E}_{mn}$ the simplest form of the equation of constraint can be put into the following form (See Appendix I).

$$C_1 J_1^2 + C_2 J_2 = C_0 \tag{10}$$

where C_1, C_2, and C_0 are positive constants and $J_1 = \mathcal{E}_{mm}$,

$J_2 = \varepsilon_{mn}\varepsilon_{mn}/2$ are strain invariant quantities closely related to the first and second invariants of the strain tensor. The variational equation can be put into the form for the system as

$$\delta\left(\int_{\omega} U(\varepsilon_{mn},\omega)\rho\, d\omega - \lambda\int_{\omega} G(\varepsilon_{mn})\, d\omega\right) = 0 \quad (11)$$

where λ is the Lagrange multiplier and $G(\varepsilon_{mn}) = C_1 J_1^2 + C_2 J_2$ is the function of constraint. From (11) the Lagrangian function $\mathcal{L}$ may be obtained and it has the form of

$$\mathcal{L} = \rho I_o + \rho K W - \lambda G \quad (12)$$

where
$$I_o \equiv I_p + \sum_{q=1}^{Q} R_q a_q^2 b_q I_q ,$$
$$K \equiv 1 + \sum_{q=1}^{Q} R_q a_q^2 b_q$$

This function of Lagrange will attain its optimum value for certain values of ε_{mn} for which

$$\frac{\partial \mathcal{L}}{\partial \varepsilon_{mn}}\, \delta\varepsilon_{mn} = 0 \quad (13)$$

Or

$$\frac{\partial (I_o + KW)\rho}{\partial \varepsilon_{mn}} - \lambda(2C_1 J_1 \delta_{mn} + C_2\varepsilon_{mn}) = 0 \quad (14)$$

For those values of ε_{mn} that $\delta\mathcal{L} = 0$, $\mathcal{L}$ must be constant. Let $\mathcal{L} = \lambda\mathcal{L}_o$, then (12) becomes

$$\rho(I_o + KW) - \lambda(C_1 J_1^2 + C_2 J_2 + \mathcal{L}_o) = 0 \quad (15)$$

Eliminating λ from (14) and (15) yields

$$\frac{d(I_o+KW)\rho}{d\mathcal{E}_{mn}} = \frac{2C_1J_1\delta_{mn}+C_2\mathcal{E}_{mn}}{C_1J_1^2+C_2J_2+L_o}\rho(I_o+KW) \qquad (16)$$

Integrating the above equation with respect to $\mathcal{E}_{mn}$ after using the equation of constraint $C_1J_1^2+C_2J_2-C_o=0$, one gets

$$\rho(I_o+KW) = C\exp\frac{\int(2C_1J_1\delta_{mn}+C_2\mathcal{E}_{mn})d\mathcal{E}_{mn}}{L_o+C_o} \qquad (17)$$

The integration constant C can be determined if for $\mathcal{E}_{mn}=0$, allow $W\rightarrow 0$ and $\rho\rightarrow\rho_o$, i.e., $C=\rho_oI_o$. Thus from (17)

$$\rho = \frac{\rho_oI_o}{I_o+KW}\exp\frac{\int(2C_1J_1\delta_{mn}+C_2\mathcal{E}_{mn})d\mathcal{E}_{mn}}{L_o+C_o} \qquad (18)$$

Through substitution the constitutive representation of the inhomogeneous polymeric system can be described by (6) with proper modifications as well as the introduction of (18). For linear elastic system C_1 and C_2 have the significance as shown below. Somewhat similar interpretations can be found in [4] .

For $\quad m \neq n \qquad W=\frac{1}{2}C_2\mathcal{E}_{mn}\mathcal{E}_{mn},$

$\quad\quad m = n \qquad W = C_1J_1^2+C_2J_2,$

where

$$C_1 = \frac{\nu E}{2(1+\nu)(1-2\nu)},$$

$$C_2 = \frac{E}{1+\nu}.$$

and ν is normally termed the Poisson's ratio and E is the modulus of elasticity of the elastic medium.

In order to get some idea about the validity of the above general formulation and to determine whether any result can be substantiated by experimental data, crosslinked vulcanized rubber has been analyzed on the basis of its microstructure. It is well known that the strength of vulcanized rubber is highly dependent upon the formation of crosslinks. This in turn affects the inhomogeneous nature of the crosslink density as well as the local orientations in the individual microstructural units. Natural rubber is essentially a polymeric chain of isoprene hydrocarbon. There exists periodically in the chain a double bond. In the process of vulcanization crosslinks are formed between sulphur and the highly reactive double bonds in the polyisoprene chain. As a result the mechanical behavior of the network material system can be calculated using the present formulation. For simplicity neglecting the effort of the side chains only the principal rubber chain and the crosslink need be considered in the analysis. For comparison the one-dimensional stress-strain relation in tension is calculated as this type of force-extension relationships is widely reported in the literature. For example, the shape of the force extension curve (conventional or nominal stress-strain relationship) for 'pure-gum' GR-S rubber at 2°C is somewhat like a reversed "S" as reported in [5].

In obtaining an analytical curve based upon the present theory, let us consider that the strain energy for any arbitrary strain tensor $\mathcal{E}_{mn}$ be given as

$$W(\mathcal{E}_{mn}) = \int_0^{\mathcal{E}_{mn}} (2C_1 J_1 \delta_{mn} + C_2 \mathcal{E}_{mn}) d\mathcal{E}_{mn} \tag{19}$$

The maximum strain energy corresponding to a maximum elastic strain that $\mathcal{E}_{33} = \mathcal{E}_m$. Since $\mathcal{L}_0$ represents a reference level of the Lagrangian corresponding to an optimum state of strains, it can be assumed to be zero for convenience. C_0 is a constant independent of the state of strains. However it can be a polynomial function of maximum strain $\mathcal{E}_m$. From (19)

$$W(\mathcal{E}_m) = (C_1 + \tfrac{1}{2} C_2) \mathcal{E}_m^2 = I_0 \tag{20}$$

Now assume
$$C_0 \sim (C_1 + \tfrac{1}{2} C_2) \mathcal{E}_m^2 \tag{21}$$

as $\mathcal{L}_0$ has already been chosen as zero. The ratio

$$\frac{W(\mathcal{E}_{mn})}{W(\mathcal{E}_{m})} = \frac{\mathcal{E}_{mn}^{2}}{\mathcal{E}_{m}^{2}} \tag{22}$$

is one of the special cases that can be considered in the constitutive representation. Other power functions of the strain ratio may also be considered.

With this information it is possible to obtain a form of the constitutive equation for this one dimensional behavior

$$\sigma_{33} = \chi \frac{\exp\left(\mathcal{E}_{mn}/\mathcal{E}_{m}\right)^{2}}{1 + \left(1 + \sum_{q=1}^{Q} R_q a_q^2 b_q\right)\dfrac{W(\mathcal{E}_{mn})}{W(\mathcal{E}_{m})}} \mathcal{E}_{33} \tag{23}$$

where $\chi \equiv 2\pi\rho_0 l^2 \eta \kappa \left(\dfrac{2\sum J_{12}\sum J_{13}\sum J_{23} - \sum J_{22}(\sum J_{13})^2 - \sum J_{11}(\sum J_{23})^2}{\sum J_{11}\sum J_{22} - (\sum J_{12})^2} + \sum J_{33} \right)$

and $\sum J_{ij} = \int_{\omega} e_i^2 e_j^2 \, d\omega + \sum_{q=1}^{Q} R_q a_q^2 b_q \int_{\omega} e_{qi}^2 e_{qj}^2 \, d\omega$.

For vulcanized rubber consider the sulpher crosslinking and the principal carbon chain

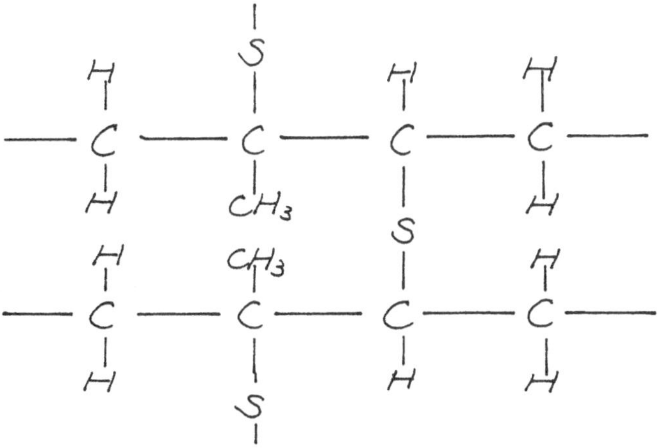

By neglecting the side chains, we have

$$Q = 1$$

$$a_1 = \frac{1.80\ \mathring{A}}{1.54\ \mathring{A}} = 1.17$$

$$a_1 b_1 = \frac{66\ Kcal/mole}{81\ Kcal/mole} = 0.81$$

which is the ratio of the band energy and $R_g = R_1 = 1$
Thus (23) becomes

$$\sigma_{33} = \chi \frac{\exp(\mathcal{E}_{33}/\mathcal{E}_m)^2}{1 + 1.95(\mathcal{E}_{33}/\mathcal{E}_m)^2} \mathcal{E}_{33}$$

The result with $\mathcal{E}_m$ chosen as 5 is shown in Fig. 1 which compares very well with the simple stress-strain curve as given in [5] when converted into the conventional stress quantities.

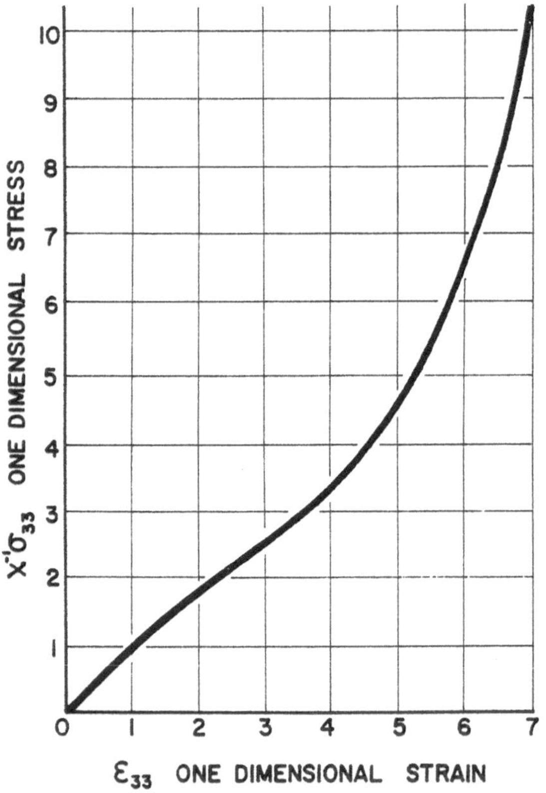

Fig. 1 Calculated one'dimensional stress-strain behavior of vulcanized rubber

References

[1] "Theory of Mechanical Breakdown and Molecular Orientation of a Model Linear High Polymer Solid," C C Hsiao, J. Appl. Phys. 30, 1492 (1959).

[2] "A Theory of Deformational Behavior of Polymers," C. C. Hsiao and S. R. Moghe, presented at the Fourth IUPAC Micro-symposium on Macromolecules, Prague, Czechoslovakia, September 1969, to appear in J. Macromolecular Science-Physics.

[3] "Orientation and Strength of Branched Polymer Systems," C. C. Hsiao and T. S. Wu, J. Polymer Science, Part A, 1, 1789 (1963).

[4] Mathematical Theory of Elasticity, I. S. Sokolnikoff, McGraw-Hill, Second Edition, (1956).

[5] The Physics of Rubber Elasticity, L.R.G. Treloar, Oxford, Clarendon Press, Second Edition, (1958).

Appendix

In our network system there exists a large number of identical microstructural units which form the polymeric body. Macroscopically we specify the deformation of the material body by allowing a spherical volume to deform into an ellipsoid under the action of a homogeneous deformation. Mathematically this means that if we consider a quadratic function $G(\varepsilon_{mn})=\varepsilon_{mn}x_m x_n$ and constrain the end point $P(x_i)$ of a line element to lie on the quadratic surface $\varepsilon_{mn}x_m x_n = C_0$ where C_0 is a constant. This gives the equation of constraint in the form

$$G(\varepsilon_{mn}) = C_0 \tag{I.1}$$

However, instead of using the above form of the constraint equation, we can take the advantage in utilizing certain alternate forms of the invariants of the strain tensor

The principal three strain invariants are:

$$\begin{aligned} I_1 &= \varepsilon_{mm} , \\ I_2 &= \tfrac{1}{2}(\varepsilon_{mn}\varepsilon_{nn} - \varepsilon_{mn}\varepsilon_{mn}), \\ I_3 &= \tfrac{1}{6}\delta_{ijk}\,\delta_{lmn}\,\varepsilon_{il}\,\varepsilon_{jm}\,\varepsilon_{kn} . \end{aligned} \tag{I.2}$$

where $\delta_{ijk}, \delta_{lmn}$ are generalized Kronecker deltas. Alternatively we can use three equivalent forms as follows:

$$\begin{aligned} J_1 &= \varepsilon_{mm} , \\ J_2 &= \tfrac{1}{2}\varepsilon_{mn}\varepsilon_{mn} , \\ J_3 &= \tfrac{1}{6}\delta_{ijk}\,\delta_{lmn}\,\varepsilon_{il}\,\varepsilon_{jm}\,\varepsilon_{kn} . \end{aligned} \tag{I.3}$$

In obtaining the proper form of the equation of constraint (I.1) should satisfy the following assumptions.

(i) Zero strain implies the equation of constraint vanishes and vice versa, i.e.,

$$G(\mathcal{E}_{mn}) = 0$$

if and only if

$$\mathcal{E}_{mn} = 0$$

(ii) $G(\mathcal{E}_{mn})$ must be a homogeneous function of $\mathcal{E}_{mn}$ of degree n (n = 1, 1, 3, ...) to effect a homogeneous deformation.

(iii) Principal strains of the system must be invariant with respect to coordinate transformation.

On the basis of the third assumption we have

$$G(\mathcal{E}_{mn}) = G(J_1, J_2, J_3) \tag{I.4}$$

Now use assumptions (i) and (ii). We can obtain for

$$\begin{aligned}
n=1 \quad & G(\mathcal{E}_{mn}) = C_1 J_1 \\
n=2 \quad & G(\mathcal{E}_{mn}) = C_1 J_1^2 + C_2 J_2 \\
n=3 \quad & G(\mathcal{E}_{mn}) = C_1 J_1^3 + C_2 J_1 J_2 + C_3 J_3 \\
n=4 \quad & G(\mathcal{E}_{mn}) = C_1 J_1^4 + C_2 J_1^2 J_2 + C_3 J_1 J_3 + C_4 J_2^2 \\
\cdots \quad & \cdots\cdots\cdots
\end{aligned}$$

where $C_1, C_2, C_3 \cdots\cdots$ are positive constants. All the constraint functions violate (i) except the one for which $n = 2$. Using this valid constraint function, we have

$$G(\mathcal{E}_{mn}) = C_1 J_1^2(\mathcal{E}_{mn}) + C_2 J_2(\mathcal{E}_{mn}), \tag{I.5}$$

where $C_1 > 0$ and $C_2 > 0$.

Therefore the simplest form of the equation of constraint becomes

$$C_1 J_1^2 + C_2 J_2 = C_0 \tag{I.6}$$

where C_0 is also a constant.

MOLECULAR BOND RUPTURE ASSOCIATED WITH INELASTIC DEFORMATION OF ELASTOMERS

R. Brown,* K. L. DeVries,** and M. L. Williams†

*Research Assistant; **Prof. of Mechanical Engineering;

†Prof. of Engineering; University of Utah, Salt Lake City

SUMMARY

Unstrained rubbers cooled far below their glass transition temperature are generally very brittle. The ductility can, however, be greatly increased by prestraining (~100%) the rubber before reducing the temperature as suggested by Andrews and Reed. Results on natural rubber and Hycar 1043 rubber are reported, showing the effects of prestrain, temperature and strain rate on low temperature ductility and primary molecular bond rupture. Molecular bond rupture was measured by electron paramagnetic resonance (EPR) techniques. X-ray diffraction in rubbers indicates that prestraining and cooling results in orientated crystallization. It is suggested that this, in effect, produces a semi-crystalline polymer with a resulting increase in ductility and rather general bond rupture (throughout the loaded sample volume) during deformation leading to fracture.

INTRODUCTION

During the last decade, the tremendous strides taken in the exploration and exploitation of space have placed new environmental demands on the engineering use of elastomers. In particular, the use of rubber polymers in cryogenic and space applications (in the binder of some solid rocket propellants and in low temperature seals, for example) requires that rubber withstand extremely low temperatures. Under these conditions, rubber can become brittle and cracks can propagate easily through the material. In the case of solid rocket propellants, these cracks can lead to uneven and uncontrolled burning. As one practical technique for increasing the low temperature ductility of rubber, Andrews and Reed [1,2] have suggested that the rubber be prestrained before freezing. They also observed hydrogen gas foaming from the extended natural rubber samples and detected EPR signals after the samples were fractured [2].

The authors have duplicated and extended these experiments to a "non-crystallizing" type rubber and have applied the techniques of electron paramagnetic resonance (EPR) spectroscopy to study the kinetics of bond rupture. A servo-controlled loading frame built around the magnet facilitated making measurements during straining rather than simply subsequent to fracture. This laboratory has been highly successful in applying EPR to nylon fracture [3], decay and other degradation of teeth and dental materials [4], and ozone cracking of rubber [5]. Zhurkov and his associates were the first to demonstrate that EPR could be used successfully to investigate polymer fracture [6,7]. Other interesting applications of the technique have been made by Campbell and Peterlin [8] and Becht and Kausch [9]. Since the molecular forces seen by a long polymer molecule vary depending on chain orientation and structure, EPR analysis can provide information on the changing morphology during failure.

BACKGROUND

EPR is a form of absorption spectroscopy in which electromagnetic radiation in the microwave region induces transitions between energy levels arising from Zeeman-type splittings in an assemblage of paramagnetic electrons. The number of electrons in the upper and lower levels are governed by Boltzmann's statistics with the magnitude of the energy gap, ΔE, between the two states given by: $\Delta E = g\beta H$ where g is the spectroscopic splitting factor, β is the Bohr magneton, and H is the magnetic field. The resonance frequency of microwave radiation necessary to induce the transactions is $\nu = g\beta H/h$ where ν is the frequency and h is Planck's Constant. The absorption of microwave energy is detectable through suitable electronics and is plotted out (for electronic conveniences) in the form of the derivative of the energy absorption spectra. A thorough description of EPR, its uses and applications can be found elsewhere [10]. In polymeric materials, fracture under mechanically applied stress involving molecular bond rupture results in the creation of free radicals and their accompanying characteristic unpaired electrons. Thus the microwave absorption intensity provides a measure of the rate and amount of bond rupture; and the shape of the spectra gives clues to which bonds are being ruptured. An EPR analysis, however, has two practical requirements: a sufficient number of free radicals must be produced, and the half-life of the free radicals formed must be long enough to record [11]. The Varian E-3 equipment used in these tests operates in the x-band region (~9.5 ghz) and has a sensitivity of about 5 x 10^{10} ΔH spins under ideal conditions. (Under usual conditions, the sensitivity is roughly 10^{12} spins.)

DEFORMATION, STRENGTH AND STRUCTURE OF RUBBER POLYMERS

Rubbers tested at room temperature are completely in their rubbery state, and their response to uniaxial loading can quite accurately be described by the equation of state from rubber elasticity [12]

$$\sigma = N_o KT(\lambda - 1/\lambda^2)$$

where σ = stress, N_o = number of cross links/unit volume, K = Boltzmann's Constant, T = absolute temperature, and λ = stretch ratio.

The rubber-like state depends on the possibility of random thermal motion of chain elements by rotation about single bonds. In any real material, such rotation cannot be completely free from restrictions imposed by the presence of neighboring groups of atoms either in the same molecule or in neighboring molecules. The freedom of rotation is a function of the relative values of the thermal energy of the rotating group and the potential barrier is only slightly dependent on temperature, the average thermal energy increases with increasing temperature. Thus, the probability of rotation is governed by Boltzmann's statistics and increases exponentially with temperature [13]. This result leads to the conclusion that, at low temperatures, the rotation and translation modes of molecular motion are frozen out and only vibrational motion is present. In this state, the rubber no longer obeys rubber elasticity and deformation is only possible by the stretching of inter- or intra-molecular bonds.

In this state, the load deformation picture is quite different [14]. Bartenev and Zuyev [15] describe the phenomenon of elastic deformation followed by a yield stress (Point A in Figure 1) and subsequent necking. The neck travels through the sample to Point B and the thinned sample stretches to failure at C. If deformation takes place at a rather low temperature (T_{br}), the yield stress becomes greater than the strength of the polymer and the resulting fracture is brittle, never reaching Point A. The translation from a brittle failure to a ductile one with a rise in temperature is illustrated in Figure 2. As suggested by Joffe [16], the slope of yield point versus temperature (Line 2) is steeper than the brittle strength versus temperature plot (Line 1). The point where the two curves intersect corresponds to the temperature, T_{br}, of brittle fracture. At higher temperatures, failure occurs in

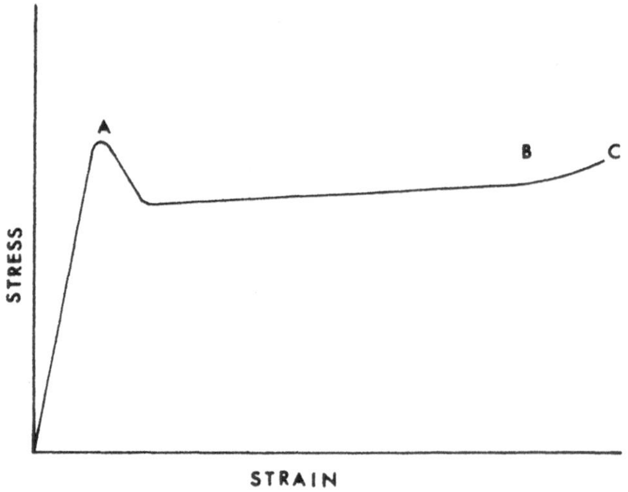

Figure 1. Stress-strain behavior of an amorphous polymer in the glassy state (after Bartenev and Zuyev).

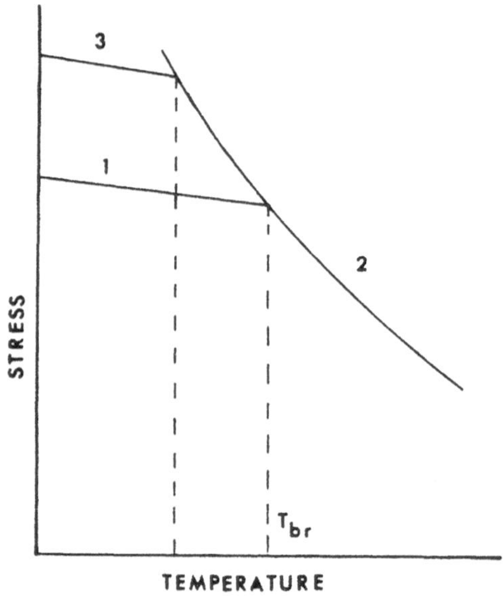

Figure 2. Dependence of brittle and ductile fracture on stress and temperature (after Joffe); 1 = brittle strength, 2 = yield point.

a ductile manner, while at lower temperatures brittle fracture is observed. If by some method the brittle strength of the polymer can be raised, say to Point 3 (Figure 2) the sample will reach the yield stress at a lower temperature, effectively lowering T_{br}. The experiments reported in this paper concern the extension of ductile behavior into temperature regions below the normal T_{br} by pre-straining the sample. The determination of brittle temperature depends to a large extent on the experimental technique employed [17]. Morris, James and Werkenthin [18] report a T_{br} of -53°C for vulcanized natural rubber and a T_{br} of -58°C for Hycar OR-20. Borders and June [19] give brittle temperatures ranging from -57°C to -15°C for various degrees of co-polymerization of Hycar.

The supermolecular structure of rubber is an important factor in the mechanical behavior. Generally, rubbers are catagorized into crystallizing and non-crystallizing rubbers. Bartenev and Zuyev [15] catalog the types of supermolecular structure found in each category as follows: crystallizing rubbers form blocks of molecules, blocks folded in ribbons and laminae, spherulites, and ribbon-shaped and laminar formations of spherulites; non-crystallizing rubbers form only blocks, and ribbons and laminae of blocks. In rubber polymers, however, crystallization is far from complete. X-ray scattering studies have estimated that the degree of crystallinity in natural rubber does not exceed 30% [20]. Natural rubbers crystallize to a greater extent than most other rubbers, and crystallization is limited by a number of factors, including molecular imperfections (branch points, cross links, chain ends) and chain entanglements. Butadiene copolymers such as GR-5, Hycar, or Buna 5 are considered as non-crystallizing due to the likelihood of a cis-trans isomer mixture in butadiene and the irregularities associated with the presence of a secondary constituent [21].

The development of crystallization in rubber depends on the temperature and the degree of deformation. The optimum conditions for crystallization are obtained when the temperature and the deformation are such that the molecules have sufficient energy and mobility to associate in an ordered way, but not sufficient to dissociate under the influence of random thermal motion. In the undeformed state, there is in general a temperature, different for each polymer, at which crystallization takes place most rapidly. For natural rubber, this temperature is about -24°C when crystallization is virtually complete in about eight hours [22]. Crystallization in stretched rubber reaches a maximum at 25°C within a very short time after extension, and is almost instantaneous at 50°C [13].

Experimental conditions also determine the morphology of the crystalline structure. In undeformed rubber, spherical semi-crystalline masses (spherulites) are generally observed, growing outward from separate nuclei until their boundaries meet and the process stops. It is suggested that these spherulites grow by a molecular ladder mechanism [22]. Pre-oriented molecules at the spherulite boundary fold back and forth upon themselves, causing the newly crystalline portion to rise as a vertical ladder from the plane of the film. When crystallization is induced by stretching, another morphology is encountered [22]. It consists of a fibrous structure oriented in the direction of extension. The size of the basic crystalline unit has been determined by electron microscope measurements to be in the order of 200Å.

EXPERIMENTAL PROCEDURE

Rubber samples of non-reinforced (gum) compounds of both the "crystallizing" and "non-crystallizing" types furnished by the B. F. Goodrich Company were used in the experiments. The rubber samples were compounded and cured as shown in Table 1.

TABLE I

Gum Compounds

Compound Number	231	235
Natural rubber	100	x
Acrylonitrile-butadiene (Trade name Hycar 1043)	x	100
Zinc Oxide	5	x
Stearic Acid	2	x
PBNA	1	x
Santocure	1	1
Sulfur	2	1.75
Total Weight	111.0	109.75
Specific Gravity	0.98	1.02
Cured 60 minutes at	145°C	150°C

The samples were carefully cut in a dog-bone shape and attached to friction grips at the end of brass pull rods (See Figure 3). The tension tests were conducted in a feed-back controlled hydraulic

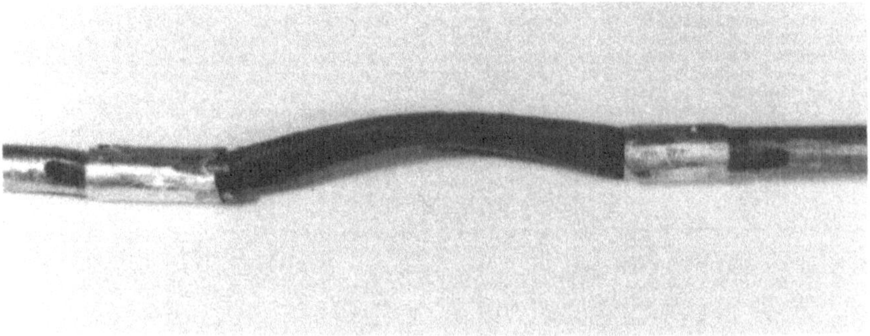

Figure 3. Rubber sample held in brass grips.

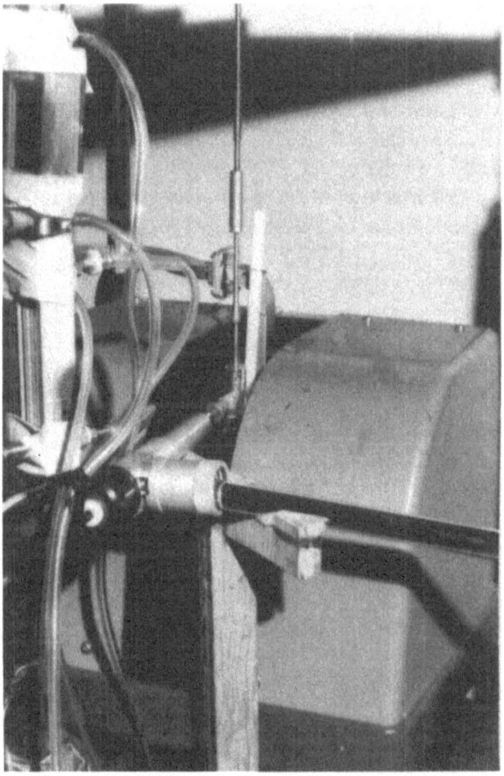

Figure 4. E-3 spectrometer with loading frame and variable temperature accessory.

Figure 5. Aluminum rack used in X-ray diffraction experiments

loading system set up with the feedback coming from the output of a linear potentiometer coupled mechanically to the hydraulic piston. This system is capable of following any programmed loading or strain history up to 20 cps but was simply programmed for constant strain rates. For testing, the sample was mounted inside a quartz tube which in turn was inserted through the microwave resonance cavity of the Varian Associates Model E-3 spectrometer. A regulated temperature controller was used to control the temperature of nitrogen gas passing through the quartz tube. Load, strain, and EPR signals were recorded continuously throughout the test. The complete set-up is shown in Figure 4.

Two experiments were conducted on natural rubber. For the first, the temperature was held at -75°C and the extension rate was 12.74×10^{-4} in/sec, prestrain was varied from 0% to 200%. In the second experiment, the strain rate was 12.74×10^{-4} in/sec, the prestrain was 100% and the temperature was varied from -44°C to -95°C. Three series of experiments were conducted on the Hycar samples. In the first series, prestrain was 100%, temperature was -65°C and the extension rate was varied between 17.65×10^{-4} in/sec and 6.85×10^{-2} in/sec. For the second series of experiments, the extension rate was held constant at 12.74×10^{-4} in/sec. The temperature was -65°C and the prestrain was varied from 0% to 162.5% of original length. In the last series, prestrain was 100%, extension was 12.74×10^{-4} in/sec and the temperature was varied -27°C and -85°C.*

In order to establish the existence of supermolecular structures in the test sample, X-ray diffraction techniques suggested by Alexander [23] were used. Using a General Electric X-ray unit, rubber samples were exposed at various stretch ratios and temperatures to a 0.025"-diameter beam of nickel filtered Cu Kα radiation (50KV, 20MA). X-ray diffraction photographs (5cm film-to-target distance) were taken at one hour exposures. The samples were

*See note on temperature at end of paper.

stretched on an aluminum rack (Figure 5) and placed in a styrofoam box during exposure. The temperature was lowered by piping nitrogen gas through a liquid nitrogen heat exchanger into the styrofoam box. Control of the temperature was possible by continuously monitoring a thermocouple fixed in the sample box and by making appropriate flow rate adjustments.

RESULTS

The results of the mechanical deformation experiments are shown in Figures 6, 7, 8, 9 and 10. Figures 6 and 7 show the stress-strain curves for natural rubber at low temperature. In every sample that had been sufficiently prestrained, ductile behavior and necking were subsequently observed at low temperatures. As shown in Figure 6, increasing the amount of prestrain before reducing the temperature resulted in increased ductility. This increased ductility was usually accompanied by a lowering of the ultimate engineering stress. Figure 7 indicates that at a given extension rate the lower the testing temperature, the greater the required prestrain to induce ductile behavior. For example, 100% prestrain causes ductility at -25°C, but the sample is still brittle at -95°C. Similar temperature and prestrain effects were observed with the Hycar specimens. In both rubbers, increasing the rate of extension tended to stiffen the stress-strain curves.

During the "plastic" deformation of the samples, readily detectable EPR spectra were produced. These spectra (see Figure 11) occurred at a g value of approximately 2.00, characteristic of organic free radicals. Free radicals once produced are inherently unstable. However, at the temperatures of interest here, these radicals exhibit long lifetimes. For example, at -75°C in a nitrogen atmosphere, no signal decay was observed, while on the other hand, the signal decayed rather rapidly at -35°C.

By suitable tuning of the spectrometer and comparison with various standards, quantitative measurements of the bond rupture kinetics are possible [11]. Figures 12, 13, 14, 15 and 16 show computer-reduced and plotted results from such measurements. The spectra began to appear after the yield point had been reached. This is consistent with the observations of Andrews and Reed who observed hydrogen gas foaming from the rubber after the yield point was reached. They attributed the evolution of hydrogen gas to free radical formation caused by main chain fracture and subsequent reactions [1]. Microscopic examination of the samples after failure occurred revealed differences in the character of the fracture surface. Micrographs of the fracture surfaces taken at 100X showed that the surfaces corresponding to ductile failure were much smoother than those for brittle fracture.

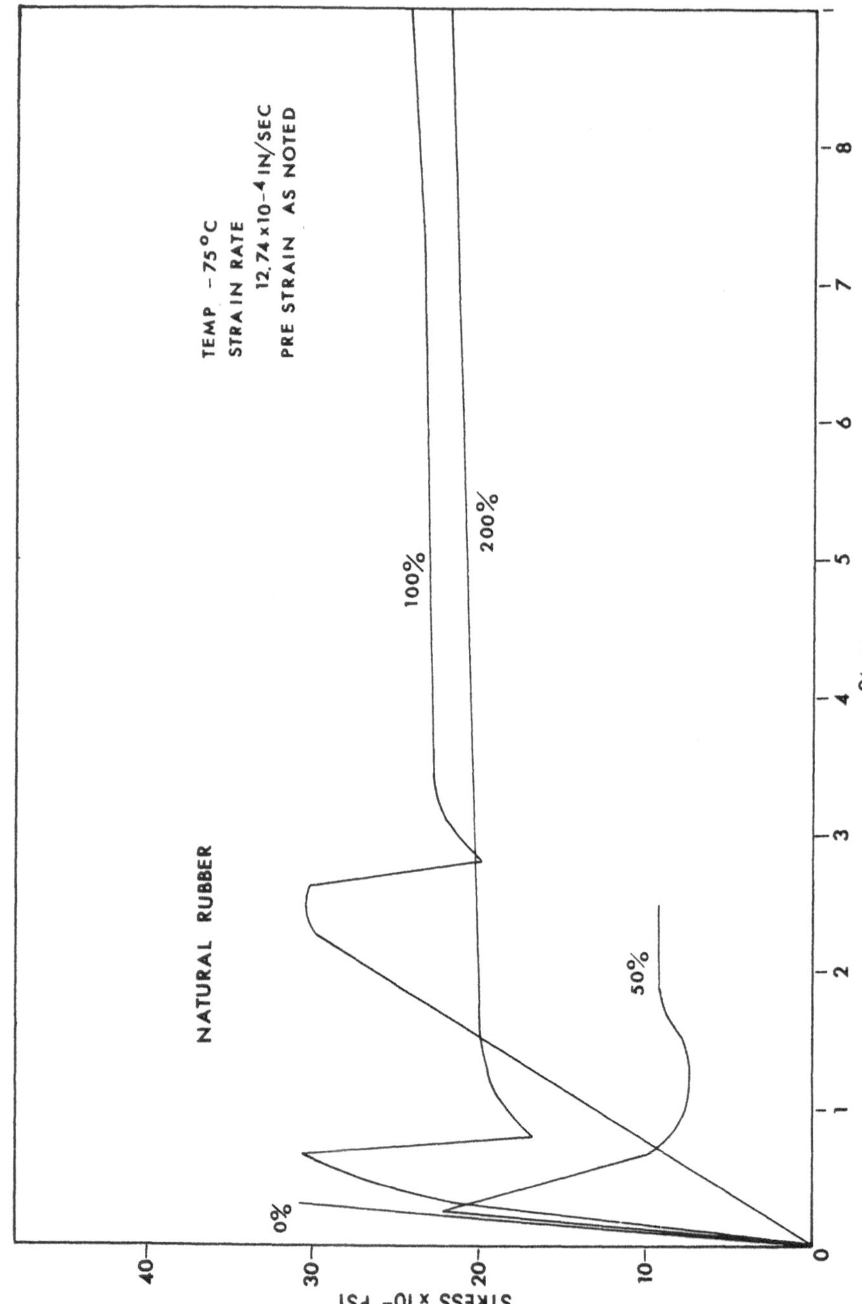

Figure 6. Effect of prestrain on stress-strain behavior of natural rubber.

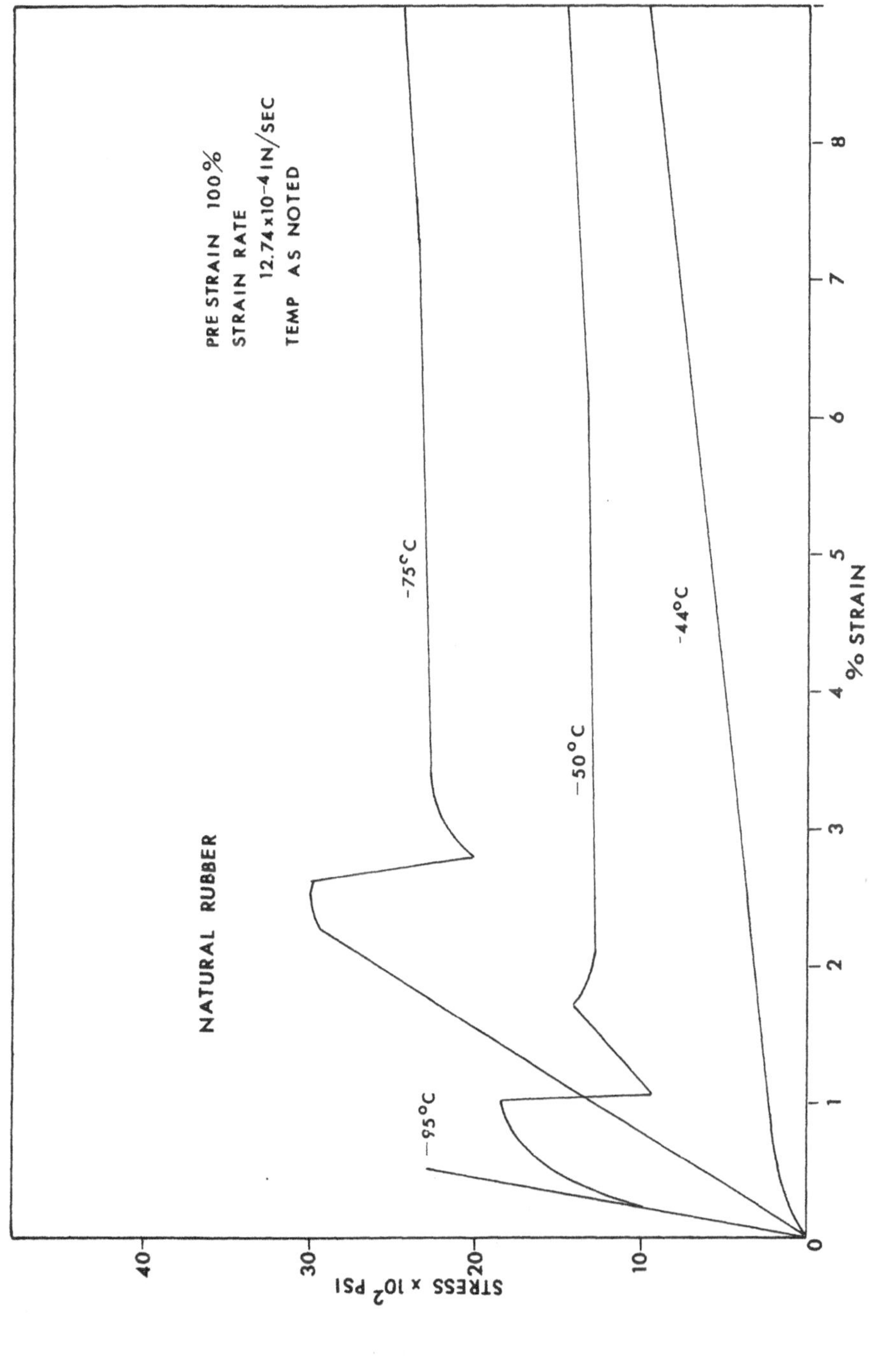

Figure 7. Effect of temperature on stress-strain behavior of natural rubber.

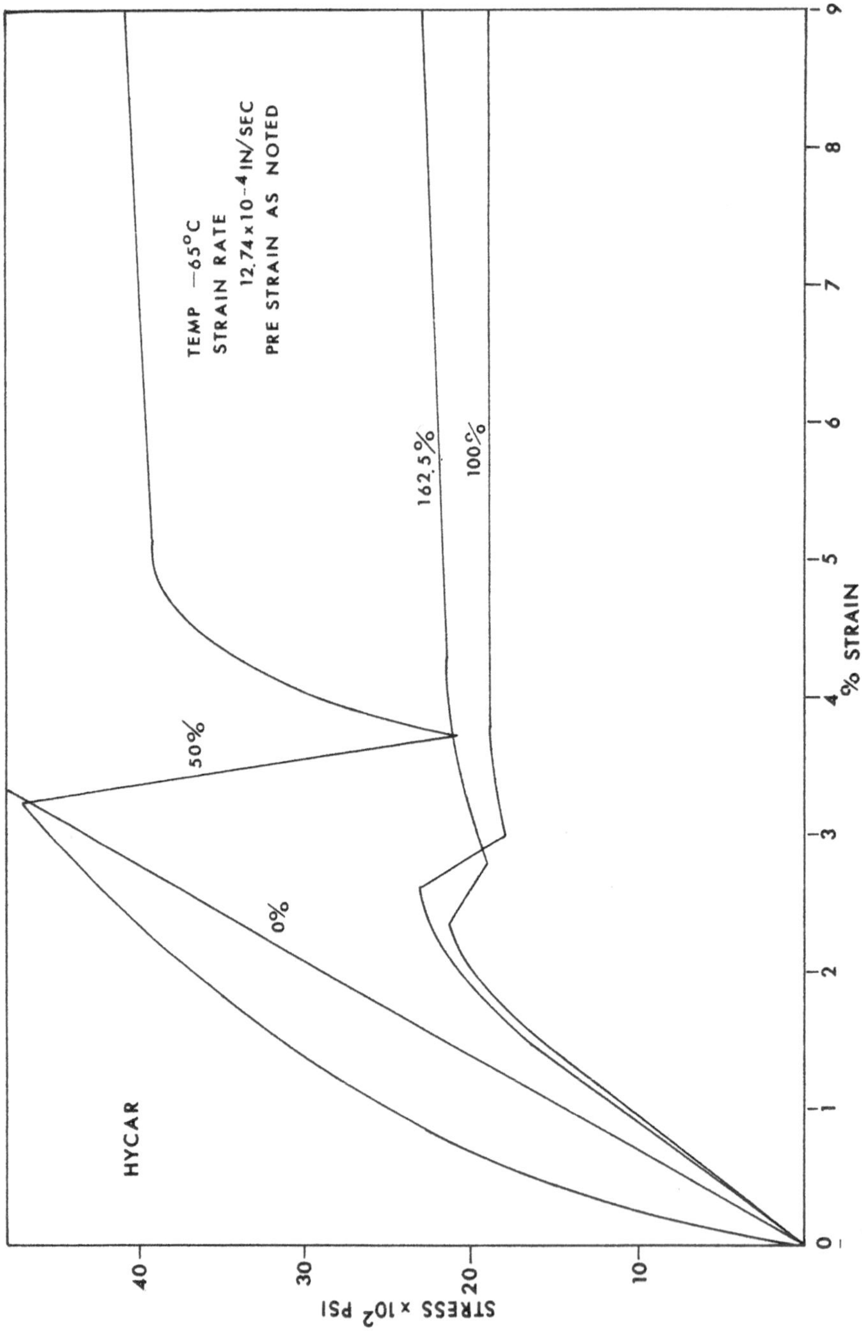

Figure 8. Effect of prestrain on stress-strain behavior of Hycar.

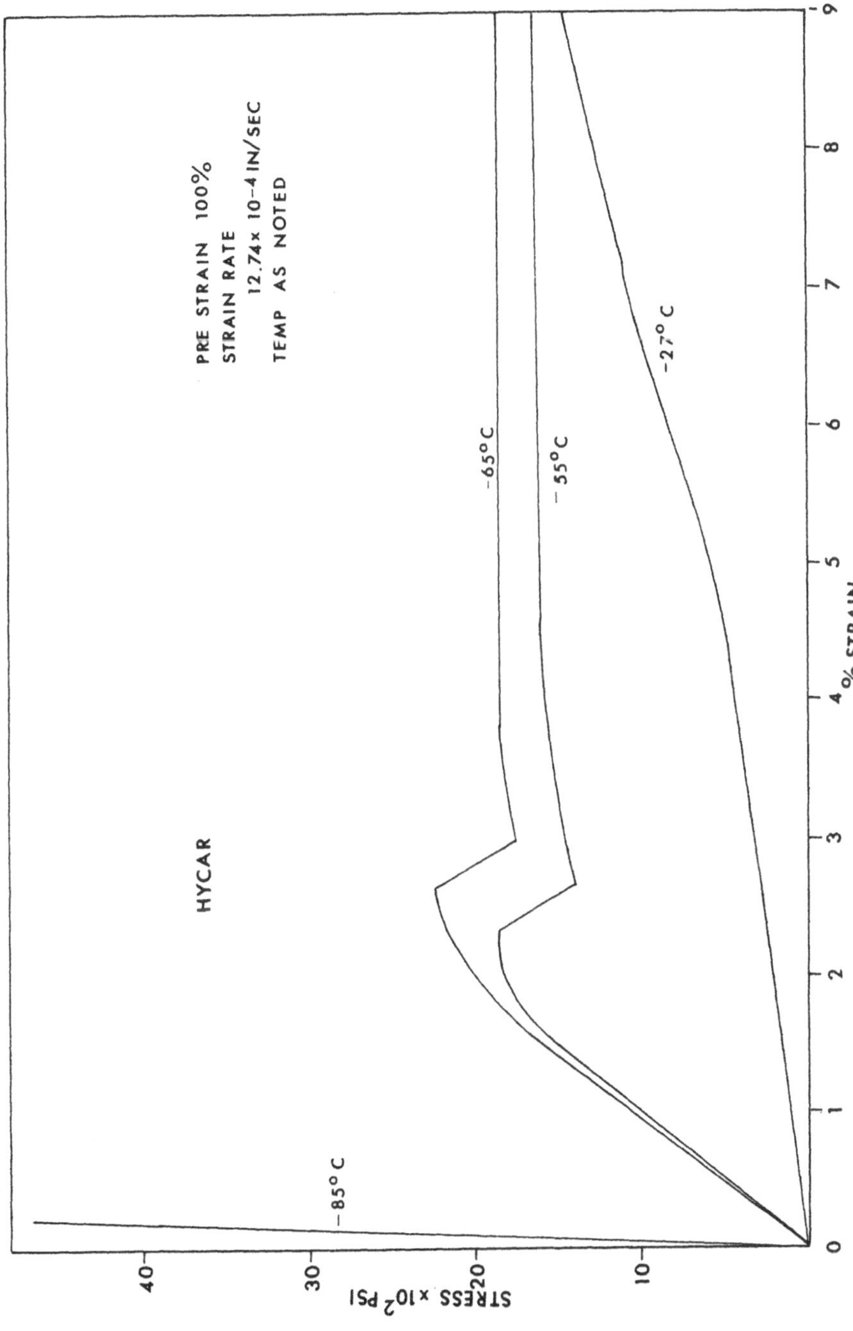

Figure 9. Effect of temperature on stress-strain behavior of Hycar.

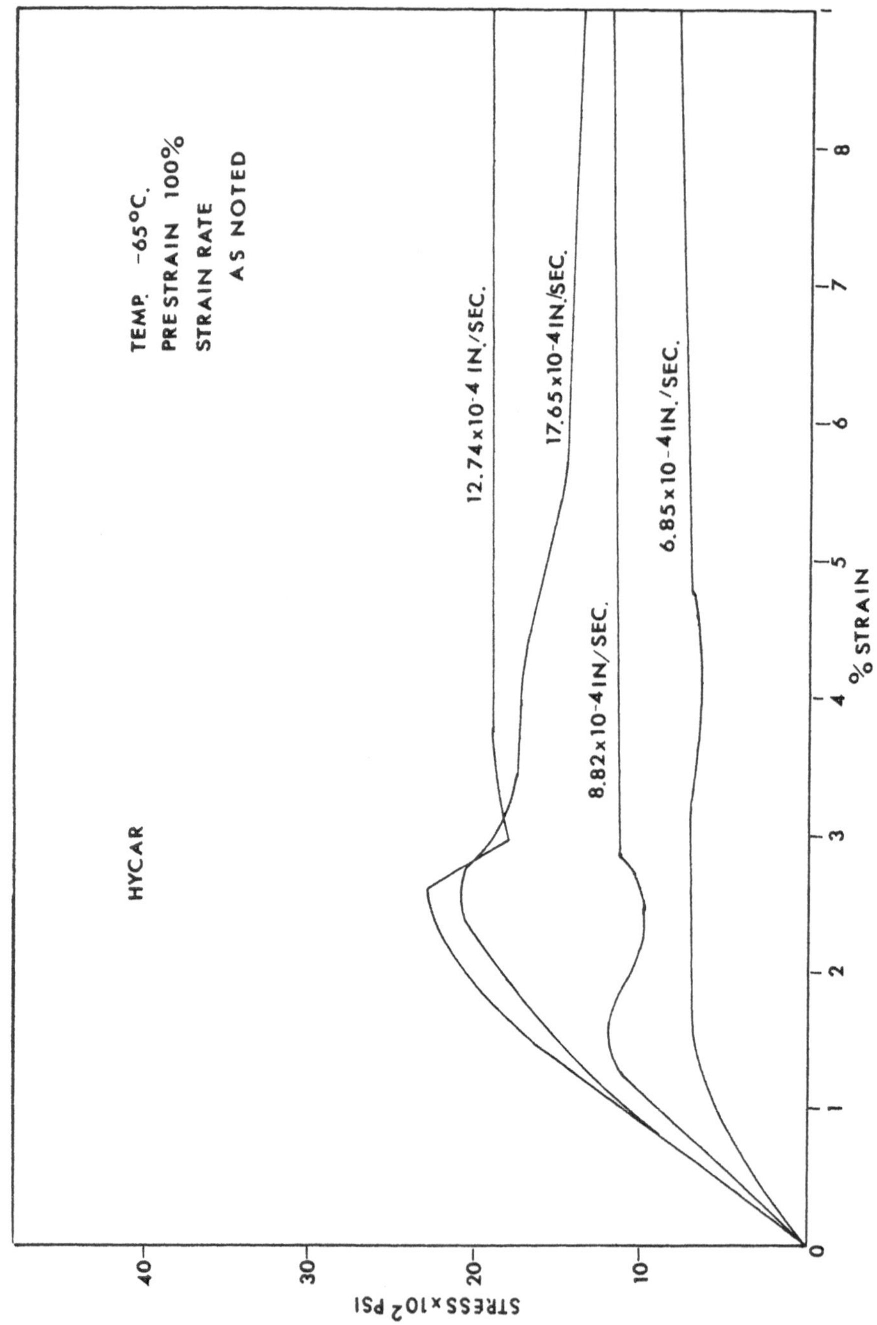

Figure 10. Stress vs. strain for prestrained Hycar at various elongation rates.

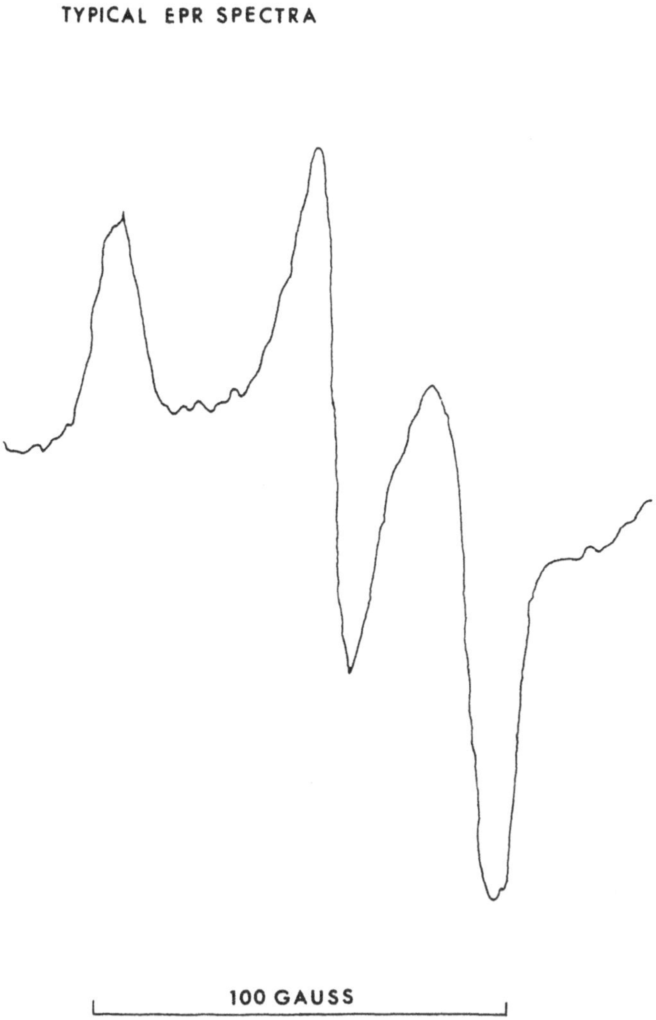

Figure 11. Typical free radical spectra.

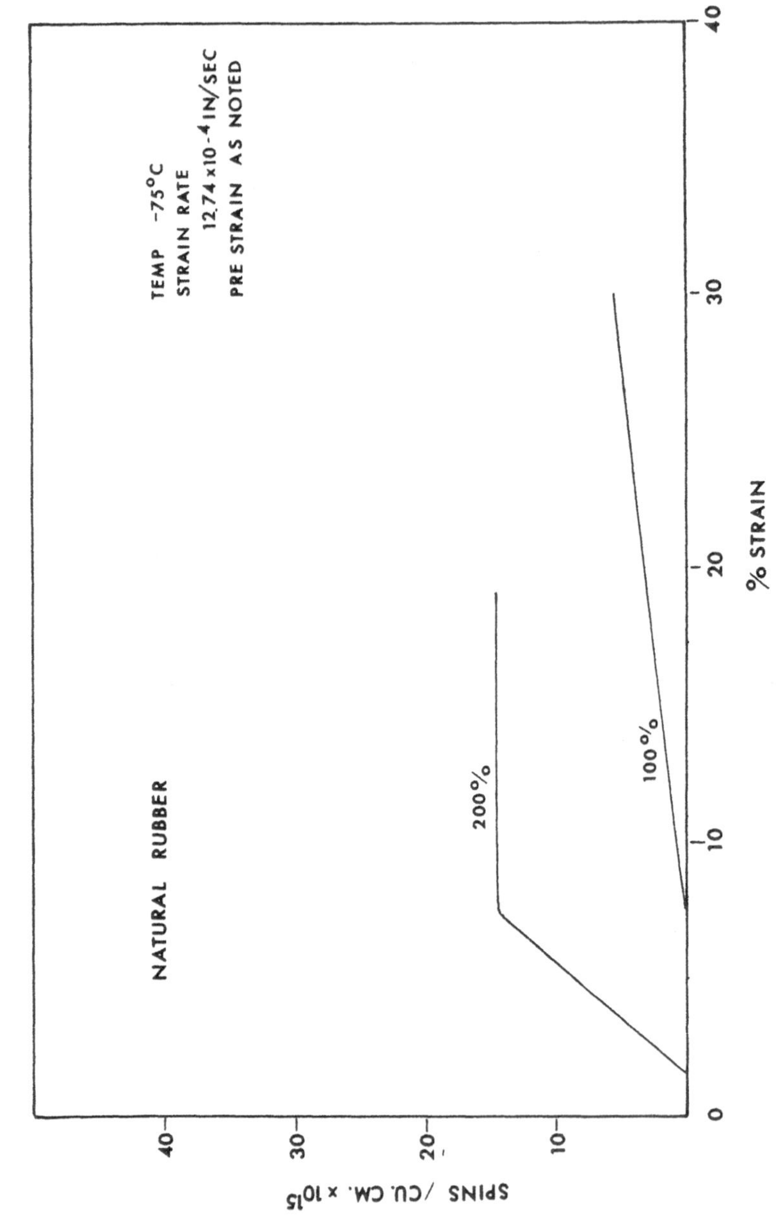

Figure 12. Effect of prestrain on free radical concentration for natural rubber.

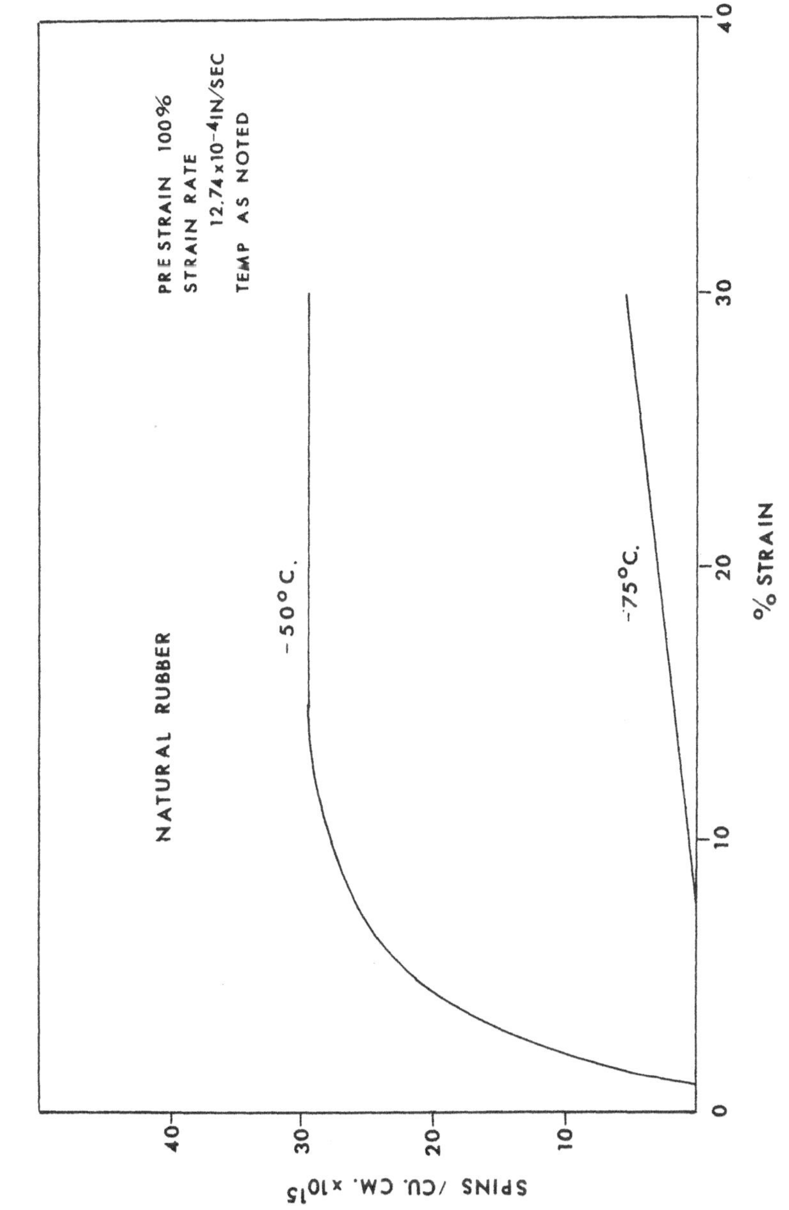

Figure 13. Effect of temperature on free radical concentration for natural rubber.

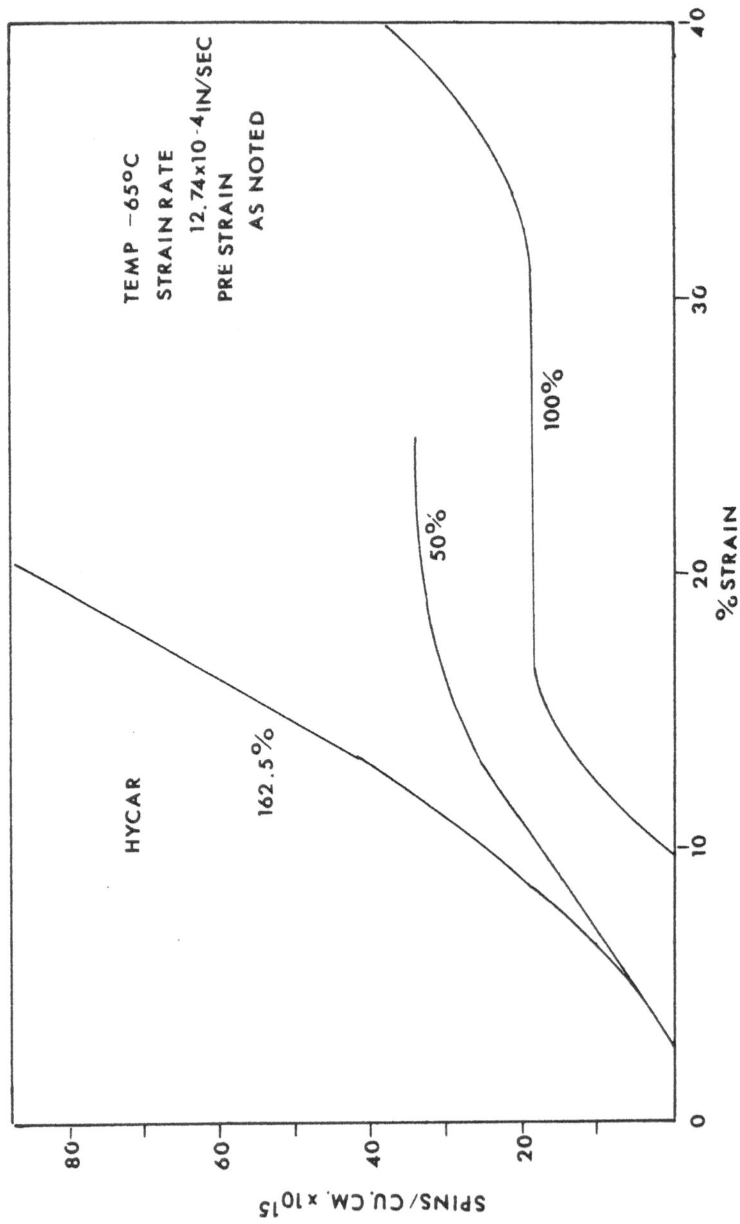

Figure 14. Effect of prestrain on free radical concentration for Hycar.

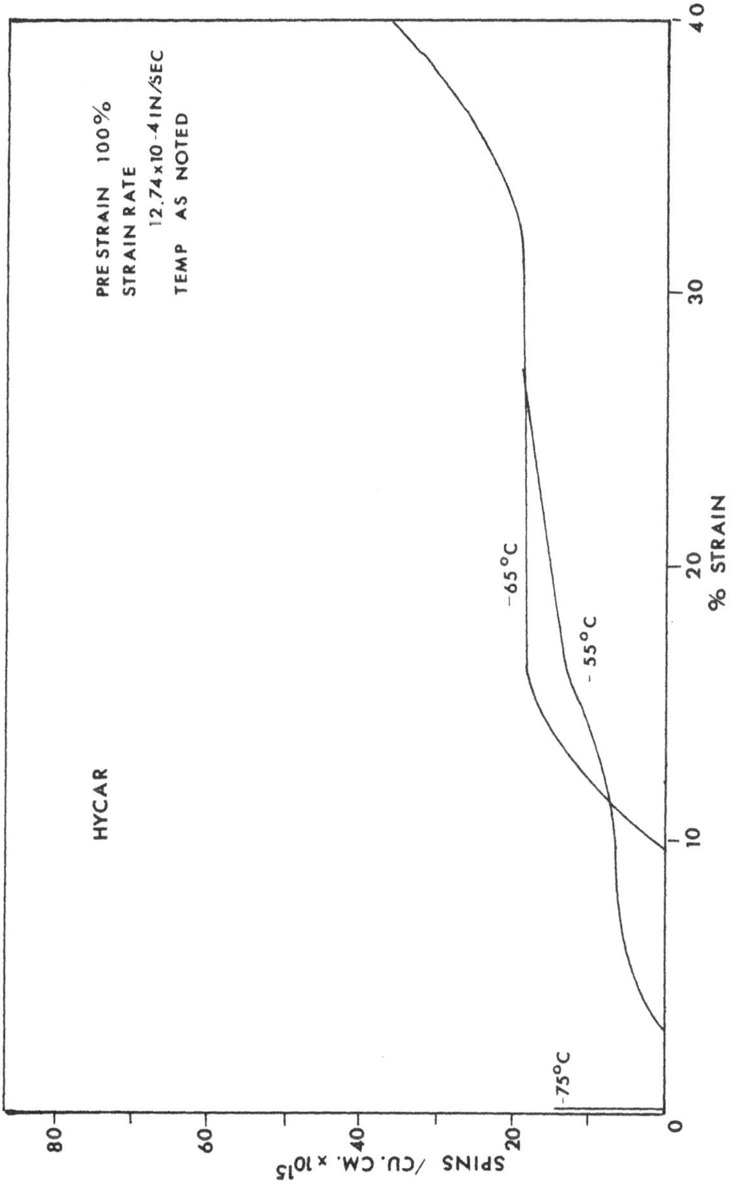

Figure 15. Effect of temperature on free radical concentration for Hycar.

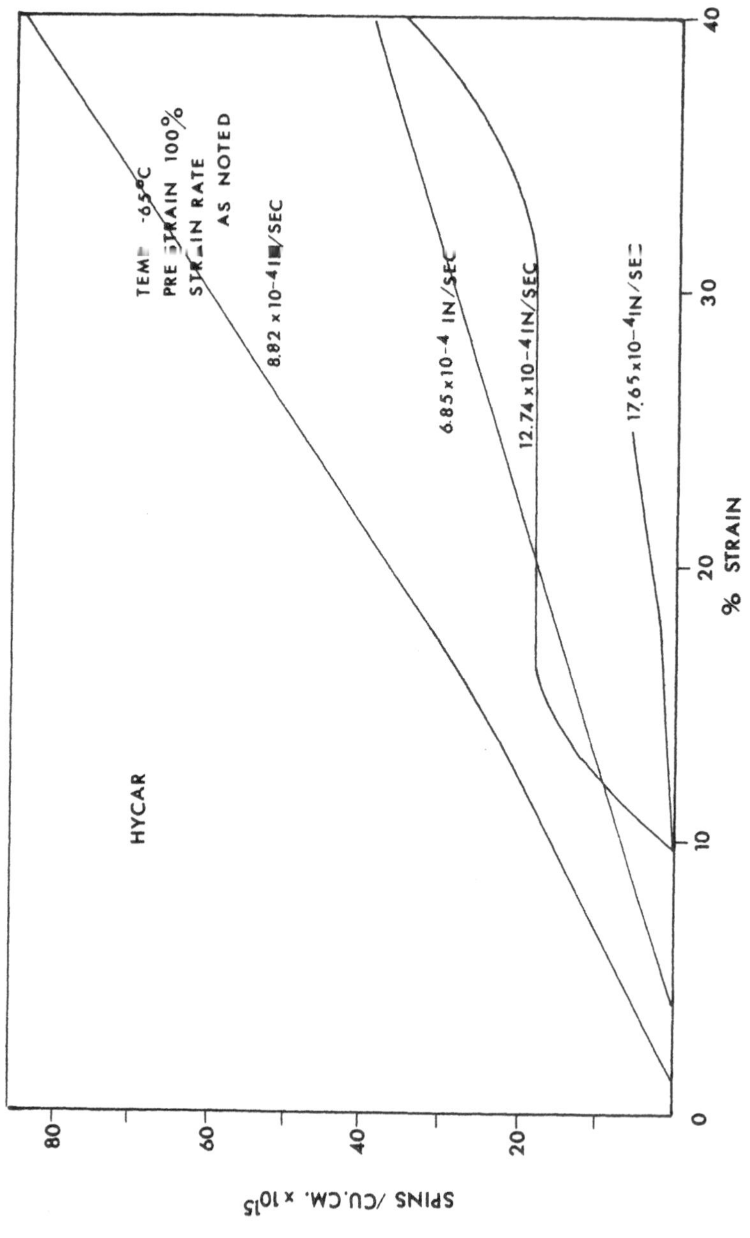

Figure 16. Effect of elongation rate on free radical concentration for Hycar.

DISCUSSION

We think the increasing ductility of the prestrained rubber can be attributed directly to the strain and low temperature induced increase in crystallinity and preferred orientation. As discussed earlier, stress crystallization of natural rubber is well established. Many authors regard co-polymerized rubbers such as Hycar as non-crystallizing. However, some authorities have found evidence of mechanical effects presumably caused by crystallization in Hycar and SBR. Mooney and Wolstenholme [24] found stress relaxations in torsion tests at low temperatures which they attributed to the formation of crystalline regions and King [25] observed the lowering of the brittle temperature in bend brittle tests of Hycar OS-20 which he related to crystallization. Beu and associates [26] used X-ray diffraction methods to show that at certain percentages of co-polymerization crystallization and preferred orientation of the polymer chains is possible.

X-ray diffraction photographs taken by this laboratory (see Figures 17 and 18) at low temperatures and at various stretch ratios show the Debye and Scherrer rings characteristic of crystallites and the transition to elongated arcs indicates a preferred orientation.

Past experience in our laboratory on highly orientated semi-crystalline polymers [11] has shown that fracture in these materials is accompanied by a large amount of free radical production (bond rupture). This has been attributed to the production of a great many "microcracks" by Becht, DeVries and Kausch [27]. These microcracks apparently originate at a great many sites in the polymer, probably in the more amorphous regions. The growth of these cracks could be arrested by the more ordered or crystalline regions. As the stress is increased, more and more microcracks become active with the associated bond rupture and free radical production, detectable by EPR techniques, until one or more attain a critical size and macroscopic failure ensues. It is proposed that something similar is occurring in the rubber samples investigated. In the prestrained rubbers at low temperatures a great many small ordered regions are produced that are comparatively impermeable to cracks. The restrictions imposed by cross-linking, etc., in the rubbers prevents these crystalline regions from becoming very large but our evidence seems to indicate that such regions are produced in very great numbers. In essence, if somewhat oversimplified, the low temperature ductility of the prestrained rubbers might be explained in terms of the temperature and strain treatment converting the polymer to an "orientated semi-crystalline polymer." More work is required before the polymer morphology can be defined or described with any completeness. These studies, as well as an investigation of the feasibility of using similar techniques to extend the ductility in highly filled rubbers, are being planned.

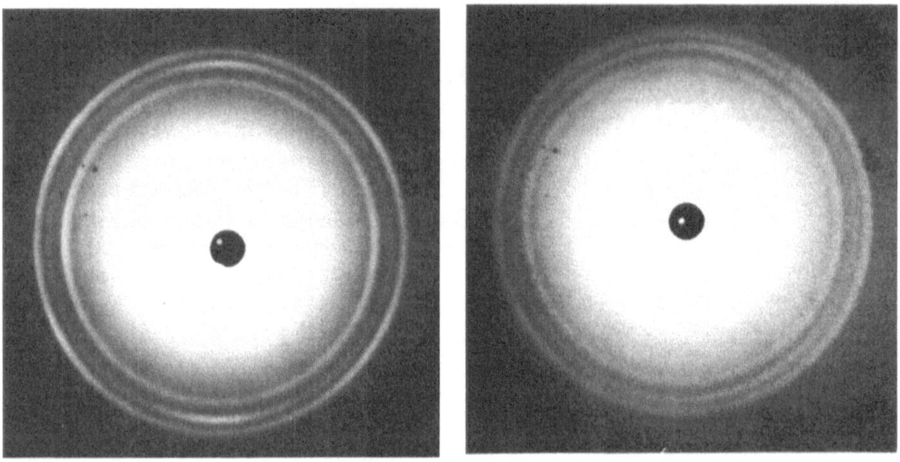

Figure 17. X-ray diffraction picture of natural rubber at -100°C. Left is unstrained; right with 300% strain; direction of strain parallel to long axis of page.

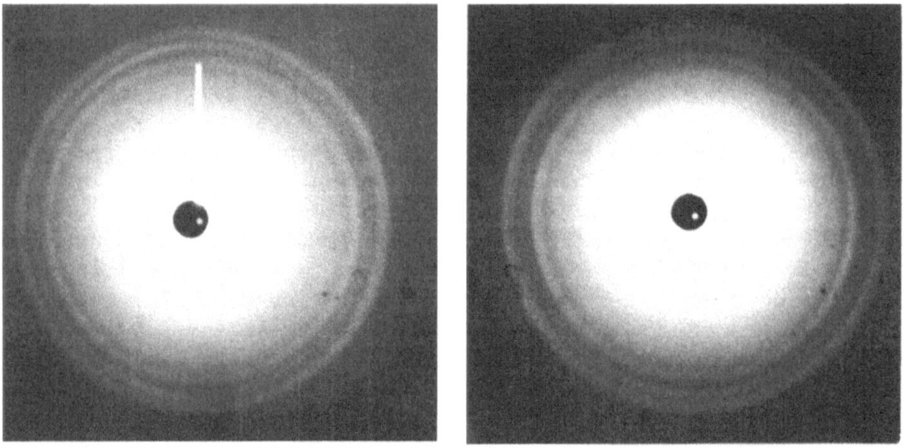

Figure 18. X-ray diffraction picture of Hycar at -100°C. Left is unstrained, right with 300% strain; direction of strain parallel to long axis of page.

NOTE

The reader may notice a difference between the temperatures reported in this paper and in *Polymer Preprints*. The difference is due to a malfunction of the variable temperature accessory gage. The temperatures reported here are correct to 5°C.

ACKNOWLEDGMENTS

Major portions of this research were supported by the National Science Foundation and the National Aeronautics and Space Administration.

REFERENCES

1. E. H. Andrews and P. E. Reed, Polymer Letters 5, 317 (1967).
2. Personal Communication.
3. D. K. Roylance, K. L. DeVries and M. L. Williams, Fracture 1969, Chapman & Hall, London, 1969, p. 551.
4. R. R. Despain, K. L. DeVries, R. D. Luntz and M. L. Williams, "Microscopic Degradation in Teeth," To be published in J. Dental Res. (1970).
5. K. L. DeVries, E. R. Simonson and M. L. Williams, J. Basic Engr. 91, 587 (1969).
6. S. N. Zhurkov, Int. J. Frac. Mech. 1, 311 (1965).
7. S. N. Zhurkov, A. Y. Savostin, E. E. Tomashevskii, Dokl. Akad. Navk. SSSR 195, 707 (1969).
8. D. Campbell and A. Peterlin, J. Poly. Sci. B 6, 481 (1968).
9. H. H. Kausch and J. Becht, Rheologica Acta 9, 137 (1970).
10. M. Baird and J. Bersohn, Electron Paramagnetic Resonance, W. A. Benjamin, Inc., New York (1966).
11. K. L. DeVries, D. K. Roylance and M. L. Williams, J. Poly. Sci. A-1, 8, 237 (1970).
12. F. A. McClintock and A. S. Argon, Mech. Behavior of Matls., Addison-Wesley, Inc., Reading, Mass. (1966).
13. L. R. G. Treloar, The Physics of Rubber Elasticity, Oxford Press (1958).

14. R. Dickie and T. Smith, J. Poly. Sci. 7, 635 (1969).

15. E. M. Bartenev and Y. S. Zuyev, Strength and Failure of Visco-elastic Materials, Pergamon Press, New York (1969).

16. A. F. Joffe, Zhrfkho Chast'fizicheskay, 56, No. 5-6, 489 (1924).

17. R. F. Boyer and R. S. Spencer, Advances in Colloid Science II, Interscience Inc., New York (1946).

18. R. E. Morris, R. R. James and T. A. Werkenthin, Ind. Eng. Chem. 35, 864 (1943).

19. A. M. Borders and R. D. June, Ind. Eng. Chem. 38, 1066 (1949).

20. J. M. Goppel, Appl. Sci. Res. A1, 3 (1949).

21. L. A. Wood, Advances in Colloid Science II, Interscience Inc., New York (1946).

22. E. H. Andrews and A. N. Gent, The Chemistry and Physics of Rubber-Like Substances, Garden City Press, Ltd., Letchworth, England 225 (1963).

23. L. E. Alexander, X-Ray Diffraction Methods in Polymer Science, Wiley & Sons, Inc., New York (1969).

24. M. Mooney and W. E. Wolstenholme, Ind. Eng. Chem. 44, 59 (1953).

25. G. E. King, Ind. Eng. Chem. 35, 9 (1943).

26. E. E. Beu, W. B. Reynolds, C. F. Fryling and H. L. McMurray, J. Poly. Sci. 3, 465 (1948).

27. J. Becht, K. L. DeVries and H. H. Kausch, "On Some Aspects of Strength of Fibers," To be published in the European J. Poly. Sci. (1970).

DISCUSSION

J. J. Bikerman (Cleveland, Ohio):

The application of electron paramagnetic resonance to the problem of fracture appears fascinating to me. My question is Can this method be rendered quantitative? Can we find out how many chemical bonds are broken during rupture? Can we say what percentage of the total work spent on breaking the sample is used up on this bond rupture?

K. L. DeVries:

EPR can be quantitative. By computer double integration of spectra such as figure 11 and comparison with suitable standards, the number of free radicals (broken bonds) can be ascertained. Some of the papers referenced do this for rubber, nylon and tooth dentins. Peterlin, Zhurkov et al. and Becht have also made quantitative measures of bond rupture. In answer to your second question, our experimental evidence (J. Polym. Sci. A-1, 7, 2125 (1969) indicates only a small fraction of the total work of fracture can be attributed to primary bond rupture. The remainder may be attributed to secondary bond rupture, plastic and visco-elastic deformation somewhat analogous to work of plastic deformation at the tip of a crack in metals.

A. Peterlin (Research Triangle Institute, N. C.):

The surface work to break enormously depends on polymer morphology. It is high in amorphous polymers particularly in the rubbery state where it may be up to 10^4 times larger than the surface energy of the newly formed fracture surfaces. The reason for that is the huge amount of plastic deformation work in front of the crack tip propagating through the sample. In glassy polymers the deformed material shows up in a permanently modified surface layer at the fracture surface. Its thickness may be up to 1μ or even more. The layer exhibits high orientation and a lower density than the undeformed material. It actually consists of loosely packed microfibrils pulled out of the bulk polymer. The crazing as this prefracture phenomenon is called is supposed to be a general feature of polymer fracture. Since the forces for plastic deformation are of the same order of magnitude as the intermolecular cohesive forces showing up as surface energy of the solid one may conclude that the difference between the work density per unit area of new fracture surface and the surface energy is a consequence of different length of path over which the forces do work. In the case of surface energy it is the length needed for physical separation of adjacent molecules, i.e., about 1Å or even less. In the crazing phenomena the effective displacement may be for instance 1 micron, i.e., 10^4 Å, thus yielding about 10^4 times larger work than the surface energy of the same area.

K. L. DeVries:

We agree with Dr. Peterlin. Most of the work of fracture in polymers goes into plastic or other dissipative mechanisms at the tip of the crack and not into rupture of primary bonds.

MORPHOLOGY AND MECHANICAL BEHAVIOR OF INTERPENETRATING POLYMER NETWORKS

L. H. Sperling, Volker Huelck, and D. A. Thomas

Materials Research Center, Lehigh University

Bethlehem, Pennsylvania 18015

SUMMARY

The synthesis and morphology of IPN's are compared to the several other methods of preparing blends of distinguishable polymer pairs. Both components of IPN's are continuous throughout, the very finely divided phase domain dimensions being controlled by the crosslink density.

When the elastic phase predominates, the IPN's behave as reinforced elastomers. As the glassy component is increased, the material becomes an impact resistant plastic. At mid-range compositions between the two glass transition temperatures, materials exhibiting leathery behavior are obtained.

As the compatibility of the two IPN components is increased, the glass transition behavior changes from two distinct transitions to one broad transition in a systematic manner. It is concluded that a broadened transition can result, even in thermodynamically compatible mixes, if the minimum volume required for independent contribution to the relaxation spectrum is subject to wide composition variation.

Introduction

There are several interesting ways of mixing two types of polymer molecules so that intimate dispersions result. The simplest involves the mechanical mixing of the two polymer melts, called mechanical blending. A refinement of this technique consists of dissolving one polymer in a second monomer that is subsequently polymerized, ususally with stirring. This type of polyblend usually contains a good deal of grafting and the resulting morphology is more complex and phase regions are usually smaller (ca. 0.1-1.0μ) than corresponding regions in mechanical polyblends. (1,2) Another important way of obtaining very finely divided phase regions involves block copolymerization. (3) In this case the dimensions of the domains are restricted by the segment length, resulting in spheres, rods, or plates having diameters or thicknesses usually of the order of 1000Å. (2,4)

The Interpenetrating Polymer Networks, (5,6,7) IPN's, may be viewed as being most closely related to the graft type polyblend. IPN's were first synthesized by Millar. (8) Both components of his IPN's were polystyrene based. When one elastomeric component and one plastic component were employed, a novel type of polyblend appeared. IPN's are synthesized by swelling a crosslinked polymer with a second monomer, complete with its own crosslinking agent and activator. After polymerization <u>in situ</u> of the second component, two continuous networks exist throughout the macroscopic material.* A greater or lesser amount of grafting may develop between the two components, depending on composition and mode of preparation. The dimensions of the phase domains are controlled by distances between crosslink sites of the individual polymers, the domain dimensions being of the order of 500Å. The present IPN's should be distinguished from the Interpenetrating Elastomeric Networks, IEN's, of Frish (10, 11, 12), which are prepared by mixing and coagulating two types of polymer latices, followed by crosslinking. Such IEN's are another novel type of polyblend, with interpenetration at the particle interfaces. Another type of IPN, termed Simultaneous Interpenetrating Networks, SIN's, may be prepared by simultaneously polymerizing two networks via independent reactions in the same container. An example of this last might involve a condensation polymerization simultaneously with an addition reaction. The SIN's, which also may have very finely divided phase domains, are the subject of current exploratory studies. (13) The morphologies produced by mixing two different types of polymer molecules via these several methods are depicted schematically in Figure 1.

Polymers are generally (but not always!) incompatible with each other, and mixing them by any of the above methods usually results in two phases. Polymer incompatibility

---

*An interesting analog of an IPN is given by George Gamow in his book "One, Two, Three, ..Infinity" where two worms eat out independent tunnels in an apple.(9)

arises because of the small entropy of mixing gained by dissolving long polymer chains in each other (14). However, simple equilibrium theory yields the conclusion that even the most incompatible polymer pairs must exhibit some degree of molecular mixing. Under favorable heat of mixing conditions, semi compatibility or possibly true mutual solutions have been obtained. (15,16)

The physical and mechanical behavior of polymer blends depends on the detailed morphology of the material: Domain size, shape and fine structure, and degree of molecular mixing. For instance, and incompatible polymer pair exhibits two glass rubber transitions, one for each polymer. After a certain degree of compatibility is reached, only one transition is observed.

The major objectives of this paper are threefold (1) to review the current status of IPN's, (2) to document the morphology and physical behavior of IPN's of various degrees of compatibility, and (3) present mechanical data illustrating the degree of reinforcement obtained for certain compositions.

Current Status of IPN's

In the first report on IPN's from this laboratory, we presented the modulus-composition and modulus-temperature behavior of polystyrene (PS)-poly(ethylacrylate) (PEA) IPN's(5). For midrange compositions, two distinct glass transitions occur at -16°C. and 75°C. Although this system is usually considered incompatible, the marked deviation from the homopolymer transition temperatures of -22°C. and +100°C. for PEA and PS, respectively, was taken to indicate a certain degree of molecular mixing. Whether or not the system was at thermodynamic equilibrium was difficult to determine, but demixing was obviously hindered.

In a second paper, we reported the transition behavior of the isomeric IPN pair PEA-PMMA. For all compositions, only one extraordinarily broad transition was found, as shown in Figure 2. The unusual breadth of the transition was confirmed via dilatometry. The question of compatibility was raised at that time. Other workers had predicted that while incompatible blends yield two sharp transitions, truly compatible blends should yield one sharp transition. (17). A single broad transition was taken to indicate semicompatibility, where a smear of composition ranges coexisted. (17) This last aspect of the problem was explored via a series of creep measurements presented in a third paper. (7) The Creep

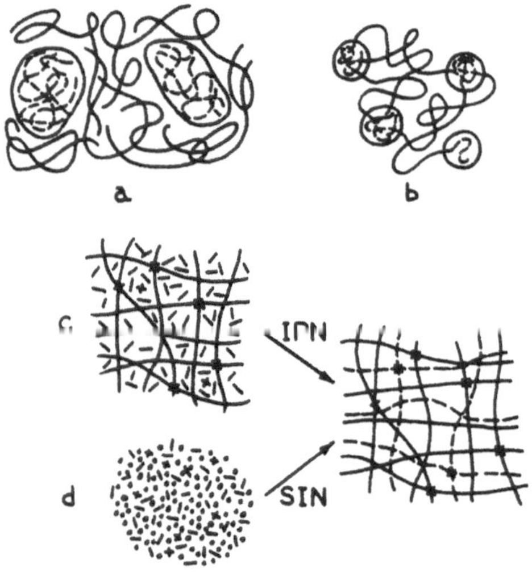

Figure 1. Schematic illustrations of molecular and phase domain morphologies in various types of polyblends.

PolymerA: – – – –
Polymer B: ————
Crosslinks:

(a) Mechanical blend. Particles of A dispersed in B.

(b) A-B-A block copolymer. Blocks of B between domains formed by end blocks of A.

(c) and (d) IPN's.
In (c), monomer A and crosslinking agent swelled into crosslinked polymer B. A is polymerized to produce IPN.
In (d), monomers A and B and crosslinking agents are mixed and simultaneously polymerized to produce similar result, designated SIN.

Master Curve relaxation presented in Figure 3 covers 22 decades of time, compared to the normal nine decades covered by simple homopolymers. We attempted to fit these data with a modified Tobolsky-Aklonis-Dupre formula

$$E_r(t) = E_1 \sum_{i=1}^{n} \frac{w_i \tau_i^{1/2}}{\tau_i^{1/2} + t^{1/2}} + R_2 \quad (1)$$

where $E_r(t)$ represents the time-dependent relaxation modulus, E_1 the glassy modulus (3×10^{10} dynes/cm^2), τ_i the shortest relaxation time of composition weight fraction W_i, t is time, and R_2 the rubbery modulus for crosslinked systems and is a constant having the value of approximately 4×10^7 dynes/cm^2 for the present polymerization formula. This equation, which allows the assumption of a broad range of compositions, fits the data better than the simple assumption of complete and intimate solution.

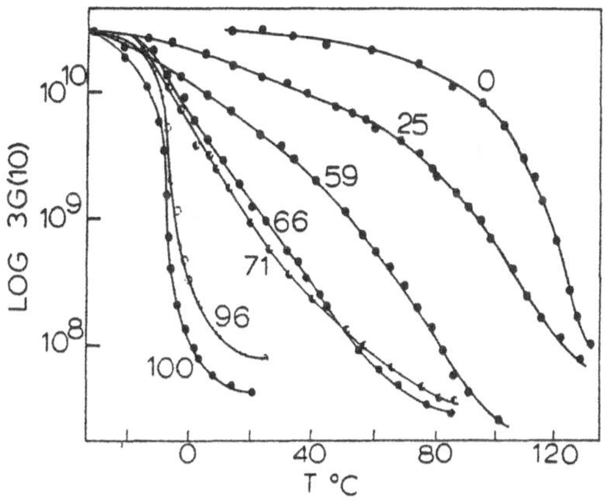

Figure 2. Log modulus vs. temperature for PEA/PMMA IPN's. Numerical values indicate wt-% PEA. All compositions were found to exhibit only one transition region. Ref. 6.

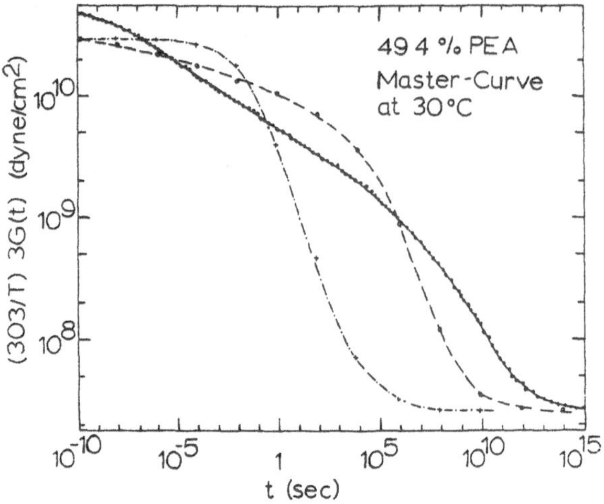

Figure 3. —●●●●— Experimental points.

-⊖-⊖-⊖- Equation (1), assuming a random distribution of compositions.

+···+·· single relaxation time assumption. (Original T-A-D theory.) The observed transition covers 22 orders of magnitude of time. Ref. 7.

However, a more subtle discrimination must be attempted. The original Rouse-Bueche theory requires approximately 50 mers to undergo coordinated motion for the glass transition relaxation to occur. A volume of approximately 10,000Å^3 is involved. Due to the size of polymer chains, however, only much larger regions, perhaps 100,000 Å^3, can be expected to have a mean composition the same as the bulk polyblend. Possible composition variations within a 10,000 Å^3 region are schematically illustrated in Figure 4. The point is, that a broadened transition will result if the minimum volume required for independent contribution to the relaxation spectrum is subject to wide composition variation, even in thermodynamically compatible polymer mixes. If this argument is valid, one cannot decide whether any polymer pair is compatible or not by observing its glass transition behavior. However, a single broad transition does say that it is either compatible or semicompatible.

Sample Preparation and Instrumentation

A series of IPN's were prepared by swelling techniques

previously described.(5,6,7) In brief, monomer solution one, containing tetraethylene glycol dimethacrylate, TEGDM, for crosslinking and benzoin for activation was polymerized photochemically for 24 hours. The recipe was 100 ml monomer, 2 ml TEGDM, and 0.3 gm benzoin. After vacuum drying to remove traces of remaining monomer, controlled quantities of monomer solution two (same general recipe) was swollen in, stored in a dessicator until homogeneous (1-3 days), and followed by a second photopolymerization and vacuum drying.

For most of the IPN's, ethyl acrylate served as monomer one, and various admixtures of styrene and methyl methacrylate served as monomer mix two. The random copolymers of styrene and methyl methacrylate formed exhibit the striking property of having essentially the same glass transition temperature, ca. 100°C., over the complete composition range.

For the electron microscopy studies, 1% by weight of butadiene was added to the ethyl acrylate solution, to enhance osmium tetroxide staining in the electron microscope portion of this investigation. Also, only 0.5 ml TEGDM per 100 ml monomer was employed for these samples.

To produce contrast between phase domain for electron microscopy the osmium tetroxide technique of Kato (1, 18) was employed. Samples were exposed to OsO_4 vapor for one week. during which time they darkened. The samples were cut to a thickness of not more than ca. 600 Å employing a Porter-Blum MT-2 ultramicrotome equipped with a diamond knife. For actual cutting operations, the samples were embedded in an epoxy resin to ensure a rigidity. An RCA EMU-3G was employed for all microscopy work.

Three times the shear modulus at 10 seconds, 3G(10), was obtained on selected samples from ca. -50°C. to + 150°C. employing Gehman torsional instrumentation. In all cases silicone oil baths were employed, the heating rate being approximately 1°C per minute. These instruments utilize samples having dimensions of approximately 4 cm x 8mm x 2mm.

IPN Morphology

One may speculate about the actual three dimensional morphology of incompatible IPN's. Where one component predominates, the minor component may be concentrated in islands, with threads of only one or a few molecules connecting them. Midrange compositions may be envisioned as intertwining rivers, containing many interconnecting branches.

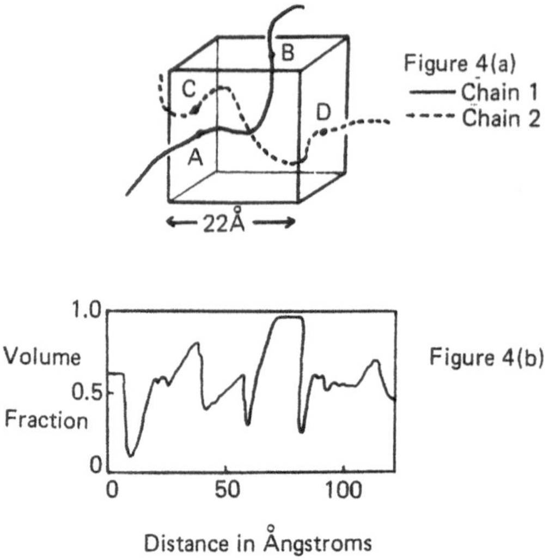

Figure 4. (a). Portions of chains 1 and 2 between points A and B, and C and D, respectively contain ca. 20-25 mers each, more or less filling the 10,000 $Å^3$ box. (Other chains are permitted in unfilled spaces.) If chains 1 and 2 are the same or different polymers because of random effects, quite different relaxation behavior in said volume would be expected.

(b). Volume fraction of type 1 chains expected in a 50/50 compatible IPN vs distance through sample. Fifty or more Angstroms (≥100,000 $Å^3$) will be required to iron out random concentration fluctuations.

However, in all cases both phases are thought to be continuous throughout, with domain sizes limited by the crosslink densities. These speculations are fortunately subject to experimentation via electron microscopy.

In Figure 5 an electron micrograph of a midrange PEA (butadiene doped)/PS incompatible IPN composition is compared with a commercial graft-type polyblend, HiPS, supplied by Monsanto Company, Springfield, Mass. HiPS contains polybutadiene (stained black with osmium tetroxide) dispersed in PS.

Surprisingly, the IPN exhibits a complex cellular structure. Similar cellular structures are evident in the polybutadiene portion of the HiPS, and in graft-type polyblends

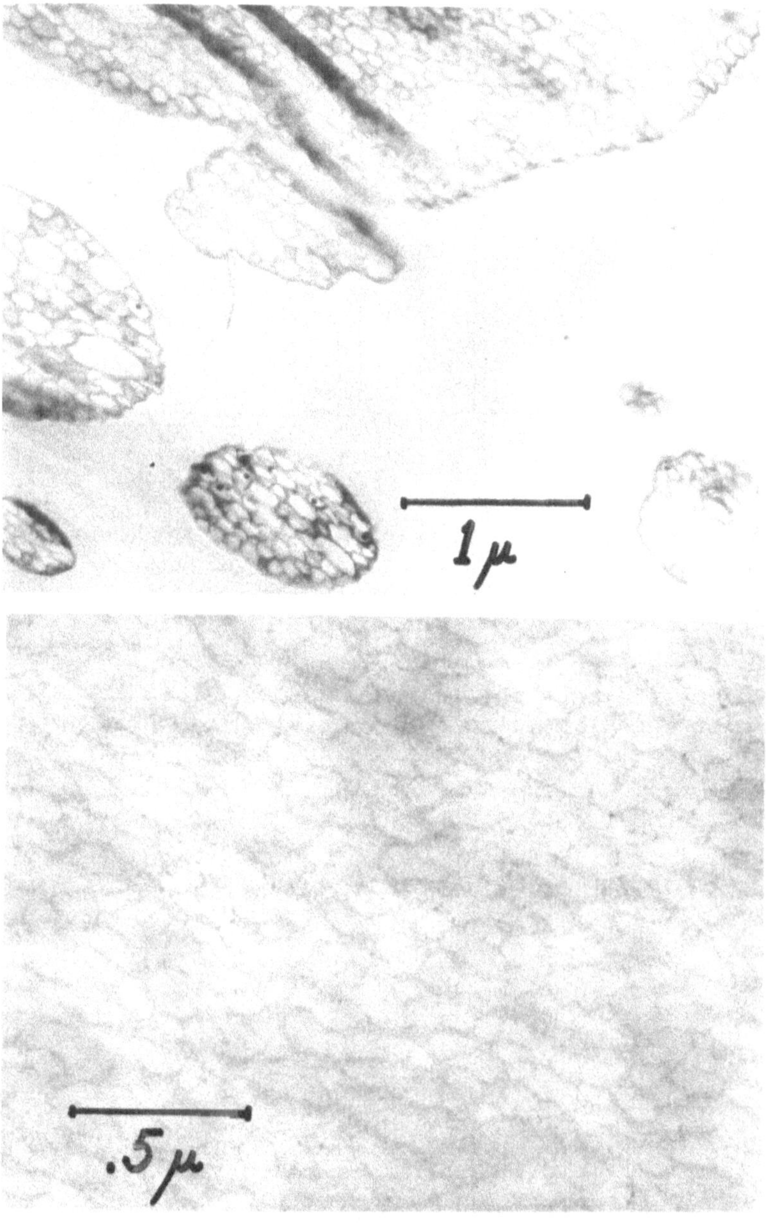

Figure 5. Graft type polyblend, HiPS, (top); IPN of poly(ethyl acrylate) containing ca. 1% butadiene monomer with polystyrene (48.8 PEA/51.2 PS), (bottom). Note similarity of cellular structures.

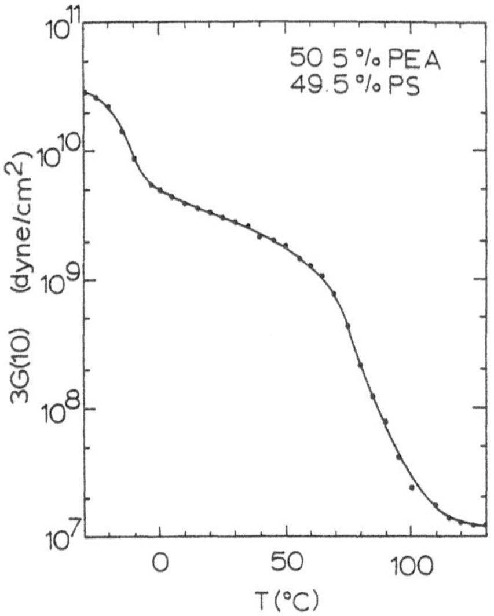

Figure 6. 50.5 PEA/49.5 PS.

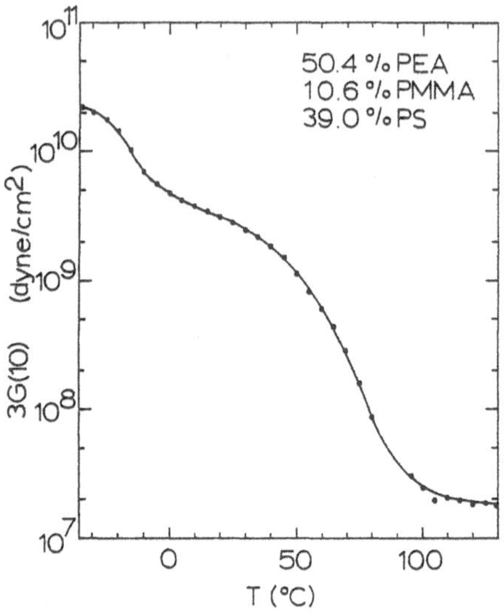

Figure 7. 50.4 PEA/poly (styrene 39.0-co-methyl-methacrylate 10.6).

Figure 6-10. The systematic replacement of styrene mers by methyl methacrylate mers does not change the glass transition temperature of the plastic component. However, the increased compatibility causes the two transitions to merge into one broad transition.

prepared by bulk polymerization techniques without stirring (19, 20). Pervading the IPN cellular structure, but on a much finer scale, interconnected spheroids may be observed in both phases. (This latter feature is not clearly visible in Figure 5; details will be published separately.)

Polyblends are normally prepared with linear polymer structures. The fine structure mentioned above undoubtedly results from the imposition of crosslinks in both components.

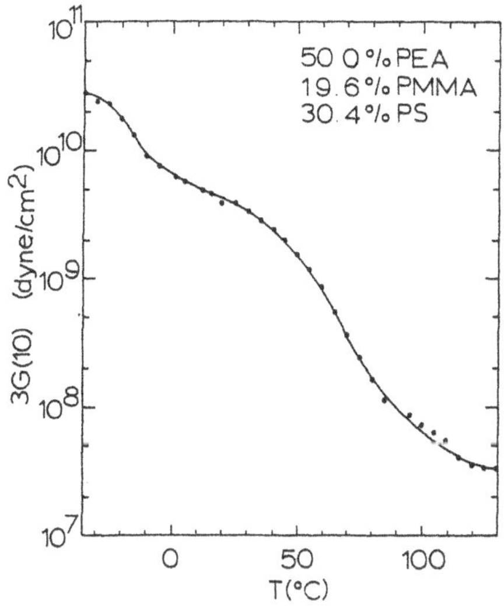

Figure 8. 50.0 PEA/poly (styrene 30.4-co-methyl methacrylate 19.6).

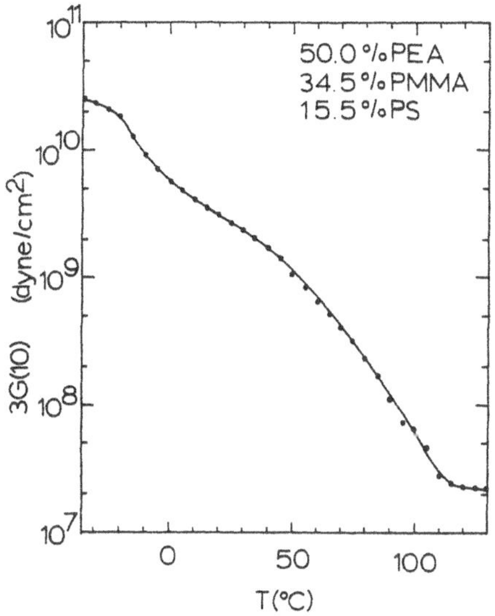

Figure 9. 50.0 PEA/poly (styrene 15.5-co-methyl methacrylate 34.5).

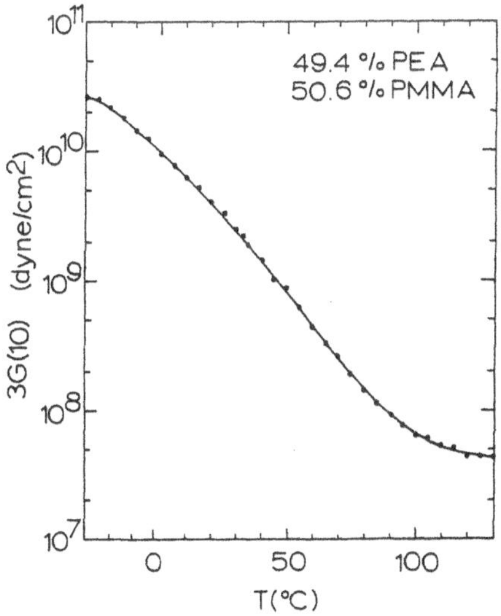

Figure 10. 49.4 PEA/50.6 PMMA. This broad transition probably indicates semicompatibility, but independent experiments will be required to verify this point, as discussed in the text.

Transition Behavior of 50/50 IPN's of Poly(ethyl acrylate) and Copolymers of Styrene and Methyl Methacrylate

A series of IPN compositions were made in the following manner: First, crosslinked PEA elastomer was prepared as component one. Monomer mixes of styrene and methyl methacrylate (containing appropriate TEGDM and benzoin), varying systematically from 0% to 100% methyl methacrylate content were prepared as component two. These solutions were swollen into component one so as to produce 50/50 w/w IPN's via procedures discussed above. Since the diffusion rates of styrene and methyl methacrylate monomers are not identical actual compositions of component two were calculated from elemental analyses on the whole material provided by Dr. George Robertson, Florham Park, N. J.

Shear modulus data obtained as a function of temperature were converted to Young's modulus equivalents through $E \cong 3G$. Values of 3G(10) vs temperature are shown in Figures 6-10. The PEA-PS material shows two quite distinct transitions. This is in accord with previous observations. (5) As methyl methacrylate mers replace styrene mers, the heat of mixing becomes more favorable to mutual solubility, since PEA and PMMA are chemically isomeric, as mentioned previously. This is reflected in the systematic merging of the two distinct transitions into one broad transition.

As with all other IPN's, the modulus becomes roughly constant (actually following $G = nRT$) at temperatures beyond the glass transition zone. This, of course, is a result of both components being crosslinked. Above ca. 125°C., these materials behave as ordinary non-reinforced elastomers. The actual modulus in the rubbery plateau region, ca. 4×10^7 dynes/cm^2, was roughly the same for homopolymers and IPN's.

Mechanical Properties of IPN's

IPN's prepared from an elastomer and a plastic material exhibit wide ranges of mechanical behavior, depending upon composition. If component one is elastomeric, addition of moderate quantities of plastic component two causes the material to become reinforced. This behavior is quite similar to the case of the thermoplastic elastomers, where the minor plastic component of these block copolymers forms spherical domains.(3) As pointed out by Kraus (21), Morton (22) and coworkers quite recently, the use of block copolymers with chemical bonds joining phases is not necessary for reinforcement. (21) All that is required is a very fine dispersion in a crosslinked elastomeric network.

As shown in Table I, PEA-PS IPN's show increased tensile strengths over simple crosslinked PEA, however an IPN where both components consist of PEA shows little if any improvement, thus the plastic phase is required.

As the plastic component is increased further, an inversion of the major and minor phases occurs. The room-temperature modulus rapidly increases about two orders of magnitude, (5) and the material becomes an impact resistant plastic. The degree of impact resistance observed in preliminary tests was quite modest, but beyond experimental error, as shown in Table II. These results are averages of several experiments.

To summarize, when the elastic phase predominates, the IPN's behave like reinforced elastomers. As the glassy component is increased the material becomes an impact resistant plastic. At midrange compositions, or at temperatures between the two glass transitions, materials exhibiting leathery behavior are obtained.

The authors wish to thank NSF for support under grant GK-13355. They also wish to thank Mr. M. Covitch for one excellent electron micrograph.

Table I.

IPN Composition	Tensile Strength, psi
(100PEA)	64
86PEA/14PS	176
81PEA/19PS	315
74PEA/26PS	1450
60PEA/50PS	1830
57PEA/43PMMA	2430
53PEA/47PMMA	2680

Table II.

IPN Composition	Impact Strength, ft.- lb./in of notch
(100PS)	0.33
67PS/33PEA	0.40
93PS/7PS*	0.26

*Millar's type of IPN.(9)

References

(1) K. Kato, Japan Plastics, 2, 6 (April, 1968).

(2) M. Matsuo, Japan Plastics, 2, 6 (July, 1968).

(3) G. Holden, E. T. Bishop, and N. R. Legge, J. Polym. Sci., 26C, 37 (1969).

(4) D. J. Meier, J. Polym. Sci., 26C, 81 (1969).

(5) L. H. Sperling and D. W. Friedman, J. Polym. Sci., A-2, 7, 425 (1969).

(6) L. H. Sperling, D. W. Taylor, M. L. Kirkpatrick, H. F. George, and D. R. Bardman, J. Appl. Polym. Sci., 14, 73 (1970).

(7) L. H. Sperling, H. F. George, V. Huelck, and D. A. Thomas, Accepted, J. Appl. Polym. Sci.

(8) J. R. Millar, J. Chem. Soc., 1311 (1960).

(9) G. Gamow, "One, Two, Three...Infinity" Bantam, 6th Edition, 1967, p. 56.

(10) H. L. Frisch, D. Klempner, and K. C. Frisch, J. Polym. Sci., B 7, 775 (1969).

(11) D. Klempner, H. L. Frisch, and K. C. Frisch, J. Polym. Sci., A-2, 8, 921 (1970).

(12) H. L. Frisch, and D. Klempner, This Book.

(13) L. H. Sperling and Robert Arnts, to be published.

(14) R. L.Scott, J. Chem. Phys., 17, 279 (1949); note especially the section titled "The Two Polymer System".

(15) L. Bohn, Rubber Chem. Tech., 41, 495 (1968); Koll.-Z. u. Z. für Polymere, 213 55 (1966).

(16) J. Stoelting, F. E. Karasz, and W. J. MacKnight, Poly. Eng. & Sci., 10, 133 (1970).

(17) M. Matsuo, C. Nozaki, and Y. Jyo, Polym. Eng. & Sci., 9, 197 (1969).

(18) K. Kato, J. Polym. Sci., B, 4, 35 (1966).

(19) H. Keskkula and P. A. Traylor, J. Appl. Polym. Sci., 11, 2361 (1967).

(20) E. Molau and H. Keskkula, J. Polym. Sci., A-1, 1595 (1966).

(21) G. Kraus, K. W. Rollmann, and J. T. Gruver, Macromol., 3, 92 (1970).

(22) M. Morton, J. C. Healy, and R. L. Dencour, Proc. Int. Rubber Conf., 1967, Gordon and Breach, 1969. p. 175.

DISCUSSION

A. J. Chompff (Ford Motor Company):

Can you give a specific example of a simultaneous interpenetrating network (SIN).

L. H. Sperling:

A condensation polymerization run simulataneously with an addition polymerization, plus appropriate crosslinking can produce a SIN. Specifically, we employed an epoxy formulation with ethyl acrylate. When a tertiary amine is used to cure the epoxy, there apparently is minimal interference with the simultaneous polymerization of ethyl acrylate.

SYNTHESIS AND PROPERTIES OF INTERPENETRATING POLYMER NETWORKS*

H. L. Frisch
State University of New York at Albany
Chemistry Department
1400 Washington Avenue
Albany, New York, 12207

Daniel Klempner
University of Massachusetts
Polymer Science Department
Amherst, Massachusetts 01002

K. C. Frisch
University of Detroit
Polymer Institute
Detroit, Michigan 48221

T. K. Kwei
Bell Telephone Laboratories
Murray Hill, New Jersey 07974

Supported by American Chemical Society Grant PRF3519C56

SUMMARY

Several partially interpenetrating polymeric networks (IPN) were made from chemically different linear elastomers

One of the physically linked networks was crosslinked by a condensation reaction, the other by a free radical addition mechanism. The IPN's were made by heat curing of films of an aqueous emulsion made by mixing the individual emulsions in various (generally equal) proportions. In two cases, the networks were cleanly separated by hydrolysis of one of the component networks to demonstrate that there was effectively no direct chemical bonding

---

*This contribution was published independently in the following journals: J. Polymer Sci. A2, 8, 921 (1970), and J. Polymer Eng. Sci. 10, 327 (1970).

between the component networks. Measurement of crosslink density showed that, in most cases, partial interpenetration does occur as evidenced by an effective crosslink density of the IPN's greater than the mean of the crosslink densities of the component networks. The swelling ratios, densities and stress-strain properties of these materials were determined. For one of the network combinations, a poly-(urethane-urea), U, and a polyacrylate-(butadiene-acrylonitrile), A, a series of IPN's varying in polymer composition was made. The swelling ratios and densities are close to the arithmetic means; however both the tensile strength and crosslink density exhibit a maximum at about 70% A. The apparent maximum tensile strength is actually significantly higher than that of either of the component polymers. The elongations all approach that of pure U, the more extensible material, except for compositions approaching 100% A, which exhibit a very low extensibility.

The phase relations in these IPN's was studied by determinations of dynamic viscoelastic properties. Two glass transitions (T_g) are found. Electron microscopy confirms that a phase inversion occurs from "U-phase particles in A-phase matrix" to an "A-phase in U-phase matrix" beyond 70% U.

CHEMICAL EFFECTS ON THE ULTIMATE PROPERTIES OF POLYMER NETWORKS IN THE GLASSY STATE

S. S. Labana, S. Newman and A. J. Chompff

Scientific Research Staff, Ford Motor Company

Dearborn, Michigan 48121

SUMMARY

Highly crosslinked polymer networks have been prepared by crosslinking low molecular weight, reactive copolymers. The dependence of the ultimate properties of these networks in the glassy state on the molecular weight of the prepolymer, its functionality, the stoichiometry of the reactants and conversion of the reactive groups has been measured. It is shown that the ultimate properties are remarkably sensitive to slight variations in chemical composition and network topology. This dependence could not be rationalized in terms of network parameters or defects on a molecular scale (as is the case in the rubbery state) but might be explained if incoherent network structures were to form.

A statistical theory is developed based on configurational restrictions, which provides a general mechanism of formation of crosslink density inhomogeneities. The theory predicts that each intramolecular reaction enhances the probability that more intramolecular reactions take place, leading to crosslinked regions which are poorly connected with the matrix or with each other.

Accordingly, carbon replicas of fracture surfaces of the thermosets before and after etching with sulfuric-chromic acid have been prepared. Electron micrographs of these replicas indicate the presence of crosslink density inhomogeneities which are 100 to 150 Å in diameter; a size range correctly predicted by theory. Thus, the present theory provides a basis for an explanation of the sensitivity of the network properties to variations in chemical composition.

INTRODUCTION

Crosslinked polymer networks are frequently considered as homogeneous three-dimensional structures; and, their ultimate properties are generally related to the properties of such a continuum. If polymers formed homogeneous continua their strengths should theoretically be about one hundred times higher than is presently observed, as pointed out by Mark[1] - as early as 1943. This indicates that polymer networks might consist of inhomogeneous network structures which, especially in the glassy state, would lead to brittle behavior. In semicrystalline polymers and composite systems, the importance of such textural features as spherulite boundaries, interface regions and other structural elements has been appreciated.

In most studies of network behavior to be found in the literature, the apparent relations between chemical compositions and properties have been emphasized especially in the rubbery state. It is the purpose of the present paper to show that amorphous, glassy thermosets are not simple solids consisting of a homogeneous continuous network. The present investigation shows that this is an overly simplistic view. Our findings indicate that ultimate properties of glassy amorphous networks exhibit remarkable sensitivity to apparently small changes in molecular topology. An attempt to rationalize variations of tensile strength on the basis of such parameters as the number of chains, chain ends, branch points, etc., is shown to be unsuccessful. For this reason, the authors have sought to identify and determine the importance of supermolecular features which may be influenced by nuances in chemical structure. The work in the present paper has only been partially successful in this direction, but it is felt that it opens a new line of investigation in understanding the behavior of amorphous, glassy thermosets.

EXPERIMENTAL DETAILS

Materials

For the prepartion of networks from reactive prepolymers, commercially available monomers were used as received. Glycidyl methacrylate, GMA, was obtained from the American Aniline and Extract Company.

Crosslinking agents were also commercially available materials and used as received. Bisphenol A (4,4'-isopropylidenediphenol) was obtained from the Shell Chemical Company, the crosslinking agent, 4,4'-methylenedianiline (MDA) was obtained from the Ciba Products Company and a caprolactam blocked triisocyanate (Isonate 123 P) was obtained from the Upjohn Company.

Prepolymer Preparation

A mixture of monomers containing an appropriate amount of 2,2′-azobis(2-propionitrile) (AIBN) was added slowly to an equal weight of refluxing toluene or dioxane. After the addition was complete, the refluxing was continued for one hour and the solution cooled to room temperature. This solution was diluted to about a 25 percent polymer concentration by adding solvent and the prepolymer isolated by coagulation in a large amount of hexane. The fine powder thus obtained was dried under vacuum and characterized for weight per reactive group and molecular weight.

For determination of weight of prepolymer per equivalent of epoxide group, the glycidylmethacrylate copolymers were titrated with perchloric acid in the presence of tetrabutylammonium iodide.[2] Hydroxy equivalents of the copolymers of 2-hydroxyethylmethacrylate were determined by the acetic anhydride-method.[3] Number average molecular weights ($\bar{M}_n$) were determined by vapor phase osmometry (Hitachi-Perkin Elmer Model 115) using 2-butanone as solvent. Weight average molecular weights ($\bar{M}_w$) were obtained from gel permeation chromatography using 2-butanone as the elution solvent. The molecular weight ratio, $\bar{M}_w/\bar{M}_n$, for all prepolymers was in the range of 2.0 to 2.1.

Sample Preparation

The reactive prepolymer and the crosslinking agent were dissolved in acetone and catalyst added. Solvent was evaporated at room temperature under vacuum to form a dry foam which was subsequently powdered and further dried under vacuum at 50°C. The resulting powder was used to obtain cured sheets (1.6 mm thick) by compression molding at 185°C under 7000 dynes/cm^2 pressure.

The extent of the crosslinking reaction was measured by infrared techniques using films of 0.025 mm thickness. Typically, the reaction progressed to 95 $\pm$ 2 percent conversion unless indicated otherwise.

Tensile Testing

Cured sheets were cut into dogbone-shaped specimens with a reduced gauge section of 0.38 cm width, parallel gauge length of 1.3 cm, radius of fillet of 0.95 cm, overall length of 5 cm and width at the ends of 0.8 cm. The specimens were clamped in an Instron Universal testing machine Table Model-S with a jaw separation of 3 cm, extended at a rate of 0.051 cm per minute and the stress-strain curve recorded. By use of an extensometer, the effective gauge length was determined to be twice the parallel gauge length of 1.3 cm. Appropriate corrections were applied to the tensile data presented in this paper. All tensile data were obtained at 25°C. The tensile strength reported was the maximum stress which for the materials described here was equal to the stress-at-break. Average values given represent a minimum of five specimens.

RESULTS

The effect of the molecular weight of the prepolymer (a terpolymer containing 30 wt.% glycidylmethacrylate) on the tensile properties of the networks formed are reported in Table I. The crosslinker used was 4,4′-methylenedianiline (MDA) in an amount to provide 1.25 amine hydrogens for each epoxy group in the prepolymer. Catalyst was used in 1 percent concentration. The average functionality $\langle f_p \rangle$ of the prepolymer chain increases with molecular weight and is included in Table I.

Both tensile strength and elongation-to-break increase with increasing molecular weight of the prepolymer: elongation-to-break increases from 1.4 to 5% and tensile strength from 41.4 x 10^7 to 69.0 x 10^7 dynes/cm² as $\bar{M}_n$ is varied from 1550 to 10,000. Above an $\bar{M}_n$ of 4000, the variations in tensile properties are small however. The tensile modulus does not show systematic trends for the entire range of $\bar{M}_n$ studied.

The effects of prepolymer functionality on tensile properties are listed in Table II, where a series of glycidylmethacrylate copolymers of number average molecular weights in the range of 6000 to 7200 were used containing 5 to 40 wt-% of glycidylmethacrylate. MDA was used in stoichiometric amounts as the cross-

TABLE I

Effect of Prepolymer Molecular Weight on Ultimate Properties.

Molecular Weight $M_n \times 10^{-3}$	Avg Functionality $\langle f_p \rangle$	Elongation-to-Break (%)	Tensile Strength (Dynes/Cm²) x 10^{-7}
10.0	21.2	5.0	69.1
6.0	12.7	4.8	68.9
3.9	8.3	4.6	65.5
2.0	4.2	2.0	58.6
1.55	3.3	1.4	41.4

TABLE II

Effect of Prepolymer Functionality on Ultimate Properties.

Glycidyl Methacrylate Wt-%	Average Functionality $\langle f_p \rangle$	No. of Effective Network Chains ν_e, (cm^{-3}) x 10^{-20}	Elongation-to-Break (%)	Tensile Strength (Dynes/cm^2) x 10^7
5	2.1	0.7	1.0	15.2
20	8.4	10.4	2.8	40.0
30	12.8	16.7	4.0	63.4
35	13.7	19.5	4.1	71.0
40	17.0	24.2	3.8	69.6

linking agent and a catalyst was added in 1 percent concentration. The number of effective network chains per unit volume have been calculated assuming that the crosslinking reaction was completed without side reactions.

The elongation-to-break and tensile strength increase markedly with increasing prepolymer functionality except for a slight decrease at the highest functionality. More specifically, the elongation-to-break increases from 1 percent to 4.1 percent and the tensile strength from 15.2 to 71.0 x 10^7 dynes/cm^2 as the prepolymer functionality increases from 2.1 to 13.7 and the number of effective network chains increases from 0.17 to 19.5 x 10^{20} per cm^3. A slight decrease in elongation-to-break and tensile strength is noted on further increase in crosslink density. The tensile modulus also increases, albeit to a smaller extent, as the crosslink density of the network increases. Thus, the tensile modulus increases from 3.0 to 3.7 x 10^{10} as the prepolymer functionality increases from 2.1 to 17.0.

The influence of the stoichiometry of amine hydrogens to epoxide on the tensile properties has also been explored by crosslinking a terpolymer containing 30% glycidylmethacrylate ($\bar{M}_n \approx 4000$) with MDA and a catalyst (see Table III).

Both tensile strength and elongation-to-break increase with increasing stoichiometry and are maximum at a stoichiometry of 1.25, i.e., when the quantity of MDA used provides 1.25 amine hydrogens per epoxy group in the prepolymer. Above a stoichiometric ratio of 1.25, tensile strength and elongation-to-break decrease. The tensile modulus shows a similar trend although its variations are small above a stoichiometric ratio of 1.10.

TABLE III

Effect of MDA to Epoxy Stoichiometry on Ultimate Properties.

$M_n \approx 4000$.

$NH/\overset{O}{\triangle}$ Stoichiometry	No. of Effective Network Chains ν_e, (cm^{-3}) x 10^{-20}	Elongation-to-Break (%)	Tensile Strength $(Dynes/cm^2)$ x 10^{7}	Tensile Modulus $(Dynes/cm^2)$ x 10^{10}
0.9	14.4	2.5	55.5	3.05
1.0	16.3	3.0	58.3	3.03
1.10	15.5	3.2	58.7	4.08
1.25	14.3	4.6	65.8	4.03
1.50	12.4	4.2	67.2	3.72
1.75	10.6	2.7	63.2	3.79
2.00	8.9	2.5	59.4	3.82

In a related study, the effects of the stoichiometry of the crosslinking agent on the tensile properties on the networks have been investigated by using 4,4′-isopropylidenediphenol (bispenol A) as crosslinking agent. The properties of this series are listed in Table IV. In this case, maximum elongation-to-break and tensile strength are obtained at phenolic hydroxyl to epoxy ratio of unity. Similar results are also obtained for networks produced by crosslinking methacrylic acid copolymers with diepoxides.[4] Properties of these networks with trifunctional branch points are analyzed in Paper No. 8 of these "Proceedings."

Compression molding of a given composition for various lengths of time yielded samples with varying degrees of conversion. The percent conversion of the epoxy group was monitored by infrared analyses of either the thin films formed by flash in the mold or the specimens ground to about 0.025 mm thickness, using nitrile absorption at 4.5 μ as an internal standard.

Variations of tensile properties as the crosslinking reaction proceeds to completion are listed in Table V. The elongation-to-break increases with increasing completion of the crosslinking reaction. Tensile strength is affected only at low conversions reaching a maximum value at about 90% conversion. Changes in tensile strength above 75 percent conversion are rather small. Above 50 percent conversion, the tensile modulus is independent of the extent of crosslinking reaction. On postcuring at 195°C for 18 hours, both the elongation-to-break and the tensile strength decrease.

TABLE IV

Effect of Bisphenol A to Epoxy Stoichiometry on Ultimate Properties.

$M_n \approx 4000$.

BPA/$\overset{O}{\triangle}$ Stoichiometry	No. of Effective Network Chains ν_e, (cm^{-3}) x 10^{-20}	Elongation-to-Break (%)	Tensile Strength ($Dynes/cm^2$) x 10^{-7}	Tensile Modulus ($Dynes/cm^2$) x 10^{-10}
0.5	3.24	2.3	69.0	3.86
0.75	6.20	2.6	74.5	4.00
0.90	7.85	4.0	84.1	3.86
1.00	8.94	4.9	80.0	3.72
1.10	7.61	2.3	72.4	4.14

TABLE V

Effect of Conversion of Crosslinking Reaction on Ultimate Properties.

$M_n \approx 6000$. Crosslinking Agent MDA 125%.

Conversion of Epoxide, %	Elongation-to-Break (%)	Tensile Strength ($Dynes/cm^2$) x 10^{-7}	Tensile Modulus ($Dynes/cm^2$) x 10^{-10}
50	1.2	63.4	3.87
56	1.6	72.4	3.97
69	2.4	71.7	3.95
77	2.9	73.1	4.00
83	3.5	75.8	3.97
86	4.2	77.2	3.87
92	4.1	77.9	4.01
100	3.1	74.5	4.01
100*	2.0	61.0	3.98

* Post cured for 18 hrs at 195°C.

The tensile strength and elongation-to-break are found to be approximately constant above 86 percent conversion when bisphenol A is used as the crosslinking agent. Again, the tensile modulus shows little variation with the extent of curing reaction.

To determine whether the observed sensitivity to chemical structure is peculiar to glycidylmethacrylate copolymers or is a general trend in glassy thermosets a series of highly crosslinked polyurethane networks were synthesized and tested. In this series a copolymer of 35% 2-hydroxyethylmethacrylate and 65% methyl methacrylate of $\bar{M}_n$ = 6500 was crosslinked with a blocked triisocyanate or various diisocyanates. Table VI shows the effect of isocyanate structure on the tensile properties. The highest elongation-to-break of 11% is obtained when a long chain diisocyanate (2,2, 4-trimethyl 1,6-hexanediisocyanate) is used as crosslinking agent. The crosslinking, in this case, was carried out at 160°C without the use of a catalyst. The absence of isocyanate absorption at 4.5 μ in the infrared spectra of the cured specimens indicate that the crosslinking reaction was completed.

The chemical composition of the prepolymer also affects the tensile properties of the networks. Effects of incorporation of styrene in copolymers of 2-hydroxyethylmethacrylate with methylmethacrylate are illustrated in Fig. 1. A series of terpolymers containing 35 percent of 2-hydroxyethylmethacrylate, 5 to 65 percent styrene and the balance methylmethacrylate were crosslinked with a caprolactam blocked triisocyanate. The elongation-to-

TABLE VI

EFFECTS OF CROSSLINKING AGENT ON THE ULTIMATE PROPERTIES OF A POLYURETHANE NETWORK

Crosslinker:	Elongation to break, %	Tensile strength (10^7 dynes/cm^2)
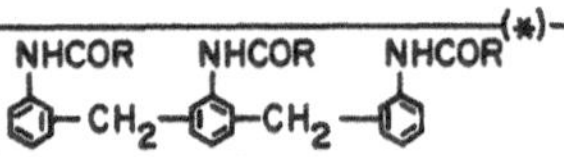 (*)	10	46.2
OCN—CH$_2$—C(CH$_3$)$_2$—CH$_2$—CH(CH$_3$)—CH$_2$CH$_2$NCO	11	50.4
OCN—C$_6$H$_4$—CH$_2$—C$_6$H$_4$—NCO	6	52.0
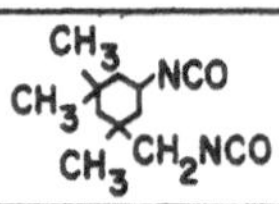	4	57.3

(*)R = 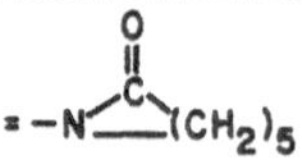

break, decreases from 10 to 2 percent as the styrene content increases from zero to 65% (Fig. 1). Tensile strength is affected to only a small degree.

DISCUSSION OF EXPERIMENTAL RESULTS

The ultimate elongation and the tensile strength of the highly crosslinked polymers prepared vary considerably with changes in prepolymer molecular weight, prepolymer functionality, stoichiometry of crosslinking agents, extent of crosslink reaction, structure of the crosslinking agent and prepolymer composition as shown in Tables I to VI and Fig. 1. The same factors have little effect on the tensile modulus of the networks in the glassy state. Although in some cases (Table VI) ultimate elongations as high as 11% are observed, many networks described here possess ultimate elongations in the range of 3 to 5%.

It should be noted that in highly crosslinked networks in the glassy state the tensile strength, ultimate elongation and modulus are roughly comparable to those of uncrosslinked polymers such as high molecular weight polymethylmethacrylate. The effect of crosslinking itself on the tensile properties of a high molecular weight glassy polymer is therefore not a large one.

Starting from a friable prepolymer, the strength properties increase, at least initially with increasing crosslink density (Tables I to V). Moreover, it may be inferred from Tables II and V that the coherence of the network is formed at a rather late

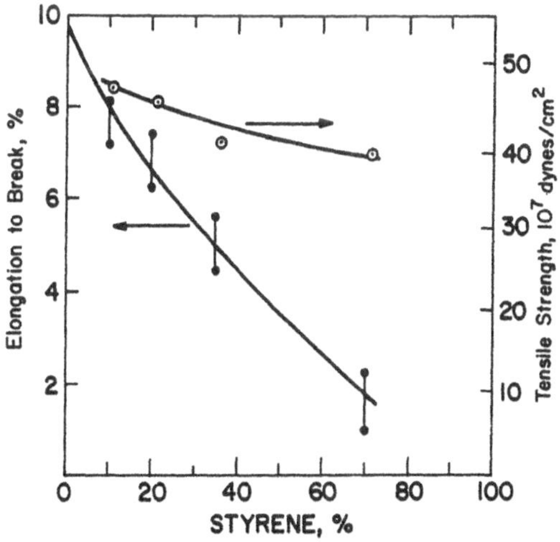

Fig. 1. Effect of styrene content on ultimate properties of a polyurethane network. Prepolymer $\bar{M}_n$ = 6500, hydroxy-ethyl methacrylate 35%, methyl methacrylate plus styrene 65%.

stage in the crosslinking process. The increase in ultimate elongation with increasing crosslink density in the glassy state is contrary to the behavior in the rubbery state.

Various factors such as prepolymer molecular weight, prepolymer functionality, chemical conversion of crosslinking reaction and stoichiometry are related to the number of effective network chains (ν_e). One may speculate that the ultimate elongation and tensile strength are entirely determined by the number of effective network chains as is the case for networks in the rubbery state. However, when all the data (Tables I to V) are plotted against calculated ν_e, no correlation is found. Similarly, the variations in ultimate elongation or tensile strength cannot be related in any systematic fashion to the number of branch points or the number of chain ends per unit volume.

Highly crosslinked polymers are expected to be coherent networks in which the whole sample is one molecule. In addition to network chains, the network may contain pendant chains, a sol fraction, fragments of crosslinked material, and an inhomogeneous distribution of effective crosslinks. No sol fraction could be extracted from the networks described in this report indicating the absence of unconnected prepolymer molecules. This is not surprising. Due to the high functionality of the prepolymer molecules, the sol fraction is exhausted well before all the reactive groups are consumed. The role of pendant chains on the mechanical properties is not clear; network defects or flaws, however, must be of a certain minimum size to have any effect on the initiation or propagation of cracks.

Although there seems to be a direct effect of chemical composition on the ultimate properties of the networks, the sensitivity of the ultimate properties to slight variations in composition cannot be directly related to structures on the molecular level. Defects or flaws of the size of a few Ångströms are not likely to influence the process of extensive plastic deformation at the tip of a propagating crack. To rationalize the ultimate properties in terms of chemical composition it is, therefore, more appropriate to search for large defects which may be a direct consequence of the chemical composition of the network components, and which are large enough to influence the fracture process. Evidence for crosslink density fluctuations in thermosets have been reported in the literature.[5-7] Crosslink inhomogeneities may be formed in many different ways. One possible mechanism is by microsyneresis or local incompatibility.[8] Data in Fig. 1 may be rationalized on the basis of microsyneresis dependent on the presence of an apolar comonomer.

The localized incompatibility, however, is not a necessary condition for the formation of inhomogeneities in crosslink density. After partial crosslinking, the network chains are no longer free to assume all possible configurations and further crosslinking may be restricted to certain regions to produce localized higher crosslink density. Such a process can be visualized by formation of a very large ring in which subsequently intramolecular reactions will be favored over intermolecular reactions. A statistical argument can be given that the network formation by crosslinking of prepolymer units is intrinsically an inhomogeneous process and the networks formed are likely to contain crosslink density fluctuations. Such an argument for the origin of crosslink density inhomogeneities has been developed and is presented in the following section.

PRINCIPLES OF THE THEORY OF CROSSLINK DENSITY FLUCTUATIONS

In this section polymer chains are considered as flexible randomly coiling freely jointed chains. Therefore, the statistics developed for randomly coiling macromolecules with no excluded volume are assumed applicable.* It can be shown that the relaxation times for rearrangement of the various polymer segments in space at the temperature of crosslinking are very small. For macromolecules of relatively low molecular weight, where no entanglements are present, these rearrangement times are assumed negligible in the time scale of the crosslinking process. Since the polymers are present in the bulk, the molecules can be considered free draining.

In the following calculations one prepolymer molecule is considered and its fate is followed on the way to incorporation into the network. For simplicity, it is assumed that the unreacted sample consists of a prepolymer of uniform molecular weight, mixed with an infinitesimal four functional crosslinking agent. Therefore, the calculations are performed on a model shown schematically in Fig. 2, a linear macromolecule of degree of polymerization P containing a number (f_P) of active groups of type A, which can react either with other A groups on the same molecule or with those on neighboring macromolecules thereby forming infinitesimal four functional branch points.

It is customary to assume that the average distribution of all segments i, or monomer units, around the center of gravity of the molecule may be represented by a Gaussian distribution function. Debye and Bueche[9] have shown that this is a very good approximation. Thus if ρ is the number of segments per unit volume

---

*Since excluded volumes frequently cannot be neglected, the results of the present calculations are correct only to an order of magnitude. The mechanism, however, leading to formation of these crosslink density inhomogeneities is not affected by this simplification.

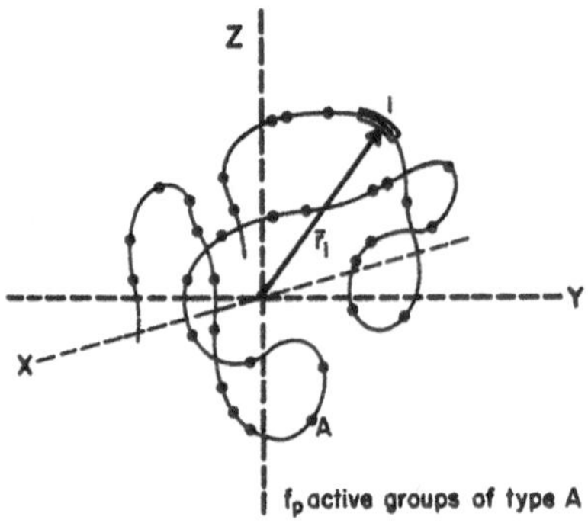

Fig. 2. Random coil model of a prepolymer molecule with active groups A.

of this particular macromolecule, one may write

$$\rho(r) = \rho(o) \exp(-B^2 r^2), \tag{1}$$

where r is the distance from the center of gravity of the macromolecule and $\rho(o)$ and B^2 are molecular constants, which can be solved from the boundary conditions

$$P = \int_0^\infty \rho(r)\, 4\pi r^2\, dr = \frac{\pi^{3/2}\rho(o)}{B^3}, \tag{2}$$

and

$$\langle R^2 \rangle = \frac{1}{P}\int_0^\infty r^2 \rho(r)\, 4\pi r^2\, dr = \frac{3}{2\, B^2}. \tag{3}$$

This yields

$$\rho(r) = P\left(\frac{3}{2\pi\langle R^2\rangle}\right)^{3/2} \exp\left(-\frac{3r^2}{2\langle R^2\rangle}\right), \tag{4}$$

and

$$\rho(o) = P\left(\frac{3}{2\pi\langle R^2\rangle}\right)^{3/2}. \tag{5}$$

The symbol $\langle R^2 \rangle$ represents the mean square radius of gyration of the macromolecule. This quantity has been considered in detail for linear and branched macromolecules by Zimm and Stockmayer,[10] and is generally defined by

$$\langle R^2 \rangle = \frac{1}{P}\sum_{i=1}^{P} \langle r_i^2 \rangle, \tag{6}$$

where $\langle r_i^2 \rangle$ is the mean square average distance of the i th segment from the center of gravity of the molecule. For linear macromolecules $\langle R^2 \rangle$ can be set equal to

$$\langle R^2 \rangle = P\, b^2/6, \tag{7}$$

where b is the "effective monomer length." It should be noted that $\langle R^2 \rangle^{\frac{1}{2}}$ is not equal to R_η, the radius of the hydrodynamically equivalent sphere as introduced by Flory and Fox.[11]

Obviously, $\rho(o)$ is always smaller than $\bar{\rho}_o$, the average number of segments per unit volume regardless to which molecule they belong:

$$\bar{\rho}_o = c\ N_A/M_o\ , \tag{8}$$

where c is the concentration in grams per unit volume, N_A is Avogadro's number and M_o is the molecular weight of a monomer unit. For instance, for bulk polymethylmethacrylate (PMMA)

$$\bar{\rho}_o = 7.1 \times 10^{21} \text{ monomer units per cm}^3 .$$

In reality, the crosslinking agent is not infinitesimal and occupies a considerable part of the space, especially in highly crosslinked systems. The average segment density $\bar{\rho}_o$ in the systems considered in this paper then reduces to approximately 6.0×10^{21} monomer units per cm^3. If b is set[12] equal to 6.9×10^{-8} cm, the value found in a theta solvent, the values of $\langle R^2 \rangle^{\frac{1}{2}}$ and $\rho(o)$ for PMMA of various degrees of polymerization P are given in Table VII.

TABLE VII

The Parameters $\langle R^2 \rangle^{\frac{1}{2}}$ and $\rho(o)$ for the Segment Distribution Function of a PMMA Molecule in the Bulk Polymer and the Corresponding ℓ_c, the Critical Number of Loops Required to Make $\rho(o)$ Reach the Value of $\bar{\rho}_o = 6.0 \times 10^{21}$

P	$\langle R^2 \rangle^{\frac{1}{2}}$ in cm	$\rho(o)$ in cm^{-3}	ℓ_c
20	12.6×10^{-8}	3.31×10^{21}	0.7
50	19.9×10^{-8}	2.09×10^{21}	1.8
100	28.2×10^{-8}	1.48×10^{21}	3.2
200	39.8×10^{-8}	1.05×10^{21}	5.5

The molecule, schematically shown in Fig. 2, is now subjected to reaction, which can be either intermolecularly or intramolecularly. The probability that a group A of this particular macromolecule reacts, either internally or externally, in a volume element dx, dy, dz located at x, y, z, is proportional to the concentration of A groups of this macromolecule in that volume element. This concentration is proportional to $\rho(r)$, where $r^2 = x^2 + y^2 + z^2$. The proposed crosslinking reaction, however, is of the second order. Therefore: the probability that this reaction occurs externally is proportional to the concentration of A groups from neighboring macromolecules, that concentration is proportional to $[\bar{\rho}_o - \rho(r)]$. The probability that the reaction will occur internally will again be proportional to $[\rho(r)]$. Summation of these probabilities over all volume elements yields the probabilities $p_{external}$ and $p_{internal}$ for the entire macromolecule

$$p_{in} = Q\int_0^\infty [\rho(r)]^2\, 4\pi r^2\, dr, \tag{9}$$

$$p_{ex} = Q\int_0^\infty [\bar{\rho}_o - \rho(r)]\,\rho(r)\, 4\pi r^2\, dr, \tag{10}$$

where Q is a proportionality constant which need not be determined.

After performing the integration one obtains

$$\frac{p_{in}}{p_{ex} + p_{in}} = \frac{\rho(o)}{2^{3/2}\,\bar{\rho}_o} \quad . \tag{11}$$

This simple result indicates that the probability of intramolecular reaction, i.e. loop formation, depends only on the value of $\rho(o)$ and not on the shape of the radial segment distribution function. This is an important consequence and will be applied to the branched structures that are formed during the crosslinking process.

It should be noted that Eq. (11) is valid only for $\rho(o) \leq \bar{\rho}_o$. Only then is a Gaussian segment distribution function for $\rho(r)$ applicable. Thus the highest value of $p_{in}/(p_{ex} + p_{in})$ that may be obtained from Eq. (11) is 0.354.

It is now necessary to consider the changes in $\rho(o)$ and in the radial distribution function during the crosslinking process. For instance, for a bulk PMMA of degree of polymerization 50 the calculated radial distribution function for one molecule is shown as curve A in Fig. 3. It can be shown[13] quite rigorously, that the radial distribution functions of the branched molecules obtained by random intermolecular reaction of this molecule with neighboring molecules of the same size are shaped like curves B and C in Fig. 3, i.e., the segment density at the center of gravity of the branched system differs very little from that in the primary molecule. Hence the ratio $p_{in}/(p_{ex} + p_{in})$ stays almost the same during the branching process. If only one intramolecular reaction occurs, however, the value of $\rho(o)$ is drastically changed as shown by curve D. Therefore, the "process of ring formation" increases the chance for more ring formation and the value of $\rho(o)$ eventually will tend to become larger than $\bar{\rho}_o$, a physical impossibility in the bulk. With the use of a normal mode

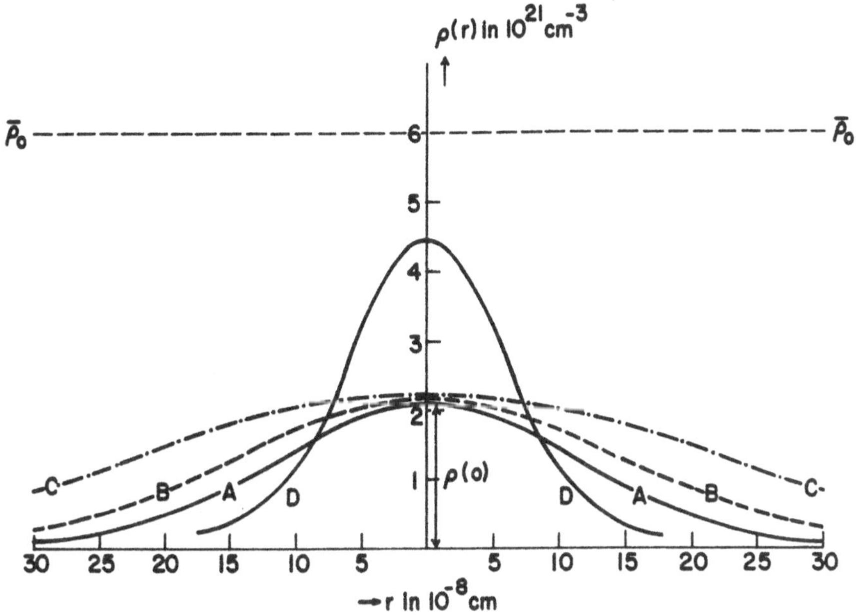

Fig. 3. Radial distribution functions for: A) one PMMA molecule of P = 50, B) and C) branched molecules obtained by coupling molecules of P = 50 at random, D) molecule of curve B) in which one cyclization occurred.

analysis developed recently,[14] the distribution functions for many ring systems have been calculated. Then for each polymer molecule the critical number of loops, ℓ_c, necessary to make $\rho(o)$ reach the limit $\bar{\rho}_o$ can be calculated.[13] These values are given in the last column of Table VII.

As $\rho(o)$ increases with increasing number of loops, ℓ, the fraction of intramolecular reactions also increases according to Eq. (11). These fractions for various degrees of polymerization, P, and number of loops, ℓ, have been calculated for PMMA thereby taking $\bar{\rho}_o = 6.0 \times 10^{21}$, and are given in Table VIII. Now the entire hierarchy of branched and ringed molecule types occurring during the crosslinking process can be calculated, even in the presence of a molecular weight distribution.

When ℓ exceeds ℓ_c a discontinuity in the above equations occurs. The effect can be described as a "nucleation" process. The segment distribution function will tend to be of a truncated type, and any further reaction within the nucleus is only intra-

TABLE VIII

Values of $p_{in}(p_{ex} + p_{in})$ for PMMA Molecules (Linear or Branched) in the Bulk Polymer Varying with P and ℓ

	P = 200	P = 100	P = 50	P = 20
$\ell = 0$	0.062	0.087	0.12	0.20
$\ell = 1$	0.13	0.18	0.26	-
$\ell = 2$	0.19	0.27	-	-
$\ell = 3$	0.24	0.34	-	-
$\ell = 4$	0.29	-	-	-
$\ell = 5$	0.33	-	-	-

molecular. Immediately, the question arises whether this nucleation leads to a process of growth similar to that occurring in crystallization. Values of $p_{in}/(p_{ex} + p_{in})$ for an assumed truncated distribution have been calculated and it can be shown[13] that the tendency of these nuclei to grow, i.e. to react intermolecularly, is even less than that shown in Table VIII. Thus their growth is retarded when compared with the formation of new nuclei.

The present mechanism for crosslinking, therefore, predicts formation of spherical entities, "gel balls," which for low molecular weight prepolymers are formed long before the gel point of the mixture is reached. The excessive intramolecular reactions within the gel balls exhausts the amount of available crosslinking agent, causing a deficiency in crosslink density in the regions between the gel balls. The resulting network structure then can be represented schematically as in Fig. 4. This figure illustrates that the coherence of the network and the randomness of the branchpoints or chain ends are two independent influences on the network morphology. Although the crosslinks and chain ends are uniformly distributed in this two dimensional model, the network is quite incoherent. A similar model has been previously suggested by Kuhn.[15]

Even for monodisperse prepolymers the distribution of ball volumes is very wide. The smaller ones may never be noticed whereas the larger ones must be visible in the electron microscope. For a PMMA prepolymer of P = 50 the number average diameter of the gel balls at the gel point is 71 Å, and the ratio of the weight average over the number average ball diameter is 1.6. Whether these

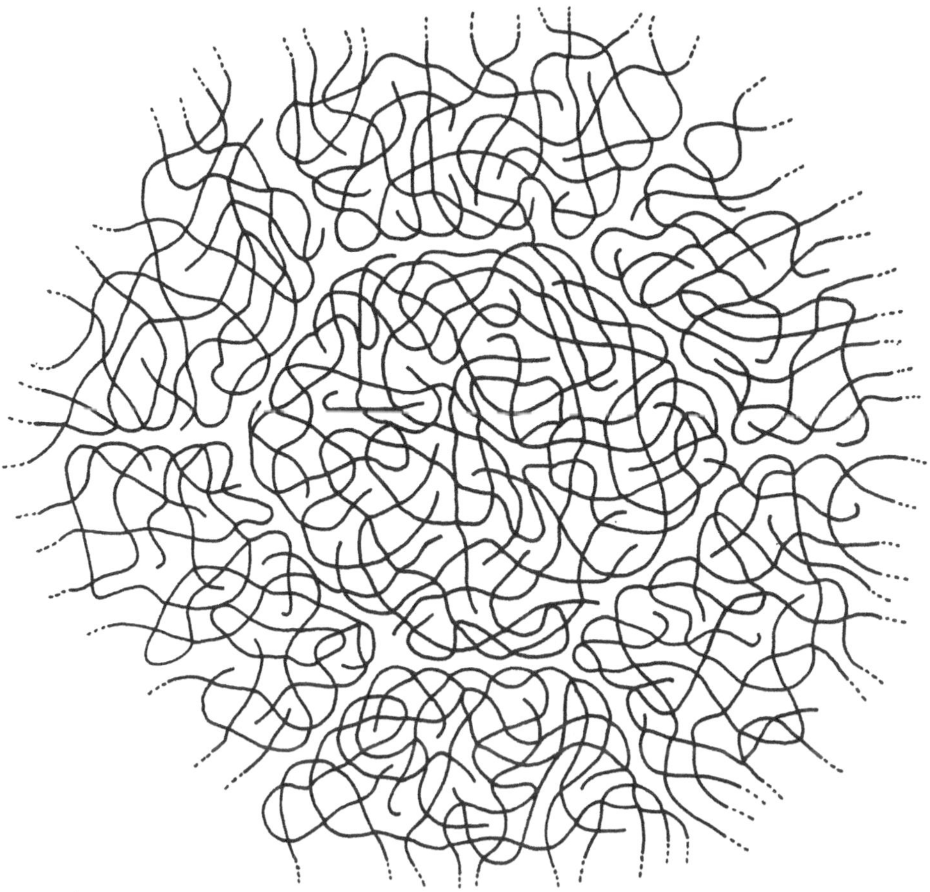

Fig. 4. Two dimensional schematic representation of crosslink density inhomogeneities. A "gel ball" at the center is loosely tied to six similar regions around it. Crosslinking in bulk requires that branch points and chain ends are evenly distributed independent of morphology.

crosslink density inhomogeneities persist, even at high degrees of cure, remains to be established experimentally. These experiments will be described in the following section.

The implications of this crosslink mechanism are manifold. The influence of the prepolymer molecular weight has already been indicated and must have an obvious effect on the fracture energy. Presence of a solvent will aggravate the formation of crosslink density inhomogeneities. This explains the poor ultimate mechanical properties of networks crosslinked in the presence of relatively small amounts of diluents. A broad prepolymer molecular weight distribution will have a similar effect, i.e., considerable portions of the low molecular weight fractions will soon be converted into indifferent solvent which tends to further reduce the coherence of the network. It also suggests why experimentally determined gel points always occur later than theory predicts.

Since the crosslink density inhomogeneities originate from the bell shaped segmental distribution function it follows that every crosslink process is inhomogeneous in principle. Completely homogeneous networks can never be made; such a structure can only be approached assymptotically.

EXPERIMENTAL EVIDENCE FOR INHOMOGENEITIES

To obtain experimental justification of inhomogeneous crosslinking described in the previous section, a network was chosen prepared from a prepolymer with number average molecular weight of 4000 and specified by the third sample in Table I. Specimens of this thermoset were fractured at room temperature in tension (the fracture was initiated internally) and the freshly cleaved surfaces were replicated by two-stage carbon replicas, prepared as follows. Surfaces were covered with a polyvinylalcohol (PVA) solution in water and dried for 24 hours. Then, the PVA replicas were removed, shadowed under 45° with platinum, coated with a coherent carbon film, and the PVA dissolved in water. Thus, a "hump" on the replica actually represents a "dip" in the fracture surface and vice versa. In the electron microscope the platinum deposits will appear as "white shadows." The carbon replicas were placed on a square copper grid with a mesh size of 60 by 60 microns, 125 microns apart. Thus, only one third of the replica is visible in the electron microscope. Electron micrographs were all taken from replicas of the same fracture surface and are representative for the morphologies encountered on fracture surfaces of this thermoset.

Of course, one may not assume that the crack would propagate neatly around the regions of highest crosslink density. Therefore, the appearance of the fracture surface may not resemble the morphology of the network before fracture since all the regions, whether lightly or highly crosslinked, will be deformed in the plastic zone preceding fracture. Moreover, the average size of these regions is probably smaller than the plastic zone ahead of the crack.

A better approach is, therefore, to utilize the difference in chemical degradation rates between lightly and highly crosslinked networks. An etching procedure was designed to bring out the desired texture. The specimens were heated in terpentine (containing some detergent) for 30 minutes at 80°C and quickly transferred to a concentrated sulfuric acid - chromic acid bath at 60°C for 60 seconds. It was presumed that lightly crosslinked regions would be preferentially oxidized. The treated samples were then washed with cold distilled water during 24 hours and dried. Replicas of the etched surfaces were selected so as to show the same areas as observed before etching.

Figure 5 shows a portion of the original fracture surface at a magnification of 40000 x. As expected, extensive local deformation preceding fracture causes the texture at this surface to obscure any pre-existing network morphology.

Figure 6 shows the same area after etching. The polymer surface is here somewhat smoother than in Fig. 5 but, superimposed on the remaining texture, a few hills in the replica have been created, which seem to have been absent before etching. From this micrograph, the diameter of the "gel balls" is estimated to be between 100 and 150 Å.

A consistent and similar phenomenon is observed throughout the entire fracture surface, as exemplified by three other areas shown in Figures 7, 8 and 9, all at 40000 x. As expected, the etching of the "gel balls" from the surface may occur at random (Fig. 7) or, frequently, in clusters (Fig. 8). Obviously, after one "gel ball" has been removed neighboring ones will be etched out more readily, leading to clusters in the replica. A region where further etching has occurred is shown in Fig. 9. Throughout the entire fracture surface, however, the etched "gel balls" are in the same size range.

A similar morphology of etched surfaces of polymer networks has been observed by Blokland[16] on polyurethane networks.

The present concept of sparsely connected "gel balls" can explain the sensitivity of the ultimate properties to slight variations in network chemistry (Tables I through V). Higher prepolymer molecular weights lead to larger "gel balls," and thus to more coherent networks (Table I). An increase in the functionality of the prepolymer causes the gelation of the entire mixture to occur at an earlier stage and, therefore, allows the "gel balls" more time to fuse together (Table II). At very high prepolymer functionalities, however, the volume fraction of crosslinking agent cannot be neglected, and to obtain a complete conversion of the reactive groups becomes increasingly difficult. Obviously, a stoichiometric amount of crosslinking agent and a high degree of conversion (Tables IV and V) are also optimum conditions to fuse the "gel balls" together. The data in Table III indicate that MDA, under the present reaction conditions, can utilize only three out of its four amine hydrogens. Consequently, optimum tensile properties are obtained with a 33% excess MDA, as shown in Table III.

Fig. 5. Replica of original fracture surface; shadowed with platinum under 45° from the right. Magnification 40000 x.

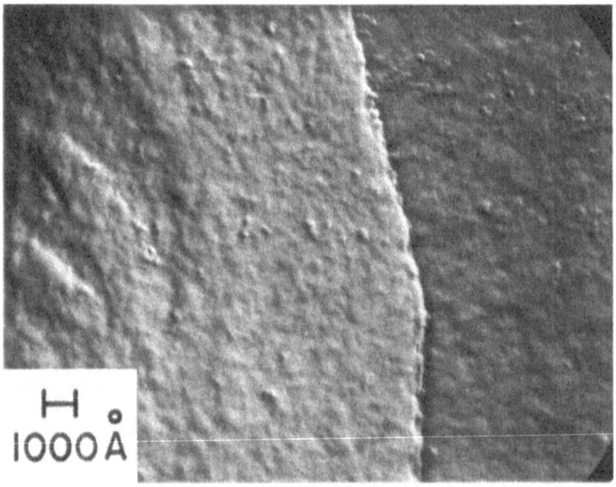

Fig. 6. Replica of same location as in Fig. 5 after etching; shadowed with platinum under 45° from the left. Magnification 40000 x.

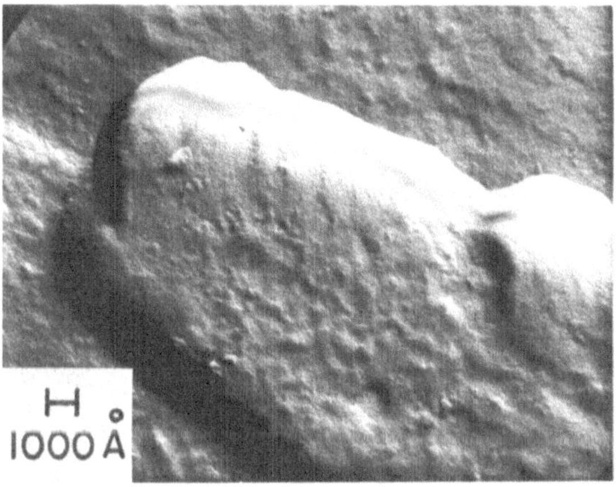

Fig. 7

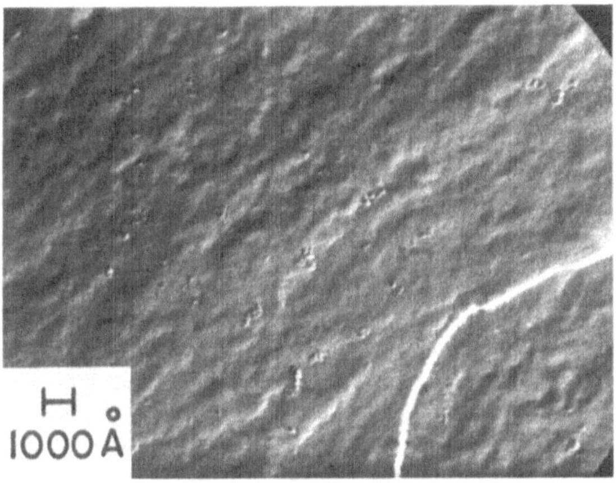

Fig. 8

Fig. 9

It has been shown that a statistical argument can be given for the formation of crosslink density inhomogeneities in homogeneous media. Unambiguous proof rests on additional studies of the effects of molecular weight, degree of cure, concentration of solvent, etc., on the network morphology.

ADDITIONAL REMARKS

It has been shown that various factors such as prepolymer molecular weight, functionality of prepolymer and of crosslinking agent, chemical conversion of reactive groups, and stoichiometry of reactive groups influence the ultimate properties of the networks. It can be predicted, however, that these factors also affect the morphology and the coherence of the network structure. It is reasonable, therefore, that in retrospect no correlation was found between ultimate properties and numbers of chain ends, branch points, or effective network chains.

Although the chemical composition is an influential factor in maximizing properties, it is important to realize that a hierarchy of structures exists from atomic dimensions to the macroscopic shape of the sample. Fracture is generally initiated at the surface of impurities (which may be as small as 1000 Å) whereas the process of crack propagation and accompanying fracture energy are dominated by supermolecular structures in the network.

For crosslinking processes in the liquid state, segmental rearrangements occur almost instantly and macromolecules may be considered at equilibrium conformations. A statistical theory has been derived to account for the segmental rearrangements which take place after each crosslinking reaction. The theory predicts that in certain agglomerates of chains, rings and branches the fraction of intramolecular reactions (favored over intermolecular reactions) will reach catastrophic proportions. This leads to formation of "gel balls" which are formed in the homogeneous phase before the gel point of the mixture is reached. The formation of these crosslink density inhomogeneities reduces the coherence of the network considerably, and are expected to exert a profound effect on ultimate properties.

Several fracture surfaces of thermosets have been etched by a subsequent swelling and oxidizing process. This treatment is designed to preferentially extract "gel balls" at the surface. Electron micrographs of replicas of these treated surfaces revealed that extractable inhomogeneities exist throughout the entire fracture surface and are 100 to 150 Å in diameter, as correctly predicted by theory.

In the past, macroscopic mechanical properties of polymer networks have usually been interpreted in terms of the properties of structural elements on an Ångström scale. It is the objective of this paper to show that supermolecular structures may and do form as a logical consequence of the crosslinking reactions, and that an explanation of network properties in terms of their morphological features is a more fruitful approach to an interpretation of mechanical properties.

ACKNOWLEDGEMENTS

The authors gratefully acknowledge the assistance of Mr. H. Plummer in obtaining the electron micrographs and Mr. Y. F. Chang for the synthesis of the prepolymers and networks.

REFERENCES

1. H. Mark, in "Cellulose and Cellulose Derivatives," ed. E. Ott, (Interscience Publishers, Inc., New York, N. Y., 1943). pp. 1001 ff.
2. R. R. Jay, Anal. Chem. 36, 667 (1964).
3. W. R. Sorenson and T. W. Campbell, "Preparative Methods of Polymer Chemistry," (Interscience Publishers, Inc., New York, N. Y., 1961), pp. 134.
4. J. Fellers and A. Golovoy, J. Appl. Polymer Sci. 15, 731(1971).
5. D. H. Solomon, B. C. Loft and J. D. Swift, J. Appl. Polymer Sci. 11, 1593 (1967).
6. R. E. Cuthrell, J. Appl. Polymer Sci. 12, 1263 (1968).
7. L. Gallacher and F. A. Bettelheim, J. Polymer Sci. 58, 697 (1962).
8. K. Dusek, Paper 11 in these "Proceedings."
9. P. Debye and F. Bueche, J. Chem. Phys. 20, 1337 (1952).
10. B. H. Zimm and W. H. Stockmayer, J. Chem. Phys. 17, 1301 (1949).
11. P. J. Flory, J. Chem. Phys. 17, 303 (1949).
    T. G. Fox and P. J. Flory, J. Phys. Chem. 53, 197 (1949).
12. P. J. Flory, "Principles of Polymer Chemistry," (Cornell University Press, Ithaca, N. Y., 1953), pp. 618.
13. A. J. Chompff, to be published.
14. A. J. Chompff, J. Chem. Phys. 53, 1577 (1970).
15. W. Kuhn and H. Majer, Kunststoffe Plastics 3, 1 (1956).
16. R. Blokland and W. Prins, J. Polymer Sci. A-2, 7, 1595 (1969).

DISCUSSION

M. Gordon (University of Essex):

I find the evidence for network inhomogeneity credible and interesting. Since the paper by Gordon, Ward and Whitney (GWW) dealt with homogeneous networks, and theoretical models appropriate to them, I wish to make some remarks to clarify the conditions leading to heterogeneity. Essentially, they amount to relatively high rates of cyclisation, especially when the ring-closures are concentrated into small neighborhoods. Any kind of polyfunctional link formation can lead to networks. The most random network formation results from polycondensation of small f-functional ($\langle f \rangle > 2$) units; in GWW it is shown that ring-formation cannot be totally avoided even there, but its effects can be quantitatively treated by the random-flight model of Gordon and Scantlebury, which represents molecules as graphs. The crosslinking of prepolymer can be simulated by first assembling the small monomer units into primary chains and then allowing the chains to crosslink. (This is a much less random process, and can be formally described as an f-functional polycondensation with a strong long-range 'substitution effect,' which prevents a chain from crosslinking before it is fully assembled). Such a process leads statistically to much heavier (intra-chain) cyclisation, and here gas-like models (i.e. based on a chain-segment distribution function) have been successfully used in the contribution by Dusek and by Labana, Newman and Chompff. The latter theory has the great merit of allowing for the effect of pre-existing rings on the rates of formation of subsequent rings in the same chain. It is worth recalling that high degrees of intra-chain ('incestuous') cyclisation were found quite early in addition polymerisations and copolymerisation [Simpson, Holt and Zetie, J. Polymer Sci., 10, 489 (1953); Gordon and Roe, ibid. 21, 75 (1956)] which are statistically closely related to the case of prepolymer crosslinking.

Note communicated later: It is doubtful that intra-chain cyclisation plays a substantial part in the vulcanisation of rubbers with long primary chains, where gelation occurs very early. As explained in the discussion of fig. 1 of GWW, cyclisation immediately after the gel point ceases to be describable in terms of random-flight type models (or equally in terms of the segment distribution function of an individual chain).

A. J. Chompff:

In the present paper, the probability of cyclization was calculated for the case that any group on a certain molecule can react with any other group on the same molecule. This probability is quite large. If, however, only end groups can react, the probability that the two ends of a linear molecule react with each other is very small. Consequently, if a polycondensation reaction is performed starting from bifunctional units and a small fraction of trifunctional units, then the chance that the ends of two long branches meet to form a ring is again very small. Moreover, formation of this ring will hardly influence the chances of forming consecutive rings and thus quite random networks will be obtained. So far we agree with Dr. Gordon.

When the fraction of polyfunctional branching units is very large, however, the chance of forming a ring is also large. Since the degree of branching is also very large, the radial segment distribution of a macromolecule is rather narrow; cyclization will make the distribution even narrower; the volume fraction of neighboring macromolecules penetrating the first will become rather small and thus the chances for external reaction will be considerably reduced after formation of the first ring. Therefore, in this case, we also predict formation of "gel balls." An indication of the presence of these inhomogeneities in polycondensate networks has been shown by Blokland.[16]

AUTHOR INDEX

SUBJECT INDEX